G000066117

ENGLISH MAPS: A HISTORY

THE BRITISH LIBRARY
STUDIES IN MAP HISTORY
Volume II

English Maps: A History

Catherine Delano-Smith
and
Roger J. P. Kain

THE BRITISH LIBRARY

First published 1999 by
The British Library
96 Euston Road
London NW1 2DB

© 1999 in text, Catherine Delano-Smith and R. J. P. Kain

British Library Cataloguing in Publication Data
A CIP record for this book is available from The British Library

ISBN 0 7123 4609 0 (cased)
ISBN 0 7123 4660 0 (paper)

Designed by John Mitchell
Typeset by Bexhill Phototypesetters, Bexhill-on-Sea
Printed and bound in England by Butler and Tanner Ltd, Frome

CONTENTS

LIST OF FIGURES

These are abbreviated titles. Full details, including approximate dimensions (vertical × horizontal), are given in the captions.

CHAPTER 5: MAPS AND TRAVEL

CHAPTER 6: MAPPING TOWNS

CHAPTER 7: THE SPIRIT OF MODERNITY: MAPS IN EVERYDAY LIFE

CHAPTER 8 THE LOTTERY OF MAP SURVIVAL

LIST OF COLOUR PLATES

following page xv

Preface

There are many ways of writing a book. That this one takes the form it does is due to one person in particular: Monserrat Galera i Monegal, for it was her invitation to contribute to her annual lecture courses in the history of cartography at the Institut Cartogràfic de Catalunya, Barcelona, in February 1996 which prompted us to take an overview of the history of English maps. Adopting the revisionist perspectives of what might be called the 'new' history of cartography, we aimed to explore the ways in which maps have interacted with society in the past, and to analyse the roles that maps have played and the uses to which they have been put, rather than to focus on the technical aspects of map construction and reproduction. We tried to take note of the many connections across the broad span of time involved, from Roman times to the twentieth century. Seeking to avoid too insular a view, we noted the influence in England of intellectual and cartographical developments in the rest of Europe. The result was not only the spoken lectures delivered within the week we were in Barcelona, but also their publication in the Institute's 'Cicle de conferències' series (*7è curs*) as *La Cartogràfia Anglesa* (Barcelona, Institut Cartogràfic de Catalunya, 1997). Three years on and despite a good deal of new material, we maintain our original approach with its emphasis on the map in its societal and cultural context, and on map-users as much as map-makers. While such an approach is not entirely novel, we believe that the present book is the first full-length exploration of the political, religious, social and economic enmeshing of maps in the history of any nation state.

In attempting to knit together ideas and studies often still in the making, we are inevitably building on the work of others. If we have misrepresented any views, that is our (much regretted) fault alone. We would like to thank many people for their generosity in a multitude of ways – from lending slides for our Barcelona lectures to reading, and re-reading, our drafts and commenting so usefully – but it becomes difficult to acknowledge in sufficient measure the debts which have accumulated over the years. While we are reluctant to single out a few when many have helped us, there are some who must be named. We are particularly grateful to our principal readers – John Andrews, Peter Barber, and Tony Campbell – and to Andrew Cook, Paul Laxton, and Laurence Worms for their substantial comments, and to David Musgrove for helping us with the bibliography and to Barbara Croucher for compiling the index. We are indebted also to Sarah Bendall, David Buisseret, Felix Driver, David Fletcher, Paul Harvey, Yolande Hodson, Richard Oliver and

Alessandro Scafi for their help over specific points. Other thanks are expressed in the appropriate end-notes. Our debt to the various British Library departments – Manuscripts, Maps, and Rare Books – is long-standing and especial thanks are due to Tony Campbell, Map Librarian, and all his colleagues. We are also grateful to the Leverhulme Trust, the British Academy, the Institute of Historical Research, University of London, and to the University of Exeter for supporting our own research in map history. On the technical side, we have benefitted from assistance from Helen Jones and Andrew Teed at the University of Exeter. Finally, we are most grateful to David Way and his colleagues in the British Library Publishing Office, for their patience and their professionalism.

Catherine Delano-Smith and Roger J. P. Kain
April, 1999

<table>
<tr><td>CHAPTER

I</td><td>Introduction:
Maps for all Seasons</td></tr>
</table>

Maps are children of their times. The economic, cultural, political and demographic factors of each period are as much a part of map history as are the more commonly-stressed scientific and technical aspects of maps and map-making. It was, for example, the surge of social and economic change in the sixteenth century that under-pinned a general and increasingly popular use of maps. The printed map quickly achieved importance in this new world of early modern England in all sorts of ways, in humanist scholarship as in the new sciences. As the print wave gained momentum, so did the manuscript map. From the sixteenth century onwards we have manuscript maps in numbers as never before, serving agricultural improvement, land management, urban planning, government and administration, fortification planning, national defence, navigation, astronomy, boundary litigation, mining, science (geology, climatology, medicine), social reform (poverty, disease) – and much else. These activities are intrinsic parts of the cultural context of maps; map use and map users are in the foreground of our history of English maps.

The range of maps which have been made and used in the past in England is compelling, and diversity in continuity is another keynote of this book. We find that the idea of maps and the practice of mapping had a place in a surprisingly wide range of contexts from early medieval times onwards. The products of modern cartography may look different from those of the Middle Ages but the underlying concepts are identical: the representation of features (places, people, phenomena, real or imagined) in their relative or actual spatial location. Precisely what defines a 'map', as opposed to a 'view', is a more open question, and we have to accept that the boundary is fluid between portrayals generally accepted as maps and those traditionally classed as pictures.[1] Many representations are placed in either class from time to time according to the investigator's starting point (art history, map history, history of science). We ourselves do not subscribe to the narrow view that drawing to a fixed linear scale is the crucial diagnostic of a map. Nor do we consider that a strict plan view is the necessary attribute of a map, although we do appreciate that an intrinsic characteristic of a map is that ground-level layout is revealed in a way that it is not in a picture. There have always been plenty of maps, particularly of towns, constructed from a non-vertical viewpoint, but even these oblique or bird's-eye perspectives are so designed as to render the whole in such a way that the relative location of each feature – the underlying plan layout – is discernible in essentials, if not in every detail. In fact, in probably the vast majority of pre-modern maps (and certainly of medieval maps), the cartographical structure is provided by plan depiction of some features, such as field boundaries, tracks and rivers, while other features, such as buildings, bridges and vegetation, are shown in profile or in oblique view. Moreover, what applies to maps in these respects also holds true for those large-scale plans of relatively small features which are sometimes excluded from map history. Plans of encampments,

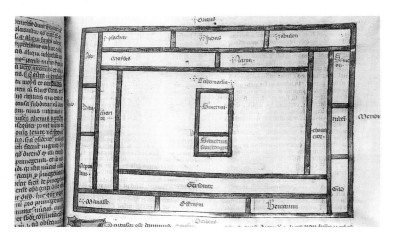

1.1 Stylised map: topological diagram of the Desert Encampment from Nicholas of Lyra's commentary on the Bible (*Postillae*, 1323–32) for Numbers 2, showing the disposition of the Twelve Tribes around the Ark of the Covenant. East is indicated at the top. 13 × 19 cm. Fourteenth-century English copy. British Library, MS Royal 3.D.vii, fol. 94r.
Reproduced by permission of the British Library Board.

forts, houses, temples, theatrical stages or even of a single room are no less cartographical in conception than a map of a wider area. Finally, it is important to bear in mind that not all maps or plans need be pictorial or made up of a spatial arrangement of graphic signs and lines; text can also be arranged to indicate spatial characteristics.

The relative importance of one map type to another over time is unpredictable. There are times and places when thinkers hold the cartographical high ground, as in the Middle Ages, when the primary function of 90 per cent of the maps which have survived or which we know about from before 1350 was pedagogic, didactic or exegetic. Such maps range from simple diagrams summarising the layout of places described in the Old and (less commonly) New Testaments, drawn as aids to understanding and teaching the meaning of obscure passages in the biblical text, to the conceptually profound and visually complex *mappaemundi* which served 'to refine and reinforce the truth of inherited tradition and to add to the sum of human knowledge'.[2] At other times or in other places the doers take over. Already, in the Middle Ages, especially between 1350 and 1500, we find increasing numbers and an ever-widening range of maps put to practical use, from property management and the resolution of land ownership disputes, to the staging of religious ceremonies and plays, as memorials to things achieved, and as tools in national propaganda and administration, a trend which continued in the sixteeenth century. Under Henry

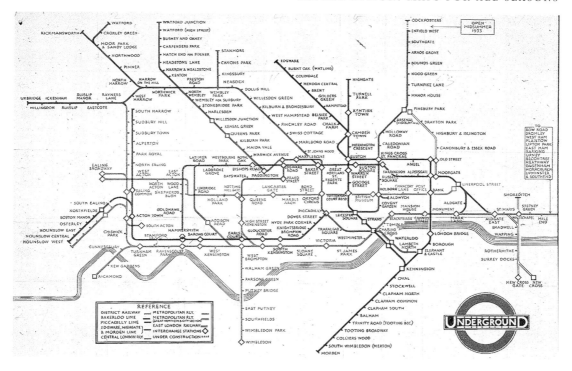

1.2 Stylised map: the first card folder edition of Henry Beck's topological *First Diagram* of the London Underground, in which the relationship of the lines, sequence of stations and interchanges is demonstrated. North is at the top. 16 × 23 cm. 1933. Victoria and Albert Museum, E.815–1979.
Reproduced by permission of the Victoria and Albert Museum, London.

VIII in the 1530s, maps began to be used in planning new-style fortifications and national defences and by the 1580s, Lord Burghley was relying on maps as tools of state business: for calculating the cost of postroads, for recording the distribution of the leading Catholic families, and for working out the capacity of southwestern beaches for enemy landings. Not that the distinction between maps for intellectual as opposed to practical functions has ever been clear or watertight; the diagrammatic style of the maps produced by medieval monks as aids to the ordering of daily life as well as in the pursuit of scholarship (Fig. 1.1) is also the salient characteristic of the most outstandingly successful practical map of all times, Henry Beck's map of the London Underground system, first published in 1933 (Fig. 1.2). Diagrammatic maps have never ceased to be important, either to thinkers or to doers.

It is also possible to distinguish maps for private or personal use from those for public use, a differentiation which takes us away from the traditional focus on surviving eye-catching individual maps, and directs attention to the implications of cartographical ephemera such as the sketch maps that were scribbled on any available space, reaching us today on flyleaves, at the end of unrelated chapters, on the covers of books, and in the margins of texts (Fig. 1.3).[3] Particularly vulnerable artefacts like early pamphlet, news-sheet and wall maps, which would have been widely seen in their day, are grossly under-represented in archives and libraries and unless a conscious effort is made to take proper account of these poorly represented groups of maps, our perception of early map history is in danger of being skewed.

Maps are secretive artefacts at the best of times. Polyfunctional and polymorphic, they are impossible to classify in a way that pleases every researcher. A map can serve many purposes, contemporary and subsequent. By the time we ourselves are confronted with some of the earliest examples, many are freighted with secondary uses and re-uses. It is also difficult sometimes to judge original function simply from content. Maps can serve a similiar purpose but look different, either because techniques of production have changed or because the values affecting presentation are different at different times. At the same time the sobering fact remains that maps have always been, and are still, very much a minority form of expression.

English maps – maps made in England or specifically for English use – have received uneven scholarly treatment. Some have been much discussed in recent decades or have a literature which goes back to the nineteenth century or even to Richard Gough's reviews of English maps in his *British Topography* (1780). Others receive here what is perhaps their first mention in a book about maps. Likewise, certain periods and themes have hitherto received more attention than others in the literature of map history. We find some striking lacunae. Even the all-important (as it is perceived) eighteenth century awaits a serious, book-length, study, as do the nineteenth and twentieth centuries. A definitive study of English town maps is needed, as is one on county mapping, and others on maps of roads, of canals, of railways; as Brian Harley once remarked, 'the vineyard is large but the labourers few.'

A book devoted to the history of English maps is long overdue. Edward Lynam's *British Maps and Map-Makers* was published as long ago as 1944. With only 48 pages of text and 30 illustrations, there was little scope for Lynam to do more than note the salient facts and factors in an overview which dealt with the whole of the British Isles and touched on the British Empire. Of more recent studies, Brian Harley's path-breaking *Maps for the Local Historian: A Guide to the British Sources* and similarly-titled books by David Smith and Paul Hindle are concerned primarily with maps as sources from which to recover information about the past, particularly for writing local histories.[4] In this book, we are able not only to present more material, to report new discoveries and updated ideas, but also to penetrate deeper into explanations in our attempt to see English maps in their place in English culture. Our underpinning theme of maps and society is made up of a number of individual strands which are woven into the fabric of our text, and which appear and reappear as we pass from topic to topic and period to period: the relationship of map literacy to literacy in general, maps and consumerism, manuscript maps and printed maps and the continuing importance of manuscript maps throughout the print age, the important distinction between professional and non-professional map users, the duality of maps as highly specialist tools and as multi-faceted artefacts, their poly-functionality, the

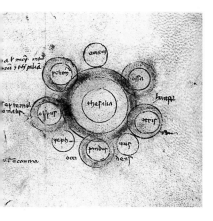

1.3 Site of the battle of Thessaly (48 BC). A cartographical gloss found in many medieval copies of Lucan's *Pharsalia* (*De bello civile*), Book VI, representing the Paulus plain and the surrounding mountains. Cardinal directions are indicated and east is at the top. Diameter 4.5 cm. Italian. Second half of the fourteenth century. Bodleian Library, MS Lat. Class.d.14, fol. 73r.
Reproduced by permission of the Bodleian Library, Oxford.

need to recognise each map's individuality and to identify specific map genres rather than just general types, the interplay between private initiative, national interest, and institutional responsibility, the importance of map style, maps as communicators of ideas as well as heuristic devices and as records of where places or people are.[5]

Close attention to the nature of medieval maps soon reveals the historical invalidity of the old neo-Darwinian model, according to which early maps are dismissed as crude and recent maps are praised because they represent the ground plan accurately from astronomical measurement. To those who would see little of relevance or importance to a history of cartography before the popularisation of scaled plan views in early modern times, we believe we have much to say. We would urge a look at what is there and some reflection on what was lost both in the Middle Ages and subsequently. We would say – do not be misled by visual style and do not heed the siren of progress and confuse change with improvement. If the past is a different country, we should not be surprised if things are done differently there, only reassured by the familiarity of certain concepts, not least those relating to cartographical representation. The basic factors of map-making do not alter much over time: surveying and production techniques, matters of style, size of output and the processes of marketing, are largely secondary to key questions such as what is mapped and why, and what for. Seen this way, what was once thought of as a Renaissance paradigm-shift in mapping turns out to have been in large part an acceleration of processes already in place.

The diversity which underpins the history of English maps from the Middle Ages onwards, and which is implicit in each chapter of this book, provides a challenging agenda. No less rewarding is the need to re-visit specific types of maps. Even seemingly well-known individual maps merit re-contemplation. In this book, we start (Chapter 2) with a survey of maps in medieval England which is at once broad-brush (to reveal the range of maps) and closely-focussed (to make some points of general applicability about the way maps from any period may be viewed). In Chapter 3, we take the story of the English printed topographical county map from its delayed beginnings in the sixteenth century through to its reshaping in the eighteenth and early nineteenth centuries. Here, for the first time in a book, we bring together regional maps of all scales to initiate an exploration of their social significance. In Chapter 4, recognising the fundamental unity of maps of landed property, we alter the lens to bring into close focus large-scale maps in which not county but field boundaries provide the formal structure, and examine land ownership mapping (estate maps), the mapping of changes in agrarian structure (enclosure maps), and maps used in the collection of tax (tithe maps). In Chapter 5, the lens is inverted to permit the first cartographical account of how people of differing professional status planned their travels and how they found their way. Finding that way-planning and way-finding are separate activities, and distinguishing routes from roads, we show that maps

had no place as land or sea travel aids in England until the late sixteenth century and the beginnings of the mass tourism with which
the former, especially, have come to be associated. In Chapter 6, a
similar story is told, inasmuch as maps of towns had their place in
the Middle Ages and after, but no strictly utilitarian role before
more recent times. We review maps of towns in a novel way,
grouping them according to the different needs they served and
recognising that many of these go back to medieval and early
modern times. In Chapter 7, we bring the story of English maps
into the modern period by drawing attention to two characteristics,
the over-riding authority of the map-making institution (the
Ordnance Survey and the Hydrographic Office of the Admiralty),
and the increasing use of maps as heuristic devices and as analytical
tools in almost every aspect of the life of the nation. Finally, in
Chapter 8 we again reflect on the fact that the maps we see from
the past are only the survivors, not all the maps there once were.
The spottiness of the historical evidence makes it doubly difficult to
take account of the full range of maps and map use at any one time
in the past. As ephemera (maps in news-sheets and political
pamphlets), maps produced expressly for the popular end of the
market are even rarer, in proportion to the numbers once in circulation, than some of the cartographical gems we hear most about.
They tend to be overlooked instead of being recognised as a specific
product, the cheap commercial map. Finally, we reflect on the
relationship between maps and books and on the possibility of
learning more about map history through the fuller records of
book history.

A question to end with – and one with which we might well
have started – is why any map is made at all. A highly developed
visual sense of expression is by no means a universal attribute, nor
does everybody easily translate mental image or real landscape into
plan. Many people are baffled rather than enlightened by a map.
One of the great paradoxes of map history is that, for all the
longevity of maps – from prehistoric times onwards – and their
variety and adaptability, maps have always been as they remain, a
minority form of graphic expression.[6] For us, it is the intimacy of
each map, and the particular configuration of the personalities and
circumstances behind its creation, that offers both challenge and
reward. The challenge of map history derives less from studying the
reductible techniques of map creation than in understanding the
human input and the motives, rational and irrational, of those who
created and used maps in the past. Maps are drawn on a blank surface, not with a blank mind. They represent points of view, not
simply a physical viewpoint.

A Medieval Flowering

Most of the maps which survive from the English Middle Ages come from a place of learning. A map in a book or kept in a library might be expected to be more successfully protected from the ravages of time than a map used in a manorial court or land agent's office or taken out into the field. Moreover, in the Middle Ages, scholarship and religion dominated culture and, consequently, map use. Religious belief underpinned every intellectual activity and every aspect of daily life. Religious works were read as literature, either privately or publicly (out loud) in secular as well as clerical society.[1] Institutions of teaching, learning and study were either part of religious establishments (monasteries and cathedrals) or, like the universities of Oxford and Cambridge, were staffed by monks, ecclesiastics and (from the 1230s) friars. Religion fired philosophers, natural scientists, biblical exegesists and chroniclers alike throughout the Middle Ages and well beyond. The ability to read was not confined to clerics and there were other types of literacy at Court and other forms of writing and reading for lawyers, merchants, civic authorities, manorial reeves and a miscellany of officials. The common skill was the ability to write and read to some degree, in English, French or Latin.

Using a pen to draw rather than to write involves a different kind of literacy, one that not all who can write can necessarily employ. Many medieval scholars had an observant eye for their surroundings and a lively curiosity about places but preferred to describe these in words, such as the sixth-century historian Gildas and later, in the twelfth century, men like William of Malmesbury, Geoffrey of Monmouth, and William Fitz Stephen. And words can be used successfully to convey quite complex abstract images; indeed, the function of the verbal *picturae* in medieval books seems to have been to help consolidate, summarise, and fix the import of a text in the reader's mind.[2] Confusingly for us today, such written geographical descriptions are sometimes called *mappaemundi*, as is, for example, Gervase of Canterbury's list of the constituent political and ecclesiastical units of Britain (*c.*1200).[3] But there *were* others who chose to draw and to map as well as to write. Towards the end of the seventh century, Bishop Adomnán incorporated architectural plans in his book on the places of the Holy Land (*De locis sanctis*). Soon afterwards, Bede studied Adomnán's plans, and copied them for his own commentary of about 702. Throughout the Middle Ages, in fact, there were many who drew maps for the books they wrote or as wall-paintings and other forms of display. There were readers, too, who sketched maps and diagrams in the margins of texts or on the covers of works they were reading to help establish the spatial relationships involved.[4]

Medieval scholars sometimes saw advantages in transferring the mapping habit to other, more prosaic, aspects of institutional life, such as the recording of property or the provision of a water supply. Why, though, we find far fewer practical maps from non-monastic contexts – from lay or crown estates, or from towns, for instance – in this period is less than entirely clear. It is important to bear in mind, when we wonder at the small number of surviving maps drawn for some local, practical purpose, how few people there were in England at the time. The total number of lords of the manor, lay or ecclesiastical, their land agents, and officers of the royal forests, for instance, would have comprised a tiny proportion of the three million or so inhabitants in England at the time of the Domesday Survey (1086). As, however, total population moved towards a peak unprecedented since Roman times (with perhaps four to five million people by 1300), as money became more important as a means of payment and recompense, as the economy became increasingly commercial and labour more diversified, and as record-keeping became more important, so maps began to be more widely used as practical tools.[5] In addition to a conceptual shift, a simple increase in the number and variety of potential users of maps needs to be taken into account in any explanation of the growth of practical, as opposed to scholarly, map use from the late Middle Ages onwards. First, though, we need to know what maps there were.

LOOKING AROUND FOR EARLY MAPS

A trawl of the archives yields richly. The haul is chronologically uneven and we shall never know for certain the extent to which the apparent dearth of maps between the late seventh and the twelfth centuries should be ascribed to actual rarity or to loss. What is clear is that while up to the mid-fourteenth century the overwhelming use of maps was in scholarship, from the middle of the fourteenth century we start to detect an increase in the numbers of maps made for practical purposes. The extent to which the different picture at the end of the Middle Ages really reflects a significant shift in the *proportion* of academic to practical maps is, however, another moot point. Even so, by the end of the period nearly every major map type familiar to us had a medieval equivalent. When we come to look more closely later in this book at the few missing types, we find that there were alternative, notably oral, ways of dealing with the same information or supplying the same need.

The Early Centuries.   There was a lively interest in late seventh- and early eighth-century England in biblical and cosmological matters. Books on cosmography were collected, read, written, and even used as currency. The Northumbrian monk Bede relates how a particularly finely-executed 'manuscript of the Cosmographers'

(*Cosmographiorum codici mirandi operis*), one of the many books Benedict Biscop, founder of the monastery at Jarrow, had procured on one of his visits to Rome, was exchanged in about 690 for the eight hides of land needed for a new monastery in Yorkshire.[6] Maps too have survived. The earliest known to us is the copy made at Jarrow of the plan in another of Benedict's acquisitions from Rome, Aurelius Cassiodorus's Codex Grandior, his copy of the Vulgate Bible, a manuscript already a hundred years old by the time it reached the Tyneside *scriptorium* in 678.[7] Cassiodorus's original manuscript and two of the three copies made at Jarrow are now lost but the third copy (the Codex Amiatinus) has survived and with it the cartographical work (albeit second-hand) of the Northumbrian monks. Spread over two large folios and almost certainly adhering closely to Cassiodorus's original, the illustration depicts a reconstruction in plan of Solomon's Temple in Jerusalem as described in the Bible (I Kings, chapters 6–7) (Plate 1).[8] The Temple is presented as a great rectangular space within which is set the Tabernacle. The perspective is 'bird's-eye', so that the outer colonnade is portrayed in profile along two sides and in plan along the other sides of the outer courtyard. All details are clearly depicted. The door to the Tabernacle is shown in profile and, within that inner sanctum, the Veil to the Holy of Holies is represented in plan by a lateral bar. Other key features, like the altar for burnt offerings, the candelabra, the Ark of the Covenant with its carrying poles, and the protective Cherubim are all rendered pictorially and in their relative spatial locations.

The vast majority of Christian and Jewish exegetical drawing during the Middle Ages (and later, in the Reformation) is accounted for by plans of the layout of the Temple and the Tabernacle and diagrams of their furnishings, and by maps relating to passages in the Bible describing the Exodus (Numbers, chapter 33), the Encampment in the Desert (Numbers, chapter 2), the Division of Canaan (Joshua, chapter 15) and the Restoration of Canaan (Ezekiel, chapter 40). Soon after the Codex Amiatinus had been completed, Bede made several references to it in his own writings. Although Bede had clearly studied Cassiodorus's plan in the process of composing his own commentaries on the Temple and the Tabernacle, he did not illustrate these himself, leaving it to contemporary and later scribes to copy or create other plans to illustrate his texts, as they did in the twelfth century at the Fenland abbey of Kirkham.[9]

The other extant late seventh-century plans come not from England but from the Scottish island of Iona. They were immediately recognised as an important contribution to biblical exegesis, however, and cannot be overlooked in the context of early English maps. The four plans came from the pen of Bishop Adomnán who, between 686 and 688, wrote a highly influential 'little book' (as Bede called it), another *De locis sanctis*, into which he incorporated Arculf's account of a visit to the Holy Land a few years earlier. Adomnán's book was based on notes taken while listening to the French bishop's narrations and may have been completed as early

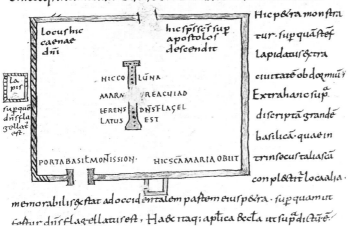

2.1 Adomnán's plan (*c*.685) of the church over the place of Ascension on Mount Sion, Jerusalem, showing the site of Christ's Flagellation. The four plans which illustrate Adomnán's *De locis sanctis* were drawn from first-hand measurements given by Bishop Arculf. All plans are introduced in the text. 4.5 × 5 cm. Ninth-century copy. Bildarchiv, Österreichische Nationalbibliothek, Vienna, Vindobonensis MS 458, fol. 17v.
Reproduced by permission of the Österreichische Nationalbibliothek.

as 686, although we hear of it only when it was presented to King Aldfrith in 692.[10] The plans represent the church at Jacob's Well, the church at the place of the Ascension, the great basilica on Mount Sion, and the complex of buildings on Mount Golgotha.[11] Everything is rendered in plan except for the column of Christ's Flagellation in the basilica (Fig. 2.1) and a chalice in a courtyard on Mount Golgotha (Fig. 2.2). The plans may not have been drawn to scale in the modern sense but they were drawn with regard to correct proportion. Adomnán relates that Arculf had taken measurements and had found, for example, the 'edge of the Sepulchre on the side [to be] about three palms', while the tomb in which the Lord's body was said to have been placed, as measured

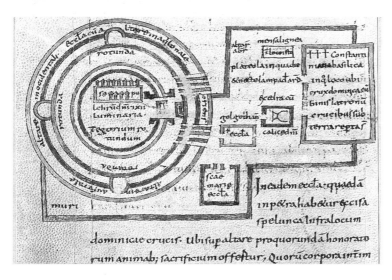

2.2 Admomnán's plan (*c*.685) of the Church of the Holy Sepulchre (the Round Church) in Jerusalem, as described and measured by Bishop Arculf. All Admomnán's plans were copied, with minor amendment, by Bede for his own *De locis sanctis* (702–3). 9.5 × 16 cm. Ninth-century copy. Bildarchiv, Österreichische Nationalbibliothek, Vienna, Vindobonensis MS 458, fol. 4v.
Reproduced by permission of the Österreichische Nationalbibliothek.

2.3 Plan of the New Jerusalem from Beatus de Liébana's Apocalypse (late eighth century), for Revelations 21. The twelve gates, one for each of the Twelve tribes of Israel, are represented topologically. Within the City, the Archangel wields a measuring rod. 12 × 12 cm. Fifteenth-century English copy. Trinity College, Cambridge, MS R. 16.2 (C.M.A. 524), fol. 25r.
Reproduced by permission of the Master and Fellows of Trinity College, Cambridge.

by Arculf 'with his hand, . . . [was] found . . . to be seven feet long'.[12] Bede made some adjustments to Adomnán's plans but no significant alterations, either to layout or presentation.[13]

The places these late seventh- and early eighth-century exegetical maps portray may have been geographically distant but they were certainly not irrelevant to those early medieval Northumbrian and Ionan monks. On the continent, the illustrated commentary on the Apocalypse (the Revelation of St John the Divine) by the late eighth-century Spanish monk Beatus of Liébana reflected a similar interest in the depiction of biblical places and was widely copied, in England as elsewhere.[14] The presentation of Beatus's representation of the New Jerusalem varies from copy to copy. It may be depicted from a high oblique angle or as if seen from the ground (in two extant thirteenth-century English copies), as a walled city, a group of buildings or, more rarely, as a single church.[15] Or it may be represented in full plan, stylised into a quadrilateral with the side views of the twelve gates, in correct topological relationship with the ground plan, arranged in threes along each side, as in a copy made in England (or possibly in northern France) about 1250 (Fig. 2.3).[16] Other manuscripts of the Apocalypse modelled their representation of the New Jerusalem on Adomnán's plan of the Round Church of the Holy Sepulchre in Jerusalem.

Mapping the whole world as opposed to a town or a few buildings required a different conceptual scale but that is not necessarily why the early eleventh-century Anglo-Saxon map of the world (see Fig. 2.22, p. 35) may be classed as a practical map rather than a purely academic map. Apart from its distant ancestry (possible a Roman geographical map), its presence in a computus manuscript associates it with the all-important practical task of deciding the date on which Easter was to fall each year – the function of such manuscripts.[17] The puzzle is why, between the clutch of late seventh- and early eighth-century maps just described and the start of the twelfth century, we can at present point only to the Anglo-Saxon map. Possibly others remain to be discovered.

The Twelfth Century.   From the turn into the twelfth century, the situation changes radically, either because many more new maps were being made in this period of medieval renaissance or because maps were starting to enjoy a better rate of survival. Of the dozen we know of from this century alone, all but four (all *mappaemundi*) are extant. We know of the lost maps through contemporary or later references. Gerald of Wales (Giraldus Cambrensis, 1146–1223) mentions something called a *Mappa Mundi* amongst books given by Hugh of Leicester in 1150 to Lincoln Cathedral.[18] Another *mappamundi* is said to have been left by Bishop Hugh le Puiset to Durham Cathedral Priory (1195).[19] For examples of surviving *mappaemundi* we have the frontispiece in a copy of Honorius Augustodunensis's *Imago Mundi*, thought to have been made at Durham for presentation to the west Yorkshire abbey of Sawley, and discussed again below (see Fig. 2.23, p. 36), and a larger map now preserved at Vercelli, Italy.[20] Both portray the world in a sub-circular form and both have east at the top. Only in the Sawley map is the Mediterranean sufficiently prominent to give, with the Danube (or Don) and the Nile, the characteristic 'T-O', tripartite, internal structure of the *mappaemundi* genre which is discussed further below.

As extant examples of practical maps from the twelfth century, we can cite the two relatively simple computus maps made in the Fenland monasteries of Thorney and Peterborough, and the ornate and highly detailed plan of the precinct of Christ Church, Canterbury, to which we also return below (see Plate 2). The Thorney computus map (1110) is the largest and most detailed of a number of diagrams in a large manuscript devoted to the calculation of Easter and related topics (Fig. 2.4).[21] Like many computus maps, it presents a circular world in abstract form, with east at the top and the traditional formalisation of seas and rivers of the T-O diagram. Thus, Asia occupies the upper part of the circular map and Europe and Africa are to the left and the right of the vertical bar respectively. On the Thorney map, the horizontal bar, which usually represents the rivers separating Asia from Europe and Africa, is labelled Jerusalem along its entire length. The centre of the map and, in this case, of the world, is marked by a cross. A series of wavy lines indicates the River Jordan and the box around the name of Mount Etna is shaped like a hill. Otherwise, the map contains only place-names.

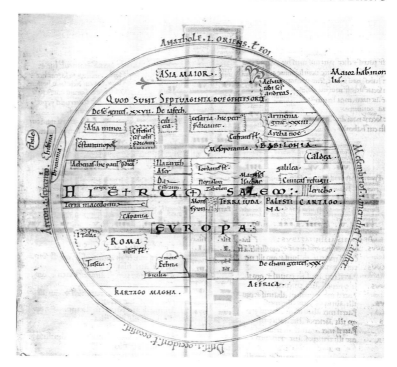

2.4 A map of the world from a computus manuscript (for the calculation of the date of Easter). The diagrammatic style echoes the zonal diagrams of Classical geographies and emphasises the peripheral location of *Britannia, Hibernia* and *Thule*. East is at the top. Diameter 17 cm. 1110, Thorney, Lincolnshire.

Reproduced by permission of the President and Fellows of St John's College, Oxford.

The Thirteenth Century. Most medieval maps cannot be closely dated or have insecure dates. Even so, it is difficult to deny the pronounced upswing from the beginning of the thirteenth century to the end of the Middle Ages (Fig. 2.5). In comparison with fewer than a score or so of maps from the entire 500 years between *c.*680 and 1200, we know of nearly 160 examples from the period 1200 to 1500, a highly notional (and possibly misleading) average of over 53 maps per century. From this point our account of the medieval map corpus of necessity becomes more selective; those maps we have chosen to review represent some well-known or important maps (chosen to serve as markers), maps for which new information is available, and a number of less familiar maps.[22]

Like those of the previous centuries, the maps produced in thirteenth-century England fall overwhelmingly into the 'thinkers' category, although three of the earliest represent the preoccupations of 'doers'. The latter include the rather stark diagram drawn *c.*1208 across half a page in a cartulary from Fineshead, Northamptonshire. This documents the distribution and dimensions of holdings of meadow and aspects of their tenancies (Fig. 2.6). No less spartan in structure is another cartulary diagram which shows a plot of land at Wormley, Hertfordshire, with the three springs which supplied Waltham Abbey, Essex, with fresh water. The lead pipes had been buried in 1220 and 1221, and the plan accompanies a detailed account of the works in the Abbey's

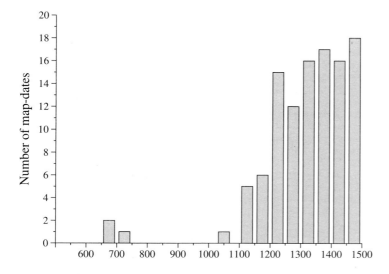

2.5 The chronological distribution of known and surviving English maps from 500 to 1500 presents a challenge. While admittedly fewer maps are likely to have been made in the first half of the period than after the renaissance of the twelfth century, the fact that only a single example is known from the four centuries between 750 and 1150 merits explanation.

cartulary, a context which indicates the importance of the document and the plan as a record for future reference.[24] Another utilitarian map, from Kirkstead Abbey, Lincolnshire, is probably a copy of an older exemplar made between 1224 and 1249 in connection with disputed boundaries.[25] The difficulty of fencing the seasonally flooded fens around the Wash in eastern England and the widespread practice of inter-commoning inevitably gave rise to disputes over land and rights. The problem at Kirkstead was to re-establish the boundary between Wildmore Fen, to which the abbey and its tenants had rights, and West Fen, which belonged to another manor. The map represents the area in question as a rectangle, across which runs an oblique line indicating the boundary, identified by the writing along it. There are no pictorial signs, only the names of places and vaccaries (seasonal dairy farms) and several squiggly lines linking the outlying vaccaries with Kirkstead

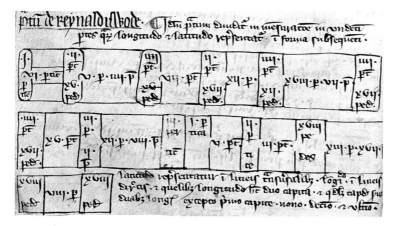

2.6 Map drawn in a cartulary showing how a certain meadow was 'divided by measure into eleven parts of which the length and breadth is represented in the following diagram . . . breadth being indicated by the horizontal lines, length by vertical lines'. The units are poles, perches and feet. 7.5 × 16 cm. Before 1208, Fineshead, Northamptonshire. Lambeth Palace Library, MS Court of Arches Ff.291, fol. 58v.
Reproduced by permission of Lambeth Palace Library.

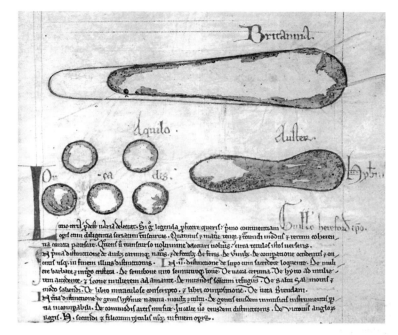

2.7 Map in Gerald of Wales's *Topographia Hiberniae* (*c*.1200), probably inserted by a scribe to show the relative positions of Britain (top), Ireland (right) and the Orkneys (left). 8 × 13 cm. Thirteenth-century copy, ?Lincoln. British Library, Add. MS 33,991, fol. 26r.
Reproduced by permission of the British Library Board.

on one side of the boundary and the barony of Bolingbroke on the other.

All other known thirteenth-century maps seem to have had a didactic or exegetic function. Of the nine or ten *mappaemundi*, at least three existed in the 1230s painted or displayed on walls, one in the royal palace of Westminster, another in the palace at Winchester, and a third in the monastery at Waltham. All three are now lost and only a fragment survives of the small *mappamundi* Matthew Paris made, he tells us, after studying 'the world-maps of Master Robert de Melkeley and Waltham'.[26] Paris went on to add that he had made another, more accurate, copy in his Ordinal, this time of the 'king's world-map, which is in his chamber at Westminster'.

There were many other maps besides *mappaemundi* in thirteenth-century England. The sharp-eyed topographer Gerald of Wales is well-known for his interest in geography, although whether he drew maps for himself or only inspired others to produce them is unclear. As a teacher, he clearly appreciated the importance of cartographical communication. The three sketch maps of the British Isles in the two copies of his description of Ireland now in the British Library may have been added to the text by later scribes (Fig. 2.7), but the more complicated map bound in the Dublin codex, between the 'Topographia Hiberniae' and the 'Expungnatio Hibernica', which shows Europe down to the Mediterranean and as far as the Black Sea, is thought to have had 'especially intimate ties with Giraldus' and to have owed much to his 'central role in its creation' (Fig. 2.8).[27] According to T.

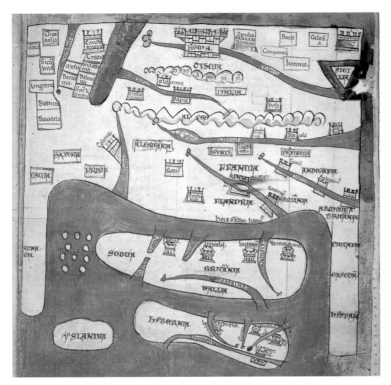

2.8 Map of Europe inserted between Gerald of Wales's two works on the topography of Ireland. The larger indentation, top left, represents the Adriatic (with Constantinople on the left); the prominent city, top centre, is Rome; the city and the triangular island, top right, are Reggio Calabria and Sicily respectively. Within England, York, Lincoln, London and Winchester are shown, together with the Thames and Severn rivers. Approx. 27 × 18 cm. c.1200, ?Lincoln. National Library of Ireland, MS 700. Reproduced by permission of the National Library of Ireland.

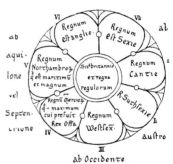

O'Loughlin, the map concerns relationships between Britain, Ireland, France and Italy, and the logic of its structure lies in the itineraries followed by those travelling to Rome from Ireland and Britain.

Itineraries were also much used by Matthew Paris in his cartographical work. Paris, a Benedictine monk from St Albans, Hertfordshire, has left us at least fifteen maps on seven subjects: two different maps of Palestine, a schematic representation of the seven kingdoms of Anglo-Saxon England (Fig. 2.9), a diagram of the roads of ancient Britain (Fig. 2.10), the fragment of the *mappamundi* mentioned above, a graphic itinerary from London to Rome, extant in four versions (see Figs. 5.5, 5.6 and 6.2; pp. 150, 151, and 182); and a map of Britain, also in four versions (see Figs. 2.29 and 2.30, pp. 44 and 45).[28] By all standards, Paris was as outstanding a map-maker as he was prolific. His duties at St Albans centred on the maintenance of the monastery's chronicle, a task which demanded that he be as well-informed on contemporary matters (hence his concern with Italy) as on English history. Duties apart, though, we get the impression that Matthew Paris was simply a remarkable person, about whose personality we should like to know more. We return to his maps of Britain later in this chapter and to his unusual rendering of an itinerary in Chapter 5.

2.9 Matthew Paris's diagram of the Anglian heptarchy shows the seven kingdoms of England, numbered in geographical order from the southeast: Kent, Sussex, Wessex, Mercia, Northumbria, East Anglia and Suffolk. Cardinal directions are given. East is at the top. Diameter 9 cm. c.1250, St Albans. As redrawn by Konrad Miller, *Mappaemundi. Die ältesten Weltkarten* (Stuttgart, J. Roth, 1895–8), vol. 3, p. 83. The original is in the British Library, Cotton MS Claudius D.VI, now fol. 10v. Reproduced by permission of the British Library Board.

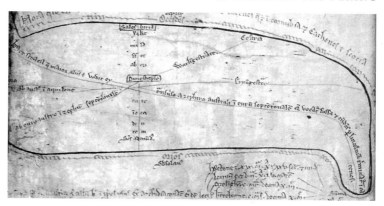

2.10 Matthew Paris's diagram of the four pre-Roman paved roads built, as related by Geoffrey of Monmouth (History of the Kings of England, c.1136), by King Belinus. Paris has given them Roman names although they do not correspond to the Roman roads: Fosse Way, Icknield Way, Watling Street, Ermine Street. West is at the top. 11 × 23 cm. c.1250, St Albans. British Library, Cotton MS Nero, D.I, fol. 187v.
Reproduced by permission of the British Library Board.

2.11 Outline of the Hereford mappamundi, to show the main geographical features of the map and the position of the marginal scenes. Dimensions of parchment 158 × 133 cm. 1290s, ?Lincoln. Hereford Cathedral.
Drawing by A. Scafi.

Less than ten years after Matthew Paris's death, the Franciscan Roger Bacon – teacher, commentator, and scholar in Paris and Oxford – produced three major works, one of which led to a now-lost map of the world (c.1268).[29] Bacon's map undoubtedly would have struck an observer as something significantly different from the usual *mappamundi* – assuming, that is, that he had structured it in the way he advocated in his book, namely, according to astronomically-defined coordinates. In contrast, what is probably the best-known medieval map, the Hereford *mappamundi* ascribed to Richard of Haldingham, is wholly traditional in concept and content. It was made, probably in Lincoln rather than Hereford, in the 1290s (or about 1300) from a variety of sources including, for the British Isles at least, considerable first-hand or local knowledge.[30] It is a large object, a single sheet of vellum in its natural pentagular shape filled to the patterned edges with a circular world and a number of flanking drawings (Fig. 2.11). East is at the top. The Holy Land takes up the centre of the map and the island-studded Mediterranean stretches prominently to the west, towards the bottom of the map. Equally prominent in a northward direction is the Black Sea which defines the northern side of the square bulk of Anatolia. The Nile, though, is shown only as one blue-coloured river system amongst all the others which, together with the towns and cities, and related inscriptions, make up the salient geographical features of the map. In addition, the map is crowded with drawings representing human history and the marvels of the natural world. At the apex, outside the frame and above the portrayal of the Earthly Paradise with the four rivers of Eden, Christ sits in majesty at the Day of Judgement. Within the two bands forming the circle around the world, gold letters spell out the word *MORS*, a reminder that man and his world are destined to pass away.

A work as complex and profound as the Hereford *mappamundi* must have taken time to produce. It certainly involved earlier drafts and the study of other maps and many texts. Other extant *mappaemundi* also testify to now-lost prototypes. The tiny, jewel-coloured Psalter *mappamundi* (c.1269), for example, with its prominently red-painted Red Sea near the top (showing the crossing of the

Exodus), is thought to be a reduced copy of a very much larger original.[31]

As has become evident from what we have seen so far, biblical exegesis accounts for by far and away the largest single genre of early medieval maps. From the thirteenth century, we can point to the maps of Canaan glossed into a student's bible.[32] The maps appear twice, close to the text to which they apply and again on a larger scale on blank folios at the end of the book. They belong to an important tradition of northern French illustrated biblical commentary which owed much to the Jewish exegete Rashi (1030/40–1105) and the Christian monks at St Victor, Paris, notably the Welshman, Richard (died 1173), and later, in the early fourteenth century, to Nicholas of Lyra. All these commentaries were avidly studied and copied in England. One of the sketch maps in the Oxford bible concerns the division of Canaan as described by Joshua. The other (Fig. 2.12) relates to Canaan Restored as described by Ezekiel and seems to have been based on Richard of St Victor's map of the same subject.[33]

The Fourteenth Century. It may be that traditional medieval map-making had reached an apogée in thirteenth-century England, at least as regards the rather specialist fields of exegetical mapping and *mappaemundi*. The nature of many fourteenth-century survivals suggests that the exuberant creativity of earlier centuries was replaced by a tendency to copy and elaborate older models. *Mappaemundi* continued to be produced – a number were made for copies of Ranulf Higden's 'universal history', the *Polychronicon* (*c.*1342), and there was the Aslake map made at Creake Abbey, Norfolk in about 1350 – but the remaining ten or so fourteenth-century Higden maps are predominantly derivative.[34] In other contexts, imagination rather than biblical or historical veracity (as we would define it) was allowed to run free when a map was needed as 'proof' of the legitimacy of a territorial claim. The late fourteenth- or early fifteenth-century map of Thanet, Kent is a case in point. One of two illustrations in a chronicle of St Augustine's Abbey, Canterbury, it was made by Thomas of Elmham. The map shows the whole of Thanet, with sea all around except at the top (east) where map and text run into each other (Fig. 2.13). Thirteen large pictorial signs represent the region's ecclesiastical establishments in profile, linked by thin lines representing, presumably, roads. Apart from a clump of trees and some small drawings of beacons and wayside crosses, a deer, and a group of boatmen on the river Wantsum, there is little else on this uncrowded map except an irregular line, coloured green, which runs across the middle of the map from north to south. The line represents the course of the running deer which, according to legend, had originally defined the territory allocated to Minster.[35] Although the Thanet map defines an estate, its purpose was to represent a historical event, perhaps against the eventuality of a challenge to the abbey's claim to its land.

2.12 Maps in biblical exegesis. Diagram of the division of the Land of Canaan, restored to the Tribes of Israelites on their return from exile, as envisioned by the prophet Ezekiel, to illustrate the commentary on Ezekiel, Chapter 45. The strips allotted to each tribe surround land set aside for the Temple and the priestly Levites. East is at the top and the Mediterranean Sea shown by wavy lines at the bottom. 75 × 65 cm. Bible *c.*1230; gloss *c.*1250, Oxford. Bodleian Library, Oxford. MS Bodleian 459, fol. 37v.
Reproduced by permission of the Bodleian Library, Oxford.

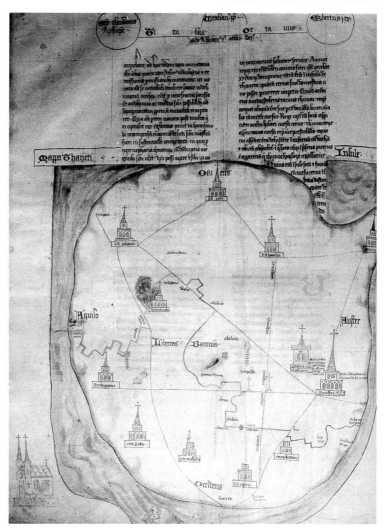

2.13 Thomas Elmham's map of the Isle of Thanet, Kent, in a chronicle from St Augustine's Abbey, Canterbury. The map illustrates a reputed historical incident, the running deer whose erratic course across the island from north to south defined the boundary of the manor of Minster. The line of the *cursus cerve* is coloured green. East is at the top. Approx. 39 × 38.5 cm. Before 1414, Canterbury. Trinity College, Cambridge, MS 1, fol. 42v. Reproduced by permission of the Master and Fellows of Trinity Hall, Cambridge.

From the maps produced in the second half of the fourteenth century we get the impression of a growing tendency to use maps in practical ways, and in secular as well as monastic and ecclesiastical contexts. These later fourteenth-century maps are on all scales, from the national to the local. The Gough map, as it came to be called from its later owner, the eighteenth-century topographer Richard Gough, shows the whole of Britain (see Figs. 2.31 and 2.32, p. 46). Dating from about 1360, it portrays not only the main topographical features of the country, with major rivers and nearly 620 settlements, but also a network of red lines. These link a number of places and are accompanied by figures giving distances. The function of these lines is not immediately clear, beyond the fact that they served a probably secular administrative function on a national scale. We return to the Gough map later in this chapter.

An entirely different type of map, to which we also return below (see Fig. 2.20, p. 29), shows nothing more than the simple outlines of a new kitchen at Winchester College and the stairs leading to it. Equally simple in both style and content is the plan from a cartulary belonging to Blackborough Priory, Norfolk. This shows a small group of domestic plots, or messuages, and 'has the air of a surveyor's or steward's memorandum, set down to clarify or record the ownership of intermingled plots' in the village of Clenchwarton.[36] Quite different in visual impact is a map of Sherwood Forest, Nottinghamshire, on which words are far more eye-catching than the few pictorial signs. The Sherwood Forest map is large (79 × 59 cm) and occupies a single sheet of vellum. East is at the top. The map, probably made for a Forest Warden early in the century, represents features within the Forest as defined in 1232 and nothing outside the boundary.[37] It is particularly interesting, as Maurice Barley has shown, for what can be learnt of the cartographer's working methods. Some of the initial pencil lines are faintly discernible. These, it would appear, were systematically inked over, starting with the outer frame. Using red ink, the map-maker ruled double lines and drew a double circle in each corner. Having marked the mid-point along each side, he used these to divide the internal space into quarters which were subdivided – again with double red lines – into eighths. The cardinal points were inscribed within double red circles over each of the mid-points and the rest of the red ink work was completed by the addition of rivers and streams. The map-maker then turned to brown ink for the realistic portrayals of the palings around Clipstone and Bestwood Parks. Finally, the map was filled with the names of some 300 settlements, roads, tracks and natural features.

While the actual use of the Sherwood Forest map can be only surmised, there is no doubt about the role of a series of illustrations in a processional, in the Old Salisbury (Sarum) rite, from Norwich.[38] The eight coloured drawings are interleaved with music for the processional chants and words for the prayers for each of the major ceremonies of the ecclesiastical year and show exactly where each celebrant was to stand once he had reached the altar. Most of the scene in each case – the altar and altar steps, and flanking pillars – are shown in profile. Other items, such as the chalice on the altar, the palms for the Palm Sunday mass, and the bishop's cope, are also in profile. Deacons and priests, however, are indicated by a plan view of their tonsured heads (Plate 3).

The Fifteenth Century. Our impression of the increasing heterogeneity in subject matter and secularisation of map-using contexts gains strength as we proceed through the fifteenth century. Only 50 per cent of the extant maps from this century are the maps of thinkers rather than doers. Amongst the former are the maps in Nicholas of Lyra's lengthy exegesis, the *Postillae*. This important

2.14 *Totius Britanniae tabula chorographica*. One of three full-page illustrations in an early fifteenth-century copy of Geoffrey of Monmouth's History of the Kings of Britain (see Fig. 6.1, p. 180). Some features of the map suggest an affinity with the Gough map (*c.*1360); the way Ely and the Isle of Anxholme are encircled by rivers, for example. South is at the top. 32 × 15.5 cm. *c.*1400. British Library, MS Harley 1808, fol. 9v.
Reproduced by permission of the British Library Board.

and extraordinarily popular work was written in northern France and took nearly a decade to complete (1323–32). From then on it was frequently copied, in England no less than on the continent, and in due course was printed, although not always with the illustrations. With its full complement of the author's (and three later) figures, the *Postillae* contain 15 architectural plans and three maps among a total of 42 illustrations (Fig. 1.1, p. 2). *Mappaemundi* also continued to be in demand for Ranulf Higden's scarcely less popular history of the world, the *Polychronicon*, of which seven copies are extant from this century alone. In 1462 a *mappamundi* of some sort was purchased, at the considerable cost of £5, for the library at New College, Oxford. However, only one entirely new *mappamundi* is known to have been created in the fifteenth century; the map from Evesham Abbey (about 1400) seems to mark the end of the peculiarly English tradition of *mappaemundi*.[40]

Also classified amongst the non-practical maps of the fifteenth century is the symbolical representation of the city of York included as an illustration to Geoffrey of Monmouth's History of the Kings of Britain (see Fig. 6.1, p. 180), and a map of 'the whole of England' (*Totius Britanniae tabula chorographica*) in the same manuscript. Both are heavily painted, the latter with the seas in a solid black wash against which the island of Britain with its variety of castle, church and city signs stands out clearly (Fig. 2.14). In the plan of York, the main feature is the circular wall, with battlements and eight towers (four looking like gateways) which surround the city, within which is a large church and through which runs a river spanned by a stone bridge. Two imaginary scenes indicate that the map was intended as a commemoration of the role York played in the conversion of Britain to Christianity. The map of Britain has south at the top although, like the Gough map, it is the outline of the eastern side of the country that is the most realistic, either because the scribe started filling in details on this side on too large a scale before he realised he had insufficient space on the page for the rest of the country, which then had to be squeezed in, or simply because he was mainly interested in eastern England.[41] In an earlier manuscript of Geoffrey's work, meticulously executed pencil sketches were added in the early fourteenth century to the margins of ten pages wherever the text refers to the founding of an English or Welsh town. In most cases, the building or buildings representing each town are portrayed in profile, but the walled city of London with the adjacent [White] Tower, is shown from a high oblique viewpoint which allows us to see the church and surrounding houses within the crenellated city walls.[42]

While these urban representations can all be said to reinforce the text they accompany, it is difficult to see in the mid fifteenth-century map of the village of Boarstall, Buckinghamshire, any function other than pure embellishment, in this case of a cartulary.[43] The map is lively and evocative. The layout of the village and its fields is clearly set out in plan, with south at the top, and trees, hedges and all buildings are depicted pictorially. The manor house with its large gateway is detailed, as is a church and a dozen houses, most of them forming two rows facing each other across the main street. The three main fields, their arable use indicated by broad lines, are surrounded by bristling hedges, as are a number of grassy closes. A large deer kneels in the woods beyond the village, and at the bottom of the map is a scene in which a kneeling man offers the king and his companions a freshly-severed boar's head. The map evidently serves as a record of a past event, the deliverance of the village from a troublesome boar. Similarly, the main

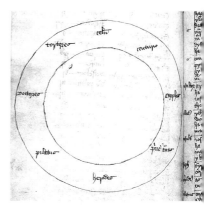

2.15   Stage plan for the Cornish play 'The Passion of our Lord Jesus Christ'. The 'station' of each of the principal characters is identified, clockwise from the top: Heaven, the centurion, Caiphas, the High Priest Ananias, Herod, Pilate, the Elders, and the Executioners. The action would have taken place in an open-air arena, like the surviving Round at St Just. Diameter 17 cm. Late fourteenth or early fifteenth century. Bodleian Library. MS Bodleian 791, fol. 56v.
Reproduced by permission of the Bodleian Library, Oxford.

function of the detailed bird's-eye view of Bristol (see Plate 14), made at the end of the century (*c*.1480) for a chronicle written by the town clerk, Robert Ricart, seems to have been no more utilitarian than a celebration of civic pride. It merits further discussion, however, and we return to it in Chapter 6.

Other fifteenth-century maps had distinctly practical uses, sometimes in unexpected contexts, like the stage plans which have come down to us from this period. The earliest are those in the 'Cornish Ordinalia', a manuscript containing three plays whose oral origins must surely go back very much earlier.[44] Unlike the more familiar mystery plays which were performed inside a church or in its precincts, the Cornish plays demanded the use of 'rounds', circular arenas outside town. Each of the three plans comprises two concentric circles between which are written the names of the eight principal characters. These plans are a guide to who stands at each 'station' on one of the scaffolds used for the action. The audience sit or stand in the centre (Fig. 2.15). A fourth stage plan comes from the eastern side of the country.[45] The play in question, the 'Castle of Perseverance', dates from *c*.1440. It also involves a circular performance area (Fig. 2.16). As explained by the text written in the space between the double circle of the plan, the two circles represent the ditch that defines the 'place' or acting area. Pieces of scenery may be represented by the drawings of a stone-built crenellated tower with, below it, a bench representing 'Mankinds Bed', and by the words 'Cupboard of Covetousness'. Outside the circles, five scaffolds, one each for the main characters, are also described in writing, and are given a compass-location. In these stage plans, mapping the places where the main characters stand parallels the illustrations in the Norwich processional. In both cases, the instructions to organisers (and perhaps participants) of ecclesiastical ceremony and secular performance had the practical function of an *aide-mémoire*.

A political intention lay behind the production of the Englishman John Harding's mid fifteenth-century map of Scotland.[46] As part of his attempt to persuade the English king to invade Scotland, Harding portrayed Scotland on his west-orientated map as enticingly fertile and well-populated. A close-packed display of boldly-styled pictorial place-signs (single towers, walled towns with great gateways, large churches, the high walls of Edinburgh) and wide rivers present a lively landscape. Thomas Chaundler's depiction of Wells, one of four full-page illustrations in a text from New College, Oxford, celebrating the good works of Bishop William of Wykeham, is gentler both in intention and (through the use of blue rather than red colours) visual effect (see Fig. 6.3, p. 184). Other maps of the later fifteenth century relate to some aspect of property management, in towns as well as in the countryside. A map showing the whole length of Tursdale Beck (Durham, *c*.1430) on two strips of vellum glued together served as 'evidence' in a legal dispute.[47] The map of southern Dartmoor and surrounding hamlets, on which the area subject to Forest Law is outlined schematically by a circle, served in conjunction with

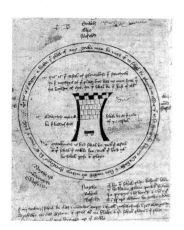

2.16 Stage plan from the earliest extant manuscript of 'The Castle of Perseverance'. The double concentric circle, probably marked out on the ground by a water-filled ditch or wooden palisade, defines the 'place', or stage. The crenellated castle contains 'Mankind's Bed'. South is at the top. 21 × 14.5 cm. *c*.1440. Folger Shakespeare Library, Washington, D.C., V.a.354, fol. 191r.
Reproduced by permission of the Folger Shakespeare Library.

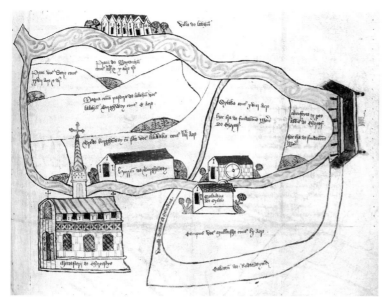

2.17 Cartulary plan of the 'island called Burway', some 400 acres of meadow and pasture lying between two branches of the River Thames in Surrey, the use of which was disputed by Chertsey Abbey and three of the abbey's tenants. Landscape features are represented in three different ways: pictures alone, words alone, or pictures and words together. East is at the top. 26 × 33.5 cm. Mid or late fifteenth century, Chertsey Abbey, Surrey. Public Record Office, E 164/25, fol. 222r.
Reproduced by permission of the Keeper of Public Records.

charters to corroborate the rights of a certain class of tenant within the Forest.[48] In the case of the individual pieces of land belonging to Chertsey Abbey, Surrey, (Fig. 2.17), Witton Gilbert, County Durham, and Staines, Middlesex, the spur to getting the land mapped was either actual or anticipated litigation.[49] A bold artistic style and the use of bright colours gives the Chertsey map a particularly striking appearance. Although land use is noted in words (meadow, common pasture, arable), the rivers, vegetation, field boundaries, bridges and buildings are depicted pictorially. Writing confirms the identity of the lane (*venella*), track (*calcetum*), the monastery of Chertsey itself with its mill and barn, and the nearby village of Laleham, presumably all relevant to the case. The Witton Gilbert map is quite different. The legal situation along the boundary between two estates, both of which belonged to the cathedral priory at Durham, had been of sufficient concern for the original (now-lost) map to have been copied sometime in the middle of the century into the cartulary as an important record. From a purely artistic point of view, the sketch is unimpressive, with a few ink lines and words defining the relevant roads, groups of *parcelles* (here called *seliones*), and properties in question. At Staines, the dispute was between the abbot of Westminster and the new tenant of a garden across which a previous tenant had dug a ditch to a mill, shown on the map by a spiked wheel together with the various ditches, tributaries of the Colne, and plots with the names of the holders/owners.[50]

There were always problems over boundaries and rights in the Fenland of eastern England. The map of Pinchbeck Fen, Lincolnshire (recently re-dated to *c*.1450) was made at Spalding

2.18 The 'Small' map of Inclesmoor, Yorkshire, may be a copy of the earlier, more elaborate and coloured but in all other salient respects identical, 'Large' map of Inclesmoor. It was drawn in a cartulary to form a permanent record of land-use arrangements agreed in the early thirteenth century, when the moor was drained for agriculture, but now disputed. North is at the top. 380 × 538 cm. *c.*1407, Duchy of Lancaster archives. Public Record Office, DL 42/12, fols 29v–30r.
Reproduced by permission of the Keeper of Public Records.

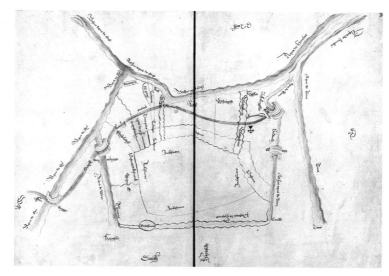

Priory to record grazing rights that had been in dispute at least since the thirteenth century, when the land belonged to the dukes of Lancaster. As usual with such maps, the selection of features shown relate closely to the matter in hand. Thus, on the map of Pinchbeck Fen, we see not only the marshy area of the fen and the local churches (drawn from life) but also the two stone crosses which served as boundary markers on opposite sides of the fen. One of the crosses was broken and its head is shown lying in the river.[51] A situation similar to that on Dartmoor and at Pinchbeck lies at the origin of a pair of maps from Inclesmoor, Yorkshire.[52] The maps, one large (63 × 87 cm) and one small (38 × 53 cm), depict the same 'island' of fen. They differ mainly as regards style. The reason for the small plain map was the need felt sometime about 1420 to ensure the map would always be available for future reference. Since the best place for a map is usually in a book, the essential information recorded in the original was copied at a reduced scale to fit on to two pages of the cartulary as a permanent, graphic, record of rights granted some 200 years earlier to the local inhabitants to graze their animals and to cut peat on the moor (Fig. 2.18). The original map is visually arresting. The rivers surrounding the moor (Trent, Don, and Aire) which flow into the Humber are boldly-rendered and coloured blue, with swirls and curls where the currents mix. Where the smaller map has words for parks and pasture, the maker of the older map painted the grass, outlined the palings, drew trees and distinguished tiled from thatched buildings in the villages.

After the liveliness of these fen maps, maps showing cultivated land in openfield parts of the country are prosaic affairs. Most show only a group of furlongs or bundles of *parcelles*. The two maps from Shouldham, Norfolk (1440–41), and the three from Tanworth in Arden, Warwickshire (1497) are all homely pen-and-

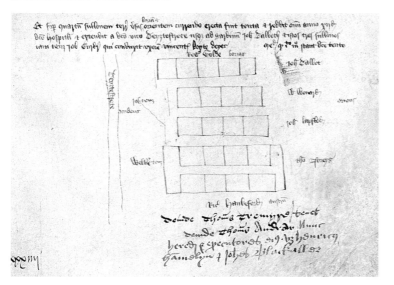

2.19 Tenter frames on an urban plot in Rack Street (Teyghstreet), Exeter. The frames were used for drying and stretching wollen cloth by tenants, whose names are given beside each rack. North is at the top. Approx. 9 × 13 cm. *c.*1420, cartulary of St John Baptist Hospital, Exeter. Devon Record Office (East Devon Area), Exeter City archives, Book 53A, fol. 34r.
Reproduced by permission of Devon Record Office.

ink sketches, like the Witton Gilbert and the earlier Clenchwarton maps.[53] For Shouldham, ruled lines define the plots or strips in question (see Fig. 4.1, p. 113), but John Archer, who was mapping his own land at Tanworth, drew freehand. Apart from names, these maps show nothing else; drawn for a specific purpose, they doubtless served it well. Similar maps were drawn for properties in the vicinity of towns. From the London area we have maps of house-plots, gardens and strips of agricultural land. For the map of the water supply of Charterhouse, on the borders of Clerkenwell and Islington, which measures over 200 × 50 cm, four pieces of vellum stitched together were needed.[54] The problem of obtaining a supply of clean drinking water for the monastery was particularly acute at Charterhouse, for the site given to the new foundation had been a plague burial pit. The map shows the solution: a pipe from a pure spring over a mile away was permitted to run across intervening properties to the cloister, whence water was distributed to each cell by lateral pipes. The plan is coloured with red, blue, dark and light green, and various shades of grey paint. The well, the pipe and its openings, and gutters are shown in detail. Vegetation surrounds the main well and hedges outline intervening properties. In contrast, the plan of a plot at Deptford, south of the Thames, and three plans of properties at Lambeth are down-to-earth in their simplicity.[55] In this respect they resemble the Shouldham and other plans of openfield property, as does the map of a plot of land at Exeter with five tenter-frames (racks on which fulled woollen cloth was stretched and dried) (Fig. 2.19).[56] The Exeter map is in a cartulary compiled about 1420 for St John's Hospital, a house for the poor. Its purpose was to record a particularly complex sequence of tenancies.

Even briefly summarised as in this section, 'looking around for early maps', the range of maps produced at different times in the

Middle Ages cannot fail to impress. We find maps of town and country, of the world as a whole, of places far away and at home, of historical places, of places observed and places imagined. The maps come from books of all sorts, chronicles and histories, exegetic commentaries, cartularies. Others were made as separate maps, to be kept rolled up or mounted or painted on a wall. A number are medieval copies of now-lost originals, or were derived from prototypes we know nothing about. Before we pursue some of the themes they raise, we have to ask: where did the medieval scholars, scribes and artists of medieval England get their ideas about maps from?

LINKING UP WITH THE ROMAN PAST

Two major areas of debate centre on the extent to which scholars in England owed their skills in spatial representation to the Romans and likewise how much of a medieval mapping tradition underlaid that of the early sixteenth century. Both raise well-worn issues of cultural continuity and the mechanisms of innovation. In both cases, available evidence flies, like common sense, in the face of earlier views of a Romano-medieval rupture or of the total novelty of the Renaissance. In fact, both medieval and Renaissance cartographical 'emergences' were deeply rooted in the preceding period.

As far as the English early Middle Ages are concerned, attention is inevitably directed to the political instability which followed the withdrawal of the Roman armies from Britain in 410 and the restless centuries of Anglo-Saxon, Viking and Danish raid, invasion and rule. At first sight, the lacunae in the cartographical evidence are seemingly insurmountable. We do not know of a single example, extant or anecdotal, of a map made in England in the Roman period, and all that we do hear about for the period after the Romans – the repeated destruction of monasteries and cathedrals, the pillaging of objects of value, and the looting and burning of libraries – would seem to put all possibility of the survival of such fragile artefacts as maps beyond discussion. At the same time, however, it is clear that steps were immediately taken to make good each important library loss, and to acquire other exemplars, imported often from Italy where libraries were being seriously neglected. We cannot ignore the extent to which classical scholarship was transmitted within Europe from Imperial Rome and late Antiquity to the Middle Ages (and on to the Renaissance).[57] The mechanism was simple; as one original manuscript of Roman date perished, so first a surviving exemplar and then, over time, some copy or other, was circulated from *scriptorium* to *scriptorium* within the British Isles or between Britain and continental Europe for copying as a replacement or as an addition to a library. Whatever the errors or modifications of transmission, we can see from texts that there was continuity of learning from Roman to medieval times. We need not be surprised, then, if some maps also appear to bear traces of a Roman inheritance.

The case for continuity does not have to apply to every type of map. In general, the Romans made considerable use of a range of maps, were excellent rural and urban surveyors, and were well accustomed to measurement in the field and to drawing to scale.[58] Although next-to-nothing is known about Roman map-making specifically in England, the British Isles are included in the itineraries of the period, including the graphic compilation known as the Peutinger map, and in the *Notitia dignitatum*, a Late Empire (395–413) list of civilian and military office-holders accompanied by maps of places mentioned. It can also be supposed that surveyors were involved in large numbers throughout Roman Britain in laying out towns and roads, siting forts, cutting canals and drainage channels and possibly in *centuriation* (government allocation of agricultural land). Perhaps something of the knowledge of these practices, if not of their products, endured beyond the end of the Roman Empire.[59] It is, though, difficult to recover even circumstantial evidence for continuity in the mapping of property, whether rural or urban, not only specifically in England but throughout the former Roman Empire.[60] Medieval rulers seem not to have followed the example of Imperial Rome, whose administrators used maps of properties as a tool for controlling the often far-flung imperial estates, for regulating land grants, and for collecting land revenue.

When we look at what we know of the chronological pattern of English medieval maps (see Fig. 2.5, p. 14), as many questions are posed as are answered. The paucity of extant maps up to the twelfth century is tantalising, particularly as we know that some scholars were accustomed to drawing maps and plans as a matter of course in their attempts to portray sometimes highly abstract intellectual problems, often introducing them in the text with an appropriate phrase. Certainly by the late fourteenth century, or at the latest by the early fifteenth century, the practical use of maps was diffusing into society at large, reaching the growing urban and merchant classes as well as estate owners and their stewards. We must suppose that the use of maps in certain – but by no means all – practical contexts reflected the role of predominantly social factors, such as the spread of literacy, book learning and the growing interest in, and acquisition of, material possessions.[61] At our present state of knowledge, though, it is difficult to be sure about the relative importance of one type of map to another. *Mappaemundi*, for example, may feature prominently in the modern literature on English medieval maps but whether they dominated contemporary output to a similar extent can be questioned. Nor can much be said about the geographical distribution of medieval map-making, and how this changed over the centuries. We can suggest a place of origin for less than half the maps we know about. Within the known distribution of medieval map-making – mostly south-east of a line running from the head of the Severn Estuary to the north of the Wash in reflection of contemporary population distribution –

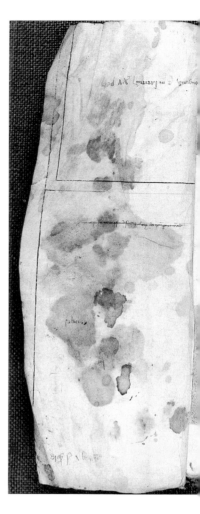

local concentrations (such as around London and in the Fenland) coincide again with regions of high population density, economic wealth, and the location of Benedictine houses and *scriptoria*.

MAP STYLE

Like maps from all periods, English medieval maps vary considerably not only in what they show but also in the way they show it. One variable concerns the angle of representation. In the case of topographical maps, landscapes are portrayed in plan; or by a mixture of plan (for rivers, roads, field boundaries, property boundaries) and side or oblique views (for hills, buildings, trees, gates etc.); or, in a few cases, entirely from an oblique (bird's-eye) viewpoint. Another variable is the degree of visual schematisation involved. Characteristic of exegetical and other didactic maps is the sparseness of content and the linear severity which results from a high degree of selective emphasis and abstraction. The simplest-looking diagrammatic representation may thus imply a greater degree of intellectual sophistication than does the essentially mimetic topographical map.[62] Style also reflects intended readership. Pictorial map signs – features portrayed from the side or from an oblique viewpoint – are easier to recognise, and lend themselves to a less cartographically sophisticated readership, than abstract or highly stylised signs. Pictorial representation can also be highly decorative.

Another variable is the proportion of drawing to writing. Some medieval maps are composed almost entirely of words (place-names, and narrative or explanatory notes), like the map of Sherwood Forest (*c.*1400) and Wildmore Fen (1224).[63] At the other extreme, some maps carry so little written information that we must suppose explanation to have been oral, as in the case of the Winchester kitchen plan (Fig. 2.20) and the map of tenter-frames in Exeter, or to have been accompanied by written descriptions as in the case of Waltham Abbey's water supply. Other maps are extravagantly rich in narrative, like the map from Canterbury. Finally, yet another variable is the amount of information conveyed in each map. Some maps seem to be remarkably comprehensive, indicating a range of landscape features, like the map from Chertsey which records a dozen categories of information. In contrast, most medieval maps (excluding the *mappaemundi*) show only features germane to the matter in hand, as in the case of the map from Fineshead, which contains only three categories of information.[64]

The extent to which precise measurement was used in medieval map-making is a much-debated point. The maps with which we started, those of the churches in Jerusalem drawn by Adomnán, incorporated Arculf's on-the-spot measurements. Many exegetical maps represent attempts to reconstruct territorial division or architectural layout from measurements found in the Bible. Indeed, a measuring rod is usually shown in the Angel's hand in both plan

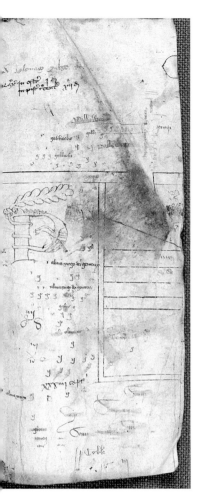

2.20 Plan of a new kitchen in Winchester College, Hampshire. Although little information is given, the ink lines are ruled and the carefully-prepared map may have served as an *aide-mémoire* for planning or costing. A few years later the parchment was cut into two for use as a book cover, with the map turned inwards. Original dimensions approx. 31 × 26 cm. *c.*1390. Winchester College Muniments 22820.
Reproduced by permission of Winchester College.

and pictorial representations of the New Jerusalem of the Apocalypse as a sign of utopian equality (see Fig. 2.3, p. 11). In general, though, it would seem that medieval topographical mapping did not involve measurement (still less anything so formal as surveying). If this really were the case, we are presented with something of a paradox, for precise measurement (in customary units) was widely practised throughout most of England and Wales from before the year 1000. Rods of local length were used in planning villages and in laying out literally hundreds of new sites and settlements after that date, not to mention the new urban developments and planned towns of the twelfth and later centuries.[65] And dimensions interested people. Matthew Paris noted the discrepancy in one account of the size of England and its bishoprics with figures given six centuries earlier by Gildas (c.570). Preferring the latter, Paris copied out the relevant passage from Gildas at least twice as if to give his map a rough scale.[66] At the same time, medieval maps remind us that accuracy can take many forms besides the mathematical and in many cases accuracy of content, contiguity, and the relative importance of specific features (usually indicated by exaggerated size) was of far greater importance. Where measurements mattered to the map user, the necessary information was imparted in words (as on the Winchester kitchen plan or on Paris's maps of England) or in figures (as for the Fineshead meadows). It could also have been conveyed by word of mouth. In short, medieval interest in absolute measurement was linked closely to the purpose of representation. It might be irrelevant to exegetic or didactic maps of scholars, perhaps of minor importance in local mapping, where a statement on the map could be matched by pacing the distance on the ground, but was essential whenever land had to be surveyed on the ground.

Map use

Even today, when we know how maps are used, they are notoriously slippery things to classify. They are simultaneously unique and polyfunctional. While each map embodies knowledge, ideas and values that are place-, time- and culture-specific, and each was intended to be used in a particular way, each type of map can serve a plurality of purposes. A portolan chart, for instance, was usually made to be taken on board ship for use as a navigational tool, but the vast majority of the richly decorated examples we see today never went near the sea. Rather, they were made for the master mariner's onshore office, or for the private libraries of a variety of people. The classification of early maps is an especially elusive business when we know so little about how maps were actually used and even less as to how they were received by contemporaries.[67] Even what might appear to be a hint on the map by no means tells the whole story. The map-maker of the Hereford *mappamundi* asked 'all who possess this history, or shall hear or read or see it' to pray for his soul – a routine injunction at the time. Most unusual (in several respects) is the remark on the thirteenth-century Ebstorf *mappamundi* that the convenience of being able to see the world's regions, provinces, islands, gulfs, lakes, seas, mountains, rivers on a single page 'is of no small usefulness . . . in that it permits [readers] to see the direction to be taken . . . and to choose for themselves which monuments they will contemplate and which routes to travel'.[68] This is but one of many texts on the map, however, and notwithstanding its apparent explicitness falls far short of indicating the main purposes of a *mappamundi*. Generally, authorial intention is clearest in the case of exegetical maps. Adomnán introduced each plan with words such as 'the accompanying picture shows the shape of this round church [so that] you may at least be able to see the way [Christ's Tomb] is set in the centre . . .'.[69] Bede used several phrases, including the comment that 'it seems a good idea to sketch for the eyes an outline of this basilica [on Mount Golgotha] as well'.[70] Otherwise, in the vast majority of cases, context is our only guide. The narrower the context, the more obvious the original function. Thus, the function of the stage plans in early manuscripts of the Cornish and other plays is more or less clear, as is that of the liturgical plans in the Salisbury Processional, where the notation of the chants, the text of the prayers, and the diagrams reminding priests where they were to stand are interleaved to form an appropriate sequence.

But the context of most extant medieval maps is far from clear. Medieval culture was itself polymorphic. Religion was the starting and finishing point for all aspects of life, and not a separate compartment, and we should expect each action and each product to have had more than one level of interpretation. The Bible itself had four: the literal meaning implied the text's direct and immediate meaning; the allegorical meaning its deeper spiritual and doctrinal significance; the anagogical, its import with regard to the goal of ultimate redemption; and the tropological, its application to the Christian life of the individual. Likewise, the construction of a map usually involves a set of several separate but integrated acts.[71] In defining context, then, in the Middle Ages or at any other time, we have to be prepared to deal with the multifarious strands of intention embedded in each map, whether we start from the map itself or with the map-maker.

Although the vast majority of maps made before 1400 (if noticeably fewer in the fifteenth century, about 80 per cent and 38 per cent respectively) came from a monastic or cathedral community, their function was by no means necessarily directly related to religion or to learning. Like the royal palaces and other lay properties, monasteries and cathedrals carried a huge burden of property maintenance and some of our maps were clearly connected with practical matters. In the early thirteenth century, a map records allocations of meadowland at Fineshead. In the late fourteenth century a map was seemingly used in connection with flood control at Cliffe, Kent. In other cases, such as the map of drains and water pipes at Canterbury, practicality is ambiguous and the map undoubtedly served more than one purpose. At the opposite end of the functional scale from the most overtly pragmatic maps are the *mappaemundi*. These represent the fusion of religion and scholarship typical of medieval culture; their study involves the unravelling of separate but inter-twined strands of geographical, historical, philosophical and biblical knowledge, belief and myth. Like maps for exegesis, they could serve as teaching aids, as mnemonic aids, or as aids to meditation.

The role of maps in medieval litigation, a highly pragmatic function, remains unclear. Traditionally (from Anglo-Saxon times at least), the boundaries of larger properties and political units were described verbally in terms of landscape features, yet a number of the local maps and plans we have reviewed and others discussed by R. A. Skelton and P. D. A. Harvey were connected with legal disputes. In such cases, they are likely to have been used as portable *aide-mémoires* rather than as legally-valid documentary surrogates for the landscape for use indoors or elsewhere.[72] That there are no way-finding maps, for either sea or land travel, is less surprising, however. Written itineraries for land travel prevailed right up to the eighteenth century, and written sailing books were the chief guide for sailors around Britain well into the sixteenth century. Nor is the lack of medieval military plans to be wondered at in view of the relatively small scale of medieval warfare and the predominance of siege tactics, in which oral description would have been equally effective and rather more discreet than any map. Yet maps can be associated with medieval military activity in other ways. For instance, contemporary records of campaigns like Edward I's Scottish incursions could explain the choice of places shown later on the Gough map or on the map in John Harding's Chronicle, just as topographical maps were used on the continent in strategic planning.[73]

2.21 The map as memorial: Thomas Chaundler's bird's-eye view of New College, Oxford, is one of several illustrations in a book celebrating Thomas Bekynton's (Bishop of Bath and Wells) endowments. 28 × 18.5 cm. Early 1460s. New College, Oxford MS C.288, on deposit in the Bodleian Library. Reproduced by permission of the Warden and Fellows of New College, Oxford.

THE IMPORTANCE OF GENRE

Obviously like should always be compared with like, an aphorism which applies no less to maps than to any other phenomena. The problem is that usually far too little is known about the circumstances giving rise to each map – the intentions of those who made it, the reasons they made it, and the function or functions it was to fulfil – for us today to be sure we know what we are looking at. Like a literary text, a map is an act of construction and, inevitably, highly specific to its context.[74] Content alone provides an insufficient basis for legitimate comparison, partly because of the peculiar poly-functionality of maps and partly because aspects other than content may be more significant. Describing a map in terms of its type alone (for example, a topographical map, a property map, a thematic map) does not take us far enough. We need to know more, not only what a map shows but why it shows it, for whom it was made or for what it was intended. The specificity of a map's context, style and function – its sub-type or genre – tells us much more than does mere content.

To demonstrate the working premise that maps superficially similar in content do not necessarily fulfil the same functions, we can take a closer look at four maps already mentioned. In many respects, they form pairs. Two maps show buildings, two show the world. Two maps are book illustrations, the other two were drawn on separate pieces of parchment. But even within the pairs, each map is also significantly different. While both building plans could be said to have lent themselves to practical use, they are stylistically so different that different functions have to be inferred. They were certainly treated differently at the time, one being carefully preserved, the other disposed of soon after use. Likewise, while both world maps are similar insofar as the basics of continental geography are concerned (continental outlines, major rivers and mountain systems, names of countries and cities), detailed analysis of their content reveals an imbalance between geographical content on the one hand and historical, religious and mythical content on the other. The four maps, in short, represent not two map types (architectural plans, world maps) but four distinctive genres: a practical building plan, a commemorative plan, a map of the world of secular origins, and a *mappamundi*.

The simplest of the two building plans is the one from Winchester College. It dates from about 1390. It is a small map (31 × 25 cm) and not much to look at (Fig. 2.20). A few straight lines drawn on vellum in black ink record the outlines of a new kitchen. However, the map was not carelessly produced. On the contrary, it would seem that a special ruling pen (hitherto either unknown or not much used in England) was used, together with a ruler and, probably, a set square.[75] The map would be difficult to decipher were it not for a note on the map that 'the site of the new kitchen is to contain 30 feet in length and 20 feet in breadth'. Clearly the purpose of the map was practical, something to do with the planning or costing of the new construction works. Twenty years or so later, it was disposed of, and the parchment on which it had been drawn was cut into two, folded (with the map on the inside), and stitched to form the cover of the College's hall-book for the year 1415–16.

The second building plan is large, ornate and (apart from some trimming) still in excellent condition despite being nearly 250 years older than the first (Plate 2). It was made sometime between 1153 and 1161 at Christ Church, Canterbury, and originally measured 55 × 66.5 cm. It shows the thirty separate monastic and ecclesiastical buildings of the cathedral precinct and part of the nearby city wall.[76] It is a highly detailed drawing, topologically accurate, and (where it can be checked) architecturally correct. Each building is portrayed in elevation but takes its proper place on a ground plan. The whole thing must have been a costly artefact to produce for, in addition to the extra large vellum and the black ink of the drawing, four coloured inks were used (red, blue, green and brown), the red serving to highlight the system of waterpipes and drains which cross and recross the open spaces and penetrate certain buildings. There are also a large number of brief explanatory texts relating to specific individuals shown in specific buildings, or describing their adventures, for which other documents provide corroboration. The conclusion is that, like the topographical views that were later drawn as a memorial to William Wykeham, Bishop of Winchester, patron of education and architecture and founder of New College, Oxford (Fig. 2.21), the Canterbury map is a memorial map, created to celebrate the architectural achievements of Prior Wilbert, the initiator in 1149 of the new buildings, waterpipes and drains represented so prominently on the plan.[77] As in the mid fifteenth-century New College views, the benefactor and his contemporaries were as important as the place and the event recorded on the map.

Our third map is the Anglo-Saxon map, also known (after its seventeenth-century owner, Sir Robert Cotton) as the Cotton map (Fig. 2.22). Its date is uncertain but it was probably made in the second quarter of the eleventh century. It occupies a full folio in a computus manuscript compiled at Christ Church, Canterbury.[78] The rectangular map shows the known world with outlines generally agreed to be 'closer than any other surviving medieval map to the Roman map of the world'.[79] The accuracy of its geographical outlines matches those on no other *mappamundi*. While it does contains some features associated with Christianity and with Old Testament history, these are modest in relation to the overall content of the map on which the boundaries of the provinces of the Roman empire strike the observer as one of the most prominent and consistently-marked features. The marvels of the east, too – another link with classical mythology – are correctly located. In overall presentation, possibly in its function, and – especially – in its (pagan) Roman antecedents, the Anglo-Saxon map stands apart from the corpus of English *mappaemundi*.

From a superficial glance, the Anglo-Saxon map seems to accompany a Latin translation of a geographical poem attributed to Priscian (*c*.600), but in fact the poem is Priscian's translation of the second-century AD pagan Dionysius's *Periegesis*.[80] Dionysius's poem itself constitutes a verbal map for, as Priscian explains, it 'is about the situation of the earth, [as] gathered by him from the writings on ancient world maps'.[81] But a graphic world portrait was also intended, for Priscian continues 'and to this work of three parts, that is to say, Asia, Africa, and Europe, there is painted a suitable map in which the location of nations, mountains, rivers, islands, and also wonders are accurately arranged'. It is by no means certain, though, that the surviving map represents the one proposed, for it is scarcely the most suitable, especially given its biblical details and the omission of many places mentioned in the poem. It looks as if what was made was different from the one Priscian intended, or that an already existing map was simply taken to illustrate the text where it called for a map.[82] While the sources used for the Anglo-Saxon map cannot be identified with any certainty, it does seem that the map-maker was looking at a secular map of world geography, perhaps one very close to, if not derived directly from, something that had come down to his day from Roman times, perhaps even from the map of the world allegedly

2.22 The Anglo-Saxon (Cotton) map of the world represents a mapping tradition going back to Roman times and still influential in the early Middle Ages. The British Isles, with Thule and the Orkneys, are in the bottom left corner. The boundaries of the Roman provinces stand out clearly, as do the Pillars of Hercules between the Mediterranean and the Atlantic. East is is at the top. Early eleventh century. 21 × 17 cm, in a computus manuscript. British Library, Cotton MS Tiberius B.V, fol. 56v.
Reproduced by permission of the British Library Board.

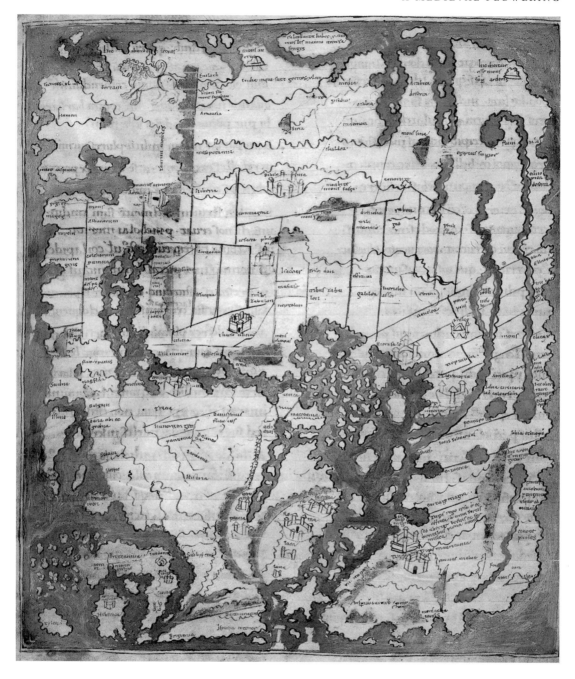

2.23   The earliest known extant English *mappamundi*: the frontispiece in the copy of Henry of Autun's *Imago Mundi*, presented to Sawley Abbey, Yorkshire, by Durham Cathedral Priory about 1200. Like the Anglo-Saxon map (see Fig. 2.22), the Sawley map draws its content from a mixture of sources; unlike the Anglo-Saxon map, however, the Sawley map has a preponderance of biblical material (including, tellingly, Paradise). East is at the top. 29.5 × 20.5 cm. Corpus Christi College, Cambridge, MS 66, p. 2.
Reproduced by permission of the Master and Fellows of Corpus Christi College, Cambridge.

made for Marcus Vipsanius Agrippa (son-in-law of the Roman emperor Augustus) at the end of the first century BC.[83]

Finally, there is the late twelfth-century map from Sawley Abbey commonly (if misleadingly) known as the Henry of Mainz map (Fig. 2.23).[84] Like the Anglo-Saxon map, the Sawley map occupies a full page in a book, the *Imago mundi* by Honorious Augustodunensis. Also like the Anglo-Saxon map, it has east at the top. Unlike the Anglo-Saxon map, the Sawley map is contained in an oval frame, leaving room in the four corners of the page for drawings of winged figures. Compared with the Anglo-Saxon map, world outlines on the Sawley map are in places almost fanciful, as in the case of north and north-western Europe. Particularly striking is the quantity of non-geographical material the map contains, most of which comes from the Bible or from classical literature.

Indeed, at the centre of the map is not Jerusalem but the pagan sanctuary of Delos. The Earthly Paradise is shown in the east at the apex of the map. By all accounts, the Sawley map is generally accepted as the earliest surviving English example of a *mappamundi*.

ENGLISH MAPPAEMUNDI

The *mappamundi* is undoubtedly the most idiosyncratic, even spectacular, map genre of all times, and was of particular importance in England. Deeply embedded in medieval scholasticism, it provided a means of communicating a rich 'narrative of Christian history cast in a geographical framework'.[85] In a *mappamundi*, space and time are equal components. That is to say, the world that the mappamundi portrays is not synchronous but one in which events take place according to spatial rather than chronological order. The beginning of time, signified by some aspect of the Earthly Paradise (the Garden of Eden, Adam and Eve, the Four Rivers), is always featured and so too, usually, is the end of time, signified by Christ Enthroned (at the top of the map), at the Last Judgement. The perennial and all-pervasive relationship of the earthly world with God its Creator may be represented by an enveloping mandorla (as in the maps for Ranulf Higden's *Polychronicon*) or by the figure of an embracing Christ (as on the Ebstorf map, where the head, hands and feet of Christ frame the map). The importance of time in a *mappamundi* reflects the unimportance of place in early Christianity, a religion of the book, in which belief was privileged over place.

Sometimes confused with elementary *mappaemundi* are the diagrams which have come to us in medieval copies of geographical descriptions originally written in Antiquity or by pagan writers or, in the case of the sixth-century encyclopaedist Isidore of Seville, by somebody taking a deliberately traditional line. Some of these early authors provided illustrations to help understand the continental structure of the world or its division into climatic zones. The typical illustration for a chapter headed *de orbe* and beginning '*orbis a rotunditate circuli dictus . . . divisus est autem trifariae: ex quibus una pars Asia, altera Europa, tertia Africa nuncupatur* [The globe is so called for its roundness . . . It is divided into three parts, which are called Asia, Europe, Africa] is a circle divided into three.[86] The earliest datable text of this sort is the geographical chapter in the History of the Jurgurthine War (*De bello Jurgurthino*) by Sallust (86–34 BC). The lines represent the Mediterranean Sea, the river Danube (or Don, denoting the boundary between Asia and Europe), and the river Nile (the boundary between Asia and Africa). Since Asia was placed at the top, the lines of the schematised Mediterranean and rivers inevitably form the vertical and horizontal bars of a capital T (Fig. 2.24).

Since the earliest extant manuscripts of these old texts are all medieval, the diagrams we see had reached the Christian medieval copyists only after centuries of transmission (in the case of Sallust, a pagan writer, nine centuries later). While scribes in Late Antiquity would have kept more or less faithfully to their exemplars, Carolingian and later medieval scribes often attempted to improve the old as they worked on the new copies. In the case of the T-O diagrams, and the climatic diagrams, this meant adding at first just one or two biblical features, such as the names of the three sons of Noah, Mount Sinai or Jerusalem, and imparting a religious significance (the Crucifixion) to the T structure (Fig. 2.25). Eventually the old diagrams were hijacked into the new context and came to carry an ever-increasing burden of extra material. The result is first, a period of transitional diagrams or proto-*mappae-mundi*, covering the centuries between about 800 to about 1100. Then, in the twelfth century, the *mappaemundi* proper emerged from the burst of learning that was affecting Western Europe as fully-fledged cosmological conceptions, freighted with history and myth, real and imagined geographies and natural history, and rich in biblical content. Given an eastward orientation, a Eurocentric view of the Old World, and a circular or near-circular shape, the T formed by the Mediterranean and the rivers between Asia, Europe and Africa tends to stand out prominently, and the Levant is more or less automatically in the centre of the map. In fact, few *mappae-mundi* were obviously deliberately centered on Jerusalem. Those that were, like the Ebstorf map (*c.*1239), the Psalter map (1260s), and the Hereford map (*c.*1289), tend to be later examples and may reflect the impact of the Crusades on Western thinking.[87]

Some two dozen English *mappaemundi* are known (excluding 19 copies of the Higden map) of which fewer than half are extant. The last great loss was that of the Ebstorf map (*c.*1239) which perished in Dresden, Germany, in the Second World War. Although made at Ebstorf, it is thought that this huge map (some three metres square) owed much to the ideas of Gervase of Tilbury, who had been at the court of Henry III five years before the mural *mappamundi* was created in the Painted Chamber at Westminster and whose Otia Imperial may have been written for the map.[88] Measuring 158×133 cm, the Hereford map is now the largest surviving *mappamundi*. The assumption hitherto has been that the map was drawn at Lincoln and then taken to Hereford, which could explain why the town sign for Hereford and the name of the river Wye seem to be 'clear additions to the map, drawn less neatly than the rest and necessitating re-writing the name of the Severn' (Fig. 2.26).[89] The map is one of the few with Jerusalem at the centre, here shown (again unusually) with a Crucifixion above it. More traditionally, the Pillars of Hercules are featured at the western extremity of the Mediterranean Sea. In the two bottom corners of the parchment, outside the circular frame of the map itself, two scenes with figures pose problems of interpretation. One is a horseman, thought variously to represent the map's author, Richard of Haldingham, or another *memento mori*. The other, which shows Augustus Caesar, relates in a slightly muddled way to the inscription around the map which describes how 'the world was first

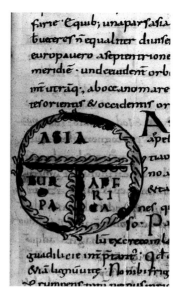

2.24 A secular diagram for a geographical textbook: the T-O diagram served originally, in Classical times, to summarise the division of the ancient world into three parts, divided by the Mediterranean Sea, the River Don (or Danube) and the River Nile. This example illustrates Isidore of Seville's Etymologies (*Etymologiae*, written 622–33), book 14, chapter 2. East is at the top. Approximate diameter 3 cm. Ninth century. St Gallen, Switzerland, Cod. Sang. 237, p. 219.
Reproduced by permission of the Siftsbibliothek St Gallen.

2.25 A Christianised T-O diagram. In this example from an unidentified ninth-century manuscript, a scribe has added the names of various Classical and Christian people and features (Paradise, Jerusalem, Galilea, Golgatha and Bethlehem). Formerly Bibliothèque Nationale et Universitaire de Strasbourg, MS C.IV, no. 15, but lost (probably in the fire of 1870) since publication by F. J. Mone in *Anzeiger für Kunde der teutschen Vorzeit* (Karlsruhe), 5 (1836), col. 113. As redrawn by K. Miller, *Mappaemundi. Die ältesten Weltkarten*, Vol. 3, p. 118.
Reproduced by permission of the British Library Board.

Left: 2.26 The Hereford *mappamundi* appears to have been altered to allow Hereford to be added. Detail showing the addition of the place-sign in a different hand, along with the River Wye, and apparently involving, as P. D. A. Harvey points out (*Mappa Mundi. The Hereford World Map*, 1996, pp. 6–7), an adjustment to the name of the River Severn, originally *ff' Sabrina* but now *Sabrina ff'*.
Reproduced by permission of the Dean and Chapter of Hereford Cathedral.

measured by Julius Caesar' with the aid of three surveyors (the Ebstorf map has a similar reference but says there were four surveyors).[90] Here is another pointer to links between medieval map-making and that of Roman times, when Vespasius Agrippa's map of the world (made between about 12 BC and 14 AD) was displayed on a portico wall in Rome. Other elements of Classical learning on the Hereford map, found also on many *mappaemundi*, are the fabulous beasts and monstrous races derived from Pliny, Solinus, Orosius and other ancient writers and portrayed at the margins of the habitable world.

The Ebstorf and the Hereford maps probably represent two 'families' within the English *mappaemundi* tradition. A possible third family group has been suggested by Peter Barber, who sees links between the Aslake, Duchy of Cornwall and Psalter maps and a common ancestry in the Westminster *mappamundi* (which was probably lost in a fire after 1263).[91] The first two maps were only recently rediscovered and their value lies mainly in what they reveal about such interconnections and about the sources used in their compilation. The Duchy of Cornwall map came to light in 1986 in the Duchy's archives. It is, though, a mere fragment of the original, which must have measured some 160 cm across.[92] Only the bottom right corner is preserved, showing Africa (complete with figures of the monstrous races) and a marginal text. The suggested date for the map is 1280 and it could have been one of the Hereford map-maker's prototypes. It may also be related to the now-lost original of the Psalter map. The latter, which measures but nine centimetres across, was drawn on the first page of a psalter in about 1260. It has Jerusalem in the centre. Small as it is, there was room for drawings of the monstrous races in southern Africa. A wedge-shaped Red Sea stands out in the southeast and, in the northeast, the wall imprisoning Gog and Magog sweeps in a great curve. Above the circle containing the world, much as on the Hereford map, is a figure of Christ between two angles, hands uplifted in blessing. On the reverse is a T-O diagram listing the provinces and cities within each of the three parts.[93]

In 1985, a fragment of yet another hitherto unknown *mappamundi* was recovered from its late fifteenth-century role as the cover of a rent book of Walter Aslake's Norfolk estates.[94] The map dates from before 1375 and had been drawn at Creake Abbey, Norfolk. Despite its plainness (it is carefully drawn but uncoloured), a characteristic which could account for its short cartographical life, the Aslake map makes a notable addition to the corpus of English *mappaemundi* for the number of place-names and coastline details which point to the use of a portolan chart (of about 1350) in its compilation, a source that would have been new to English map-makers.[95] The map is greatly faded but it can be seen to portray Africa, the Atlantic islands, part of Arabia and the Red Sea. It also marks the virtual end of new *mappaemundi* in England, for only the Evesham map (*c*.1400) is known to have followed (Fig. 2.27).

By the middle of the fifteenth century, as scholarship changed and the world began to be viewed differently, the genre was losing its relevance. As four-dimensional maps, maps with a frame of reference going beyond the human environment, the *mappaemundi* had been uniquely positioned to accommodate the contradictions which so concerned medieval thinkers. The mapping of the Earthly Paradise, for example, could not be – and was not – fitted into any other type of map. The three factors which saw the removal of the Earthly Paradise from world maps as the fifteenth century turned into the sixteenth – a shift in medieval scholasticism away from seeking a rational, that is, geographical, location for the Earthly Paradise; the failure of travellers to the east to return with confirmation of its existence; and the rediscovery in western Europe of Ptolemy's Geography with its mathematical approach to mapping – were precisely those which left no place for the *mappamundi* with its dominance of belief over first-hand observation.[96]

MAPS OF ENGLAND

England occupies a very small space on a *mappamundi*, squeezed with the rest of the British Isles into an inadequate space on the bottom left edge of an east-orientated map, a peripheral position remote from the foci of Christian affairs in Rome, Byzantium and the Holy Land.[97] Although the British Isles are presented remarkably realistically on the Anglo-Saxon world map, this was not always the case on *mappaemundi*. Only the Psalter map might be said to match the Anglo-Saxon map for its detailed realism. Other *mappaemundi* show various degrees of generalisation as on the Hereford map, where Britain is represented summarily as a rounded oblong, albeit with appropriate indentations (Fig. 2.28). However, it was not the prime function of a *mappamundi* to be explicitly geographically realistic, except perhaps for the mapmaker's own region (thus, Britain on the Hereford map, and north Devon and the Severn valley on the Evesham map, include details of purely local interest) and the outlines of Britain remained schematic right through to the Evesham map (see Fig. 2.27). To present in detail the real shape of the British Isles, a different cartographical approach was needed, something closer to the geography of Roman times, not the cosmology of the *mappaemundi*.

The making of a geographical map imposes certain constraints on its maker, and a fundamental question concerns the manner by which a region could have been mapped accurately in the Middle Ages, that is, with regard to its mathematically-determined size and shape. What grid or system of coordinates was available to structure the map? Claudius Ptolemy's Geography (written in Alexandria in the second century AD) contains all that was needed: Book 1 describes the idea of a systematic global graticule and instructs the reader how to plot each place according to its unique pair of coordinates, and Book 2 lists the latitude and longitude for

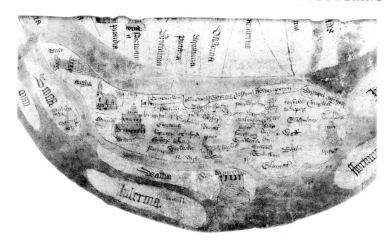

2.27 The British Isles on the Evesham *mappamundi* are wholly out of proportion with the rest of the map, occupying roughly a fifth of the map. Thus, although peripherally located right at the bottom of a rather elongated map, England is visually prominent, and Evesham and surrounding places are also readily picked out. East is at the top. Whole map 94 × 46 cm. *c.*1400, Evesham. College of Arms, MS Muniments 18/19.
Reproduced by permission of the College of Arms.

places in the British Isles. Once available in western Europe, the Latin edition of the Geography left an almost indelible mark on the outline of Britain for much of the sixteenth century (notably in the shape of Scotland) but it was not to be had in any language in Europe until, in 1295, the Greek monk Planudes found a copy in Constantinople. Even then, it was well over a century before a Latin translation was completed (Florence, 1409) and longer still before any copy reached England.[98] Another of Ptolemy's mathematical works, however, was circulating widely in the West from the twelfth century onwards. This was Ptolemy's Almagest (the Mathematical Syntaxis) which although earlier than the Geography contains the essence of his cartographical ideas.[99] More specifically, it articulates his principle that topographical maps should be plotted from tables of latitude and longitude. Translated from the original Greek into Arabic in 827, the Almagest was then translated into Latin by Gerard of Cremona in 1175. However, unlike the Geography, the Almagest could have been read much earlier in its Arabic form, for by the early 1150s Arabic was studied at the English court in Normandy, if not already in England itself.[100]

Exactly when a copy of the Almagest reached England, and into whose hands it fell, is uncertain. The time would have been ripe, for between 1250 and 1350, the notion of quantification as an absolute rather than a relative measure, was being established.[101] By about 1265, it had occurred to Roger Bacon, Franciscan friar and teacher at the University of Oxford, that the system described in the Almagest could indeed be used in making a map of the world.[102] Bacon's description of a coordinate system based on parallels and meridians is found in his *Opus maius* (Part 4) but his map, which must have looked very different from any *mappamundi*, is now lost. The importance of Bacon's mapping procedure, David Woodward points out, 'lay not only in its innovative way of ordering terrestrial space geometrically but also in its practical importance in providing a systematic inventory of places on the earth, as a tool for the expansion of political power'.[103] In the event, Bacon's ideas had little impact, partly due to the shortage of astronomically-determined values for the latitude and longitude of English towns and places. The Almagest does not provide the long list of coordinates that is found in the Geography, nor were any given in the *De spherae* of Sacrobosco (John of Holywood, Yorkshire, 1220/1230), another well-read and much-glossed text.

It was not that there was a lack of interest in scientific data at this time. Tables of latitude and longitude had existed since classical times. The first Arab astronomers adopted Greek and other works in order to re-calculate entire handbooks of tables according to new meridians.[104] In Spain, the Arab astronomer Arzachel was able to draw on such tables for his own compilation (the Alphonsine Tables, 1080, updated *c.*1272 as the Toledan Tables) while western scholars translated works into Latin; Adelard of

2.28 The outline of the British Isles on the Hereford *mappamundi* is generalised but realistic, albeit distorted to accommodate the curve of a circular map. Whole map 158 × 133 cm. East is at the top, ?Lincoln. Hereford Cathedral. Reproduced by permission of the Dean and Chapter of Hereford Cathedral.

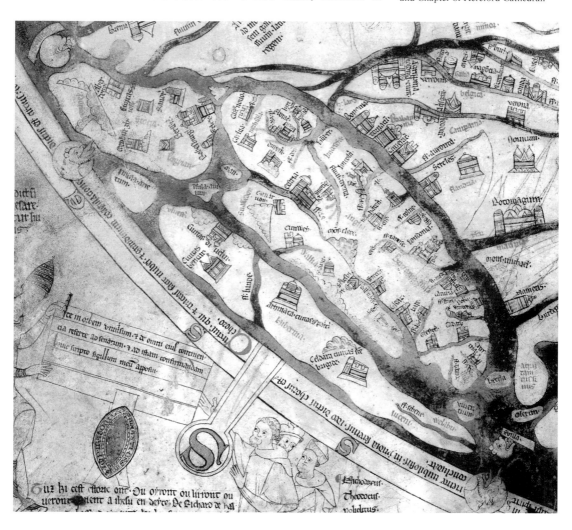

Bath translated the Khorazmian Tables in 1162.[105] In England, Gildas's figures for the dimensions of Britain were challenged in the thirteenth century by some scholars (for example, Henry of Huntingdon) and supported by others (for example, Matthew Paris, and, following Bede, William of Malmesbury).[106] By 1300 at the latest, coordinates for London had became available from a French source and in 1340 those for Oxford had been calculated by William Rede, fellow of Merton College.[107] Calculating longitude was always a problem but by the early fifteenth century, figures of some sort were being suggested for an increasing number of English towns, such as Lincoln, York, Canterbury, Colchester, and Hereford, as well as for London and Oxford.[108] Books like the *De spherae* described the methods and instruments needed for astronomical observation.[109] An Arabic treatise on the astrolabe, an instrument necessary to such calculations, is thought to have been translated into Latin in England in 1147, possibly by the Spanish Jew, Abraham ben Ezra.[110]

Despite a lively interest and the efforts of individuals, it seems that it was not until western thought and imagination had become 'mathematised', until philosophy (pure mathematics) fused with metrology (practical calculation), and until Ptolemy's Geography was accessible in Latin, that Bacon's theory could be turned into practice.[111] Then, at last, the new data found in the Geography – coordinates for no fewer than 120 places in Britain (Albion), 34 in Ireland (Hibernia) and eleven for various islands – could be combined with long-recognised principles. So it happened that the English *mappaemundi* bear no trace of any coordinate scheme. Nor, despite numerous illustrated treatises dealing with Euclid's geometry, was what we now recognise as geographical accuracy (mathematically-defined size, shape and direction) of consequence to medieval map-makers; selected features were included or exaggerated on maps in relation to their perceived importance. Nor do the medieval topographical maps of England (or Britain or the British Isles) owe anything to either Ptolemy or Bacon. However, this is not to say that the best of these topographical maps had not been carefully structured; they had been, but from material of a different nature.

The earliest maps focussing specifically on Britain or the British Isles are topological in structure and diagrammatic in style – those illustrating the various manuscripts of Gerald of Wales's topographies of Ireland, the most detailed of which has been likened to 'a sketch drawn on a blackboard during a lecture to act as a visual aid to the audience' (see Fig. 2.7, p. 15).[112] Interestingly, Gerald is said to have read the topography of Wales to students at Oxford in about 1186, affording them 'great entertainment'.[113] One wonders if he also drew for them the sort of simple sketches which eventually found their way into some copies of his works, either in his day or later – in the thirteenth century, for example – by courtesy of map-conscious scribes. Gerald himself may not have been a great thinker or an original scientist, but he does seem to have been a sharp and largely accurate observer of natural features and

2.29 Matthew Paris's first map of Britain. The influence of an itinerary as a source of information about places and rivers between Dover and Berwick is clearly seen. North is at the top. 35 × 25 cm. Late 1250s. British Library, Royal MS 14 C.VII, fols 2–5, fol. 5v. Reproduced by permission of the British Library Board.

phenomena, with an interest in cosmography (on which he wrote, describing the primordial universe as the *vetus globus*).[114]

Gerald of Wales had spent thirteen years in Paris before returning to Wales in 1175 and then retiring from clerical duties to Lincoln. He was well-travelled. In contrast, medieval England's most prolific map-maker, Matthew Paris (*c.*1200–1259), seems to have travelled little apart from a voyage to Norway and visits to London. Yet his visual and geographical sense was outstanding and his cartographical output was as imaginative as it was varied and prolific. Where Paris's ideas came from is not clear, but he was well-known at the court of Henry III and in the highest ecclesiastical circles and must have been integrated into a lively network of

2.30 Matthew Paris's second and most elaborate map of Britain. The influence of the north-south itinerary is still predominant but Paris has been able to revise his coastal outlines to produce a more realistic shape, although both the Wash and Cardigan Bay are missing. North is at the top. 34 × 23 cm. Late 1250s. British Library, Cotton MS. Claudius 14. D.VI. fol. 12v.

Reproduced by permission of the British Library Board.

scholars and informants. The title of his first map of Britain, *Britannia nunc dicta Anglia*, refers to the post-Roman change of name for the island and has been interpreted as one of the indications that, like the Anglo-Saxon map, Paris's map was derived from a Roman prototype that was still available in some form or other in the middle of the thirteenth century.[115]

Some of Matthew Paris's maps exist in several copies. There are four versions of the map of Britain, all probably drawn between *c.*1255 and 1259. P. D. A. Harvey has demonstrated how important it is to consider the different versions together, for it is only when they are seen together that the way the map was constructed is revealed.[116] Figure 2.29 shows what is thought to be the earliest, together with one of the most striking aspects of all of them, the alignment of places along a vertical axis running from Berwick on

Tweed on the Scottish border, to the English Channel port of Dover. Along this line, the sequence of places and the distances from one to the next are remarkably correct. The more important river crossing points are indicated although the river network is incomplete, a feature consistent with the use of a written itinerary in the creation of the map. To discover Paris's second main source, however, a different version of the map has to be studied. This one has a number of details along its eastern margin (Fig. 2.30), including notes referring to neighbouring lands and an indication of the territories across the North Sea. On the other side of the map, the north-western coast of Scotland is compressed into a marked curve, a feature which adds to the impression that Paris was now working from a circular or oval source map, yet other details of his outlines suggest that he was using a map like the Anglo-Saxon map, namely a map of the world of Roman secular origin.

Matthew Paris's procedure for the construction of a geographical map of Britain thus seems to have involved at least three stages. First, he took the island's outline from a world map. Then (like the Hereford map-maker) he used an itinerary for the sequence of places and rivers between Dover and the Scottish border at Berwick. Finally, he filled in with the hundred or so additional names of towns, regions, mountains and other features. Paris's comment, written on the map, that 'had the page allowed, the whole island would have been longer', should not be taken, Harvey stresses, to mean that Paris was worried about conforming to mathematical scale but that he would have been able to make the outline closer in *shape* to the original he was working from had he had more space. Once the map had been completed in draft, however, Paris must have seen yet another source map, one which showed the river network in full as well giving a better outline overall, and so he went on to an improved version. A fourth version was finished for him after his death by his assistant.

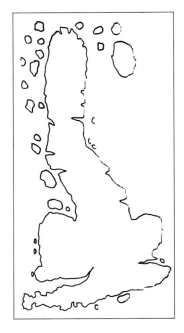

2.31 The Gough map of Britain: outlines. Here shown with north at the top, although the map has east at the top. Whole map 115 × 55 cm. *c.*1360. Bodleian Library, Oxford, MS Gough Gen. Top. 16.
Drawing by A. Scafi.

2.32 East and south-east England on the Gough map. East is at the top and the area shown is from the Wash to the Channel. Ely is encircled by rivers. London, at the head of the Thames estuary, is marked by the largest place-sign. Lines link a selection of places, with distances noted alongside. Bodleian Library, Oxford, MS Gough Gen. Top. 16.
Reproduced by permission of the Bodleian Library, Oxford.

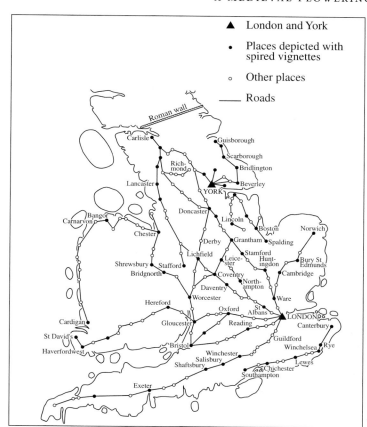

2.33 The rationale of the route network shown on the Gough map is unclear. The total absence of routes in the northeast, apart from the mainly coastal Beverley-Guisborough connection, is particularly striking, especially as a London-Carlisle route is indicated. York seems to be the central node of a virtually independent system. In the absence of any link between London and Dover, the coastal route between Southampton and Canterbury is wholly isolated. Redrawn from Philip Beale, *A History of the Post in England* (Aldershot, Ashgate, 1998), Fig. 3.1.

We do not know what other attempts were made to represent England during the next hundred years for we have only one map of this sort from the fourteenth century, the Gough map of about 1360 (Figs. 2.31 and 2.32).[117] Virtually nothing at all is known about it, its precise date, origin, patron or first owner, or function. Internal evidence (place names which changed in the course of the fourteenth century, and specific details such as the construction of Coventry's town walls) point to the middle of the century for its date. Like Matthew Paris, the maker of the Gough map must have relied on whatever information he could obtain; other maps (perhaps even a portolan chart) and information from his own travels or the firsthand knowledge of others around him.[118] For information about the interior of the island, the maker of the Gough map must have had access not to one itinerary, like Paris, but to several. Again like Paris, the Gough map-maker also appears to have started with the outline and then filled in with place-names. If, however, the places and distances taken from the itineraries fit the outline so well, this was not because the map had been composed around the itineraries but because it was an extraordinarily well-constructed map, based not only on widely-gathered information but also on astronomically calculated coordinates for a handful of towns.[119]

The second distinctive aspect of the Gough map is the network of ruled lines linking some of the places shown. These lines have been described as roads, of which some 2,940 miles (4,727 km) in three classes (main, secondary, local) have been identified, connecting the 600 or so walled and unwalled settlements shown on the map.[120] However, it might be better to see these lines, with their accompanying distance figures, as indicating a selection of routes (Fig. 2.33). The network is highly idiosyncratic. There are, for instance, no lines at all in Scotland, nothing in the North East, and no link between Dover, or any part of the south coast, and London. Instead, the towns of Lincoln and York are linked in a virtually independent system, unrelated to the five arterial routes which reach out from London to the rest of the country. Equally eye-catching is the circular route from Gloucester around the coast of Wales and back to Gloucester, and the strangely isolated route along the south-east coast of England from Southampton to Canterbury. The presence of the east Yorkshire-Lincolnshire network has led to the suggestion that this particular copy of the map was prepared specifically for use within that area and that the two other copies thought to have formerly existed would have focussed on different parts of the country; but it has to be admitted that the map remains an enigma. The impression is certainly gained that the red lines define clear-cut operational areas, though for what purpose is impossible to guess. Whether made for government or ecclesiastical use, the Gough map, and its possible companions, became well-known, for its outline was used nearly two centuries later by map-makers as far afield as Germany (Sebastian Münster, 1540) and Rome (George Lily, 1546). Not only was the Gough map bettered by no other medieval topographical map in originality or quality, but there were no serious competitors until well into the sixteenth century.

Two points may be made by way of a coda to this review of English medieval maps. First, although there is still a need to set the maps we know about more deeply into their context, there is now a considerable cartographical literature to turn to for new light on the question of cartographical continuity from Roman times to the early Middle Ages, in particular on the transmission of texts from Late Antiquity and on cross-Channel contacts. Hitherto unknown Classical texts, Jewish and Arabic learning, new practical techniques, and new ways of thinking about the natural world stimulated the medieval renaissance, and must have had an impact on map-making. Second, the variety of surviving medieval maps to which attention has been drawn in this chapter challenges a number of assumptions about the use of maps not only in the Middle Ages but also at other times.

Mapping Country and County

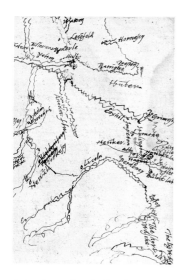

3.1   A notebook sketch: part of John Leland's map of the Humber estuary in the manuscript of his *Itinerary*. Whole map 25 × 25 cm; as shown 25 × 13 cm. 1530–1540s. Bodleian Library, MS Top. Gen. C.3., fol. 203r. Reproduced by permission of the Bodleian Library, Oxford.

In some respects, the sixteenth century saw a radical change in the cartographical situation in England. In other respects, little had altered. We need to look carefully to see precisely where novelty lay, and what was well-rooted in the past. One transformation is obvious. Whereas medieval maps used to be the intellectual or pragmatic specialist tools of small groups of users, by the end of the sixteenth century the cartographical image was fast moving towards the ubiquity and widespread familiarity it enjoys today. Maps were alluded to in Renaissance plays and poetry.[1] They featured in portraits, on playing cards, in news-sheets and in books.[2] They stood as symbols without need for further explanation.[3] They decorated not only the walls of palaces, or the lodgings of university teachers and students, leading statesmen and prelates, but now also the relatively modest private residences of successful merchants and *arrivistes* amongst the landed gentry.[4] In the new Protestant Bible from John Calvin's Geneva, maps helped the reader understand the written Word.[5] Woven into tapestries or painted onto leather screens, maps were an element of domestic furniture.[6] Maps and globes showed the inhabitants of these islands hitherto undreamt of new worlds and the distant achievements of English explorers and traders. Individuals continued to sketch maps as a form of note-taking, as did John Leland in the 1530s (Fig. 3.1) and as did the court official responsible for the planning of Queen Elizabeth's Progress into East Anglia in 1578.[7] Now, though, printed maps were taking their place in the middle ranks of society alongside the manuscript maps of the specialist user and the traditional cartographical treasures of the rulers. As the total population of England rose rapidly from the 1540s to double over the next one hundred years or so, the proportion of those who could read also increased, reaching an unprecedented 25 to 30 per cent of the total population by the mid-seventeenth century. To appreciate the role of maps in sixteenth-century England, attention has to be paid to changes in the map-using classes.

As far as map-users in general are concerned, it was the printed map which lay at the core of the cartographical transformation of England. The wealthy might imitate the ruler and order exotic artefacts such as tapestry maps (Figs. 3.2 and 3.3) but, apart from the increasing demand for maps of the nation's defences from rulers and ministers and the growth, towards the end of the century, of the fashion for a customised estate map, most people seeing maps for the first time would have come across them in books and bibles or in newsletters and political pamphlets. From the 1530s, maps printed on separate sheets were being acquired by the poorer student as well as the rich house-owner who could also afford the costly sets of printed sheets which had to be assembled and mounted on rollers before they could be displayed as wall-maps.[8] For nearly half a century, however, all these maps had to be imported. The breakthrough in home production of both book and sheet maps came only in the 1570s.

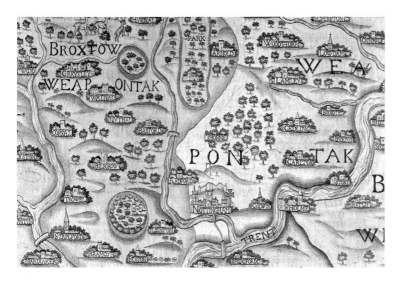

3.2 Detail from the tapestry map of Nottinghamshire made for Mary Eyre of Rampton, Notts., probably by redundant weavers from the Sheldon tapestry works in Warwickshire. Local information has been added from a variety of sources, including Saxton's and Speed's county maps and Speed's 'Battles' map. 1632. Whole tapestry 2.77 × 2.21 metres; as shown 75 × 122 cm. City of Nottingham, Museum of Costume and Textiles.
Reproduced by permission of City of Nottingham Council.

The question is not just how did the factors of change in six-teenth-century England affect the popularity and use of printed maps in England, but why topographical mapping was so tardy in its introduction in England compared with the continent. All the types of maps (except *mappaemundi*) found in England in the Middle Ages continued to be made and used throughout the six-teenth century, although not all were immediately translated into print. Indeed, most specialist maps have remained in manuscript. In the sixteeenth century, maps from overseas trade and exploration, for national defence and military planning, for boundary disputes and other legal matters, for land reclamation, of fortified towns, of buildings, of forests and woods and, especially in the last decades of the century, of manors and estates, were produced in ever-increasing numbers but almost exclusively in manuscript.[9] The earliest printed maps were book illustrations, where the map was an integral part of the text. In Italy and Germany, where book maps were also produced, and where separate topographical maps were being prepared for Ptolemy's Geography, printers had been quick to seize the marketing advantage of producing the 'first' printed map of a region, province or country. In England, prospective purchasers were obliged to get their printed topographical maps, showing other countries or the world as a whole, from abroad.[10] It was only in the early 1570s that Christopher Saxton was instructed to proceed to a survey of the whole kingdom of England and Wales, 1574 before the first printed county map of England became available, and 1579 before all English and Welsh counties were represented in print and bound into a national atlas.

The history of the mapping of England falls into three periods, each with its distinctive characteristics.[11] The first period is that of the small-scale county map produced in the second half of the sixteenth century and the early decades of the seventeenth century. Christopher Saxton's country-wide survey provided the first

3.3 Detail from a manuscript map of the Trent valley, showing the castle and bridge over the Trent at Newark, Nottinghamshire. Before 1540. West is at the top. Whole map 45 × 122 cm; as shown 45 × 76 cm. British Library, Cotton MS Augustus I.i.65.
Reproduced by permission of the British Library Board.

national coverage, and this was followed by maps made by John Norden, William Smith (whose printed maps were for long known only as 'anonymous') and John Speed. While each of these contributed something of importance to the format of county mapping (for example, key, graticule, and, in Norden's case, some roads), their seventeenth-century successors were not map-makers so much as map-publishers who added little to Saxton's, Norden's, Smith's or Speed's cartographical achievements to justify their claim of a 'new' map. The second period saw new county surveys and the production of large-scale county maps (from one inch to three inches to one mile) in the eighteenth century. The distinctiveness of these maps goes beyond mere scale, for they are outstandingly the scientific products of the Age of Enlightenment in which concern for accuracy of measurement gave the maps – notwithstanding their rococo decoration – a clinical precision, even though most landscape features were represented by map signs which Saxton and his contemporaries would have recognised. The final period brought a profound change to the organisation of the topographical mapping of England and Wales, through the Ordnance Survey's use of sheet lines rather than county boundaries (the latter being used only for Kent, 1801; Essex, 1803; Devon, 1809) and, above all, the replacement of private initiative by institutional provision.

Towards regional mapping

Most of the factors encouraging a wider recognition of maps in sixteenth-century England are those of the European Renaissance in general. While England's engagement with topographical mapping materialised over half a century later than in countries such as Italy, Germany, and the Netherlands, this was due less to ignorance of continental cartography than to English protectionist policies (affecting the book trade), English politics and reactions to the Reformation, and the idiosyncrasies of England's rulers.

A new map market

At Westminster in 1480 or 1481, William Caxton produced the first scientific book to be printed in English. This, the first printed edition of the *Myrrour of the Worlde*, a prose translation from the French of a thirteenth-century poem, was illustrated with a number of woodcuts including (in some copies) a simple T-O diagram (Fig. 3.4).[12] Other English books published before 1536 included woodcuts of an essentially cartographical nature among their illustrations, like the liturgical plans of the processional from Norwich and Hieronymus's stylised cosmological diagram.[13] There may also have been copperplate printing before the end of the fifteenth century. A copperplate map of London reported as in the library of Christopher Columbus's son Ferdinand in 1497 was said to have been printed in London.[14] But despite such promising beginnings, English printing and book illustration persistently failed to match the standards of the best continental printers in Paris, Frankfurt and Antwerp. High-quality books and maps were imported, and for much of the sixteenth century English book illustration involved either woodcuts of relatively inferior local craftsmanship or material originating abroad. When, in 1535, a map was to be included in Miles Coverdale's translation of the Bible, the beautifully designed woodcut map of the Holy Land may well have been drafted in England by Hans Holbein the Younger, but it had to be cut and printed on the continent.[15] The first maps known to have been printed in England, half-a-century after Caxton's diagrams, are three rather clumsily executed woodcuts illustrating the battle between the English and the Scots at Pinkicleugh in 1547 (Fig. 3.5).[16] Between 1548 and the end of the sixteenth century, though, over 100 book maps (mainly woodcuts) were published in England, together with some 50 separate (mainly copperplate) maps, the vast majority of which were produced in the second half of the century, after 1570.[17]

Virtually nothing is known of English copperplate printing until the last decade of the sixteenth century. Even then, some regarded copper engraving as a comparative novelty.[18] The mid-century situation is confusing. It is possible that Thomas Geminus was producing engraved prints on a press at Blackfriars in the 1540s and that in 1555 he reworked the plates for the map of Britain made by George Lily in Rome ten years previously and brought to

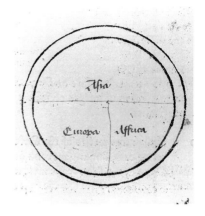

3.4 The first map printed in England: the T-O diagram from William Caxton's *Mirrour of the World*. The names of the three parts of the world were added by hand. Diameter 7.3 cm. 1481. British Library, IB.55040, fol. 35r.
Reproduced by permission of the British Library Board.

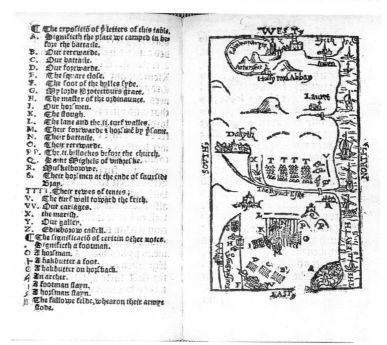

3.5   One of three maps of the battle of Pinkicleugh in William Patten's *The Expedicion into Scotla[n]d* (London, Richard Grafton, 1548), printed from woodblocks thought to have been cut in England. The map is keyed to letterpress on the facing page. 11.5 × 7 cm. British Library, C.12.d.10. Sig.G viii. Reproduced by permission of the British Library Board.

London on his return from exile.[19] The multi-sheet 'lost copperplate' map of London of 1558 or 1559 was undoubtedly made (for or by whom is not known) from accurate local information but again appears to been engraved abroad, probably in Antwerp.[20] When, finally, copperplate printing was permanently, or successfully, established in London, it was reserved for printing maps on separate sheets of paper, to be inserted into books, sold separately, or assembled into multi-sheet wall maps. For maps in books, especially those printed on the same page as text, woodcuts continued to be regarded as cheaper to produce and technically more convenient to print in the same run as letterpress.[21]

The greatest single impact of printing on cartography was in the way the new technique transformed a hitherto relatively little-used type of map, the topographical map, into a marketable commodity. There had been, as we have seen, topographical maps in the Middle Ages, both small-scale regional maps (like Matthew Paris's maps of England and of the Holy Land) and a number of relatively large-scale maps. The significant difference between those medieval examples and the printed sixteenth-century topographical map is the way the latter immediately became a multi-purpose map, made to satisfy a diverse, largely unknown and impersonal market and available for a multitude of uses. It would not be stretching the point too much to suggest that the sixteenth-century printed topographical map is a genre of its own, whose florescence put the possibility of map ownership within reach of an ever-increasing number of people from an ever-widening social spectrum.

The existence of a printing industry, and a book- and map-trade, presupposes not only demand but also the means of acquisition. The new wealth came from several sources. The inflow of precious metals and other commodities from the New World increased the supply of money in England, as in Europe in general. The English wool and cloth industry dominated European trade and financed a flourishing import trade. More particularly, Henry VIII's dissolution of the monasteries in the late 1530s enriched many people directly while inflating the land market to the general advantage of social aspirants – liberated in the process of England's precocious breakdown of feudalism and early transition to a capitalist society – as well as established landowners. The new levels of wealth encouraged a consumerism which extended, through education and travel, to books and maps, particularly the new-style topographical map. A product, usually, of systematic field surveying, the printed topographical map has distinctive characteristics. In general, its scale is sufficiently large to depict towns, nucleated settlement and the country houses of the gentry, rivers, notable bridges, forests and hills. The area selected for the map usually comprises a single major administrative unit (county or province) or a group of them. The map includes a linear scale from which measurements could be taken for route planning, and it shows the cardinal points, also useful for travellers and administrators. The printed topographical map was, as it still is, a generalist map for a wide range of uses and users.

Politics and religion

English politics also helped change attitudes towards maps in the early sixteenth century.[22] The proximity of the continent of Europe has always been an immediate and powerful factor in English social, economic and political life (Fig. 3.6). Since Roman times, neither trade nor cross-channel cultural relations have been seriously interrupted for any length of time. Even the political isolation following Henry VIII's divorce from Katherine of Aragon and the ensuing break with the Pope and the Church of Rome in the 1530s was short-lived. The seas surrounding the British Isles are narrow and have always been heavily used. The corollary of external political hostility is fear of invasion. The very real threat from France in the 1530s and 1540s and Spain in the 1580s came to be reflected first in Henry VIII's and later in Lord Burghley's wholehearted use of maps in military planning. English expansion overseas, primed by international rivalry, also increased the need for maps.[23] The theologically-based *mappamundi* was outmoded not so much by the new geography as by Protestant hostility to the products of the Catholic church. Instead, printing offered a new portrait of the world, the 'universal' or world map. But even these, like the maps of European countries, had to come from abroad, directly or through importing booksellers. Production of separate sheet world maps, suitable for display like the '*mappa mundi*' in John Checkyn's room in Pembroke College, Cambridge, in 1535/6, remained exclusively continental throughout the sixteenth century (and most of the seventeenth century). Maps of the world printed in England before 1600 were to be found only in books.[24] The first surviving separately-printed English world map was John Speed's *A New and Accura[t]e Map of the World* of 1626, but even this was published first as an atlas map.[25]

Maps were given a significant role as aids in the reading and understanding of religious texts by leading continental reformers such as Martin Luther, Philip Melancthon, Johannes Oecolampades and, above all (for the prominence he gave to maps in Genevan editions of the Bible from 1559 onwards) by John Calvin.[26] Equally relevant is the way the fickle tides of Reformation politics swept Catholic and Protestant refugees back and forth across the Channel between England and the continent. The English government's political sensitivity kept a whole cross-channel traffic of exiles, spies, court envoys and diplomats on the move. One way or another, or for one reason or another, contacts with continental Europe (and further afield) helped keep England fully supplied with new ideas, techniques, materials, craftsmen and, of course, strategic information.[27] They also fuelled a state of nervousness. Both Tudor and Elizabethan governments felt the need to be constantly informed and in a state of readiness. As Peter Barber has pointed out, by the 1540s a whole generation of statesmen had grown up under the influence of books exalting the use of maps in government.[28] Gradually the exhortations to learn about maps, and from maps, were applied to a wider reading public. An increasing number of surveying treatises and, above all, publications for self-improvement such as M. Blundeville's *Exercises, containing eight Treatises . . . necessary to be read and learned of all young Gentlemen . . . desirous to have knowledge* (1597) and, later, Henry Peacham's *The Complete Gentleman* (1622) emphasised a knowledge of maps as a desirable accomplishment. Publications like these reflected the degree to which the map-using epicentre had moved out from the traditional centres of learning (monasteries and universities) to embrace a wider, less discriminating and, no doubt, sometimes not wholly comprehending audience.

From playing an important but relatively passive role on the wall of a medieval monarch's audience chamber, maps came to be pressed into active government service. Henry VII may have benefited from the use of maps in diplomacy but it was his son on whom maps made the most discernible, and enduring, impact. The young and extrovert Henry VIII, the first of the English rulers to benefit from the wealth sweeping Europe as gold and other precious minerals poured in from the New World, clearly enjoyed his spending power.[29] Imitating long-established practices at the Burgundian and Italian courts, Henry's initial enthusiasm was for pageants, processions and *tableaux vivants*, occasions of celebration, in which maps had a place.[30] Major European artists like Hans Holbein and Vicenzo Volpe were involved in creating the decorations. In 1521, under John Ratzell's guidance, the voyages of exploration and discovery in the New World were fêted, and both

3.6 Manuscript map of the English Channel, probably by Jean Rotz. A remarkable view which brings out the proximity of England and France as forcefully as any modern aerial photograph. *c.*1542. Note the scale bar, top right. West is at the top. 90 × 44 cm. British Library, Cotton MS Augustus I.ii, fols 65, 66.
Reproduced by permission of the British Library Board.

maps and surveying instruments were displayed on stage. In 1527 the king's German astronomer Nicholas Kratzer, creator of a number of what were described as 'great' maps on ceilings and walls, also produced maps for the Greenwich festival. Other contexts provided an excuse for cartographical display: maps of the heavens were said to have been painted on the ceilings of tents and other temporary structures at the meeting of Henry VIII with Charles V in Calais in July 1520. By definition, such maps were short-lived, and we learn of them usually only through contemporary documents and letters. At the time, though, their impact would have been considerable, providing many with their first sight of a map.

Slightly less public was Henry's display of tapestries. These often huge wall-coverings were made in the Low Countries and had for long been a favourite form of cross-channel dynastic or diplomatic gift. By the end of the fourteenth century, eleven tapestries had been received by Richard II and his uncles.[31] By Henry VIII's time, the royal collection of Flemish tapestries had grown to be rich and varied but Henry's own acquisition of tapestries as a form of political ostentation seems to have been a function primarily of imitation rather than (as in the case of maps for military planning) creative initiative. By his death in 1547, the walls of his palaces were covered with 'painted cloth' (a cheaper version of tapestry decoration) containing maps and cartographical depictions of some sort.[32] The twelve tapestries celebrating Charles V's North African campaign of 1535 – one of which is a remarkable cartographical representation of Tunis as seen from Barcelona – were manufactured in the Netherlands and destined for Spain. They were first exhibited in England, on the occasion of the marriage of Charles's son Philip to Henry VIII's daughter Mary in Winchester Cathedral in 1554.[33]

Rarely, in English history, can the personality and private life of individual monarchs have had so much impact on the culture of the country at large as in Tudor England. The personal interests of a ruler, however absolute, are one matter; those of his ministers, responsible for the daily routine of government and foreign affairs, are another. Map-use among the echelons of what we would call the civil service was slower in coming. Although maps had probably been referred to in high level decision-making since the later Middle Ages in England, as far as the use of maps in routine government in the early sixteenth century goes, both king and country were 'rather backward by German and Italian standards'.[34] The situation was changing in the 1530s. Henry VIII began to take a personal interest in the use of maps in planning national defence. War with France was an imminent possibility and the need not only to preserve Calais, England's sole remaining cross-Channel possession, but also to protect the English coasts presented Henry with a specific problem for which maps provided no small part of the solution: the modernising of the nation's fortifications. Henry rose to the occasion, studying maps carefully and critically and ordering new surveys. The English settlement of Ireland too, called for maps, both general maps, to inform Queen Elizabeth's Privy

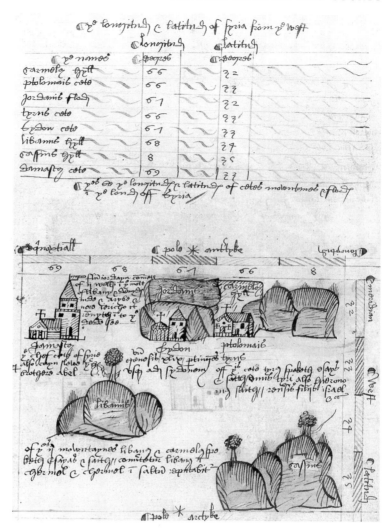

3.7 Anthony Ascham's map of Palestine, plotted from coordinates taken from Ptolemy's *Geography*. One of 19 maps in his well-illustrated manuscript commentary on Sacrobosco's *De Spherae*. 38 × 27.5 cm. *c.*1526. Yale University, Beinecke Library MS 337, fol. 55v.
Reproduced by permission of Yale University.

Councillors, and maps of the new 'seignories' – large estates – for the benefit of colonists hoping to claim their rewards.[35]

Henry's military maps were all manuscript maps. A map drawn to show defence installations and other strategic matters is hardly a public document. However, any suitable printed map can also be pressed into military service and later in the century Christopher Saxton's newly-printed topographical maps were used in this way by Queen Elizabeth's chief minister, William Cecil, Lord Burghley. In Henry VIII's time, though, there were, as far as we know, simply no printed maps of the country as a whole, or even part of it, suitable for such use. Tudor military needs, and the ruler's espousal of maps in the service of the state, were factors propelling statesmen and cartographers alike towards a national topographical survey.

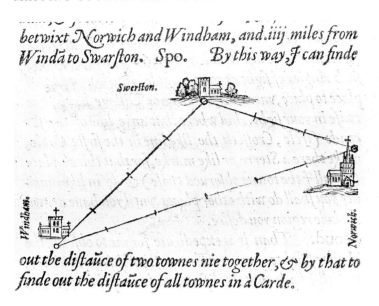

betwixt Norwich and Windham, and .iiij miles from Windã to Swarston. Spo. By this way, I can finde

out the diſtaũce of two townes nie together, & by that to finde out the diſtaũce of all townes in à Carde.

3.8   In *The Cosmographical Glasse* (1559), William Cuningham provided the first English-language description of Gemma Frisius's system of triangulation. He said nothing, though, about the calculations involved in measuring either base-line or angles, and in fact few people in England would yet have had the necessary arithmetic, despite the recent publication of Robert Recorde's *The Grounde of Artes. Teaching the Worke and practise of Arithmeticke* (London, 1543). 7 × 12 cm. British Library, G.6583, fol. 140r. Reproduced by permission of the British Library Board.

Methods of map-making

Before 1570, there were a number of now-lost manuscript maps of England known from inventories such as those compiled for Henry VIII's possessions, but few would have been sufficiently detailed to be really useful as a source of the sort of information needed in administration and planning at either national or county levels. Reasons for such a state of affairs could not have included either ignorance about the principles of topographical map-making or of a lack of potential volunteers, for there was expertise in abundance. When the sixteenth century opened, three methods of making topographical maps were known in western Europe. There was the well-tried tradition of using itineraries for the internal structure of maps, as in the Middle Ages. There was also, at least in theory, the idea of building up a map by plotting places in their mathematically correct location, that is, according to astronomically-determined coordinates of latitude and longitude. This too was a long-known idea if little used, hampered by the unavailability of a sufficient number of observations and perhaps also by a lack of widespread interest in this type of map. In 1526 or 1527, though, a young Cambridge student was already demonstrating something of his potential as a future royal tutor. Anthony Ascham has left us his translation of Sacrobosco's well-illustrated treatise on the globe (*De Spherae*, 1220/1230) to which he had added nineteen maps constructed from coordinates of latitude and longitude taken apparently from Ptolemy's Geography (Fig. 3.7).[36]

The third method of map-making was new both in England and on the continent in the first decades of the sixteenth century. Triangulation, a mathematical technique based on ancient Greek geometry of the Euclidean type, was rediscovered in Western

3.12 The anonymous manuscript map *Angliae figura* is the earliest known surviving non-Ptolemaic map of the British Isles from the sixteenth century. Degrees of latitude and longitude, calculated from a western prime meridian (probably that of the Azores) are given in the margins. Between 1534 and 1546. 64 × 47 cm. British Library, Cotton MS Augustus I.i.9.
Reproduced by permission of the British Library Board.

opposition to the Protestant powers dominating England in Henry VIII's last years, Lily would anyway have found access to modern surveys difficult, and he certainly had no interest in recording the conversion of formerly Catholic bishoprics to Henry's new church. His map accordingly represents little advance on the late medieval Gough map, except as regards Catholic Scotland, for which he was evidently supplied with recent material.

Many, then, spoke of making a map of England. There were yet others who would have had the necessary skills (Fig. 3.14). John Elder, for instance, was another Catholic under the Protestant Henry VIII but he had the better fortune to be a Scotsman. In 1543 he presented his map of Scotland to Henry VIII, describing it as a map 'wherein your Highness shall perceive and see not only the description of all the notable townes, castles, and abbeys set forth . . . but also the coast . . .'.[46] It is possible that John Elder was the mysterious 'friend' who supplied Gerard Mercator with the manuscript map of the British Isles which Mercator engraved and printed in Duisburg in 1564.[47] No other extant maps by Elder are known. Contemporary with Elder was the middle-aged Yorkshireman John Rudd (1498–1574). Again, none of Rudd's maps is known to have survived. However, records show that in 1561 he had requested a two-year leave of absence from his clerical post in Durham to travel and to prepare a map of England to replace one he had apparently already made – under considerable difficulty, he said – in the 1550s.[48] Rudd was certainly keenly interested in map-making if not yet an active map-maker. He talked about a map of the Holy Land made while he was imprisoned on suspicion of Protestant leanings, and he may have been responsible for two surviving manuscript maps, one of the diocese

3.13 England and Wales on George Lily's map of the British Isles. West is at the top. 1546, Rome. The copper plate was brought to London where it was reworked by Thomas Geminus before printing in 1555. Whole map 53.5 × 74.5 cm. British Library, K. Top. 5 (1).
Reproduced by permission of the British Library Board.

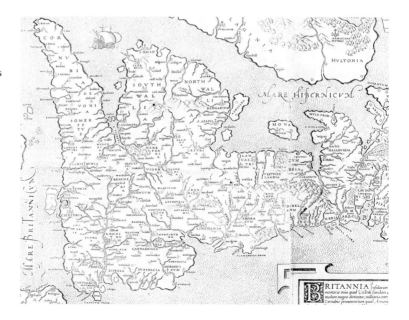

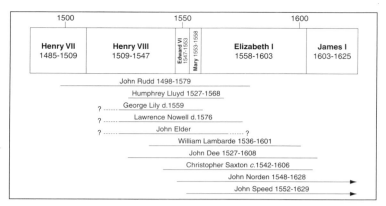

1500		1550		1600	
Henry VII 1485-1509	**Henry VIII** 1509-1547	Edward VI 1547-1553 / Mary 1553-1558	**Elizabeth I** 1558-1603	**James I** 1603-1625	

John Rudd 1498-1579
Humphrey Lluyd 1527-1568
? ...... George Lily d.1559
? ...... Lawrence Nowell d.1576
? ...... John Elder ...... ?
William Lambarde 1536-1601
John Dee 1527-1608
Christopher Saxton c.1542-1606
John Norden 1548-1628 →
John Speed 1552-1629 →

3.14 There was no shortage of map-makers in England by the middle of the sixteenth century who could have undertaken a national survey or who had already expressed an interest in producing a map of the country suitable for practical use. In the event, the task fell to Christopher Saxton, but not until the 1570s. Military surveyors are excluded.

of Durham (*c*.1569–1570), characterised by a scale bar sur-mounted by dividers, and one of Northamptonshire, also with a scale.[49]

Other map-makers who could have undertaken the production of a modern large-scale survey of England included the Englishman Robert Lythe, known mostly for his work in Ireland between 1567 and 1570, and the highly-competent Welshman Humphrey Lluyd, whose small-scale maps of England and of Wales were post-humously engraved for Abraham Ortelius and printed in 1573 (Fig. 3.15).[50] There was also the well-placed John Dee, the out-standingly able mathematician and cartographer, whose surviving map work is readily recognised by an italic hand said to match, in its neatness and precision, his personality.[51] There was Lawrence Nowell, not Dean of Lichfield (as generally stated in the older liter-ature) but a cousin of the same name, an antiquarian and philologist, and a member of William Cecil's household (Fig. 3.16). Nowell's position would have given him access to material and information useful for his map-making; certainly his surviving

3.15 South-east England on Humphrey Lhuyd's map of England and Wales, posthumously engraved and published by Abraham Ortelius (*Additamentum*, Antwerp, 1573). Coordinates are lacking. Whole map 38 × 47 cm; as shown 12.5 × 21.5 cm. British Library, Maps C.2.c.8 (3).
Reproduced by permission of the British Library Board.

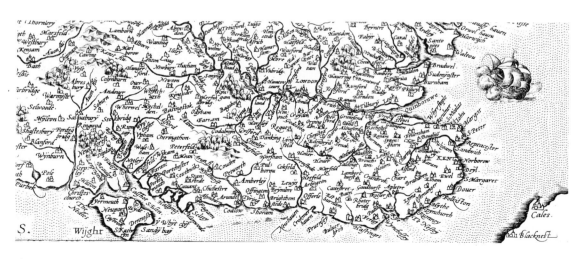

3.16 South-east England and Calais on Lawrence Nowell's manuscript 'A General Description of England and Ireland with the costes adioyning'. The margins show degrees of latitude and longitude calculated (like those in *Angliae figura*, see Fig. 3.12) from an Atlantic prime meridian. The seated figure may be a self-portrait. *c.*1564. Whole map 21 × 31 cm; as shown 9 × 14 cm. British Library, Add. MS 62540. Reproduced by permission of the British Library Board.

work reveals a first-rate cartographer.[52] In June 1563, Nowell, having already worked in Ireland (1560) and Wales, volunteered to produce a map of England and Wales.[53] He was asked to prepare a trial map but for some reason his scheme failed to materialise. Before he disappeared abroad (he was last heard of in Leipzig in 1570) Nowell left a very small copy (21 × 31 cm) of a map of the British Isles which was at least provided with, if not constructed on, a graticule of latitude and longitude and which Lord Burghley is said always to have carried with him.[54] Then, finally, there was the lawyer turned topographer and antiquary, William Lambarde,

Nowell's former pupil and friend to whom Nowell had handed over his map work before leaving England on his ill-fated trip. However, all Lambarde's known maps are in books, and it is unlikely that he ever embarked on anything cartographically more ambitious. The original of the map of the beacon network in Kent, a printed copy of which which Lambarde inserted into the second edition of his *Perambulation of Kent*, 1596 (see Fig. 7.11, p. 230), is thought to have been drafted by Philip Symonson from another now-lost map.[55]

While the names just mentioned are probably a fair representation of the outstanding cartographical talent of mid-sixteenth century England, there were yet others with known map-making aspirations. For instance, Queen Elizabeth's printer and bookseller, Reyner [Reynold] Wolfe (who died sometime between 1573 and 1579) had 'spent much time on maps', as he himself put it, and who (possibly inspired by Leland's *Itinerary*) was preparing a *Universall Cosmographie*, to include maps of England and the 'provinces', as a companion volume to Raphael Holinshed's *Chronicles*.[56] Whatever Wolfe's own skills, the point is that by the 1550s the means of making a new map of England according to the most up-to-date standards were present. There were plenty of highly competent map-makers. There were engravers willing to work in London – even if most were not English. There were instrument makers. And there was a press for printing copper plates. More significantly, the need for a national map was acutely felt in both educated and government circles. So the question is: why did nothing happen about the systematic mapping of England until the 1570s?

THE FIRST SURVEY: SMALL-SCALE COUNTY MAPS

Christopher Saxton's county maps

The answer to the above question may lie in the way factors and circumstances come to combine at one particular moment. In the second half of the sixteenth century, there was William Cecil, later Lord Burghley, who has been described as 'the most cartographically minded statesman of his time' (we should not, though, overlook Francis Walsingham, about whose now widely scattered map collection so little is known).[57] Burghley's interest in 'lineal descriptions', as opposed to verbal, is well-documented.[58] Burghley not only made sure he was provided with maps but that he made full use of them, often sketching his own. The British Library now has his 'atlas', in which – interleaved between his proof copies of Saxton's maps – are two non-Saxton printed maps, eighteen manuscript maps, and various lists and itineraries. Some of the manuscript maps are in Burghley's own hand, like the sketch-map of Liddesdale in the Scottish Borderlands (*c.*1561) and many of the annotations on the printed maps betray his personal scrutiny. The margins of Saxton's maps of Northumberland, Westmorland-

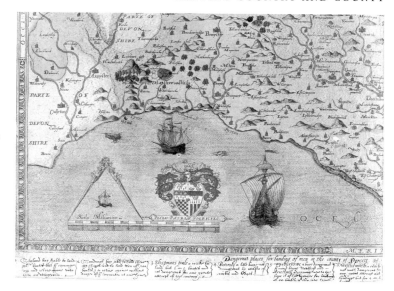

3.17 William Cecil, Lord Burghley, made considerable use of maps in government. Here he has annotated a proof copy of Christopher Saxton's map of Dorset with details of 'Dangerous places for landing men'. 1575. Whole map 37.5 × 54 cm; as shown 13 × 27 cm. British Library, MS Royal 18.D.iii, fols. 13v–14r.
Reproduced by permission of the British Library Board.

Cumberland, Lancashire, Worcestershire, and Montgomeryshire-Merionethshire have lists in Burghley's hand of the names of leading families and their residences. The map of Northamptonshire has names of the justices. On the map of Essex, Burghley has penned a jingle relating to the houses of the great. In the margins of the printed map of Devon, he has jotted down military details such as the location of gunpowder stores in local towns and a 'A Breefe Note of ye places of Descent [landing] [that] are most dangerous and require greatest regard and assistaunce' and made similar notes on the map of neighbouring Dorset (Fig. 3.17).[59] Burghley must have found the absence of coordinates or any sort of grid on Saxton's maps no small inconvenience. When it came to the question of getting hold of an adequately large-scale, accurate and up-to-date map, or set of maps, of the whole country, a prime mover had to be William Cecil, Lord Burghley.

Even when the intention to commission a map exists, some means of turning the idea into reality is needed. William Cecil's interest in maps may have been long developing, but it could be argued that it was only at a certain point in his personal or ministerial career that he found himself in a position to take responsibility for a project of this size and nature and to put it into action. The precise circumstances may never be known, although certain elements may be guessed at. In 1570, Cecil embarked on his second term of office as Secretary [of State] to Queen Elizabeth. In 1571 he was awarded a peerage (and took the title Lord Burghley). At the same time, another event, further afield, must also have been relevant. In 1570, Abraham Ortelius's *Theatrum orbis terrarum* was published in Antwerp. Sold as separate sheet maps as well as a bound atlas, Ortelius's Theatre was an immediate success throughout Europe and not least among English scholars.[60] It could have been one of the factors which triggered the English government into action and to sponsorship of the first national survey of England and Wales. Yet another prompt could have been the publication in 1568 of Philip Apian's remarkable survey of Bavaria.[61]

To a limited extent, some sort of topographical mapping had been proceeding in England before Christopher Saxton embarked on his survey in 1574, albeit in piecemeal fashion. The distinction of Saxton's work, however, is that it was based on systematic field work – surveying, observation and sketching. It covered the entire country, and the resultant maps were engraved on copper plates, which meant that they could be run off in multiple copies and reprinted (as they were, with alterations where necessary) on many occasions for years after. There is as yet no evidence which sheds light on the initial arrangements but by 1573 Thomas Seckford had selected the 30-year old Yorkshire estate surveyor to survey and produce maps of all the English and Welsh counties. Like Burghley, Seckford held prominent posts in Elizabeth I's government. He was, or was soon to become, Master of the Queen's Requests, Surveyor of the Court of Wards and Liveries, Steward of the Court of Marshalsea and Porter and Keeper of Prisoners in the Marches of Wales. He was also sufficiently wealthy to undertake the financing of the entire project, in which he was obviously co-operating with Lord Burghley. Quite why he selected Saxton remains unclear. Saxton had been apprenticed in surveying and cartographical draughtsmanship to John Rudd, although none of his early estate survey work is known.[62] Nor is anything definite known in detail about Saxton's methods of work in preparing the county maps. There was at one time considerable discussion as to whether he used a method of traversing (taking lateral measurements from a road or track) or triangulation.[63] He certainly worked fast, completing his field work in five summers. It may be assumed that he used as much ready-made material as was available and that he did in fact practise a form of Gemma Frisius's triangulation. He surely would have made use, too, of his official pass to gain access wherever possible to suitable vantage points for his theodolite sightings, including not only church towers and steeples but also the beacons which had formed a critical system of intervisibility across the kingdom since the Middle Ages for national defence.[64]

As Saxton completed the field work and the drafting of each map, it was engraved and printed, and a proof copy sent immediately to Lord Burghley. Saxton's first maps (1574) were probably those of Norfolk, on one sheet, and the three counties of Oxfordshire, Buckinghamshire and Berkshire on another. Since the last counties to be done were those in Wales (either Pembrokeshire or Glamorgan, 1577) Saxton must have worked first mainly in southern England, then in the midlands and the north and finally in Wales. A number of engravers were employed on the maps. At least six of them signed their work, accounting for 22 of the maps out of the total of 34 county maps. Although all worked in London, only one is known to have been English. Augustine Ryther (responsible for five maps) was certainly English but it is less

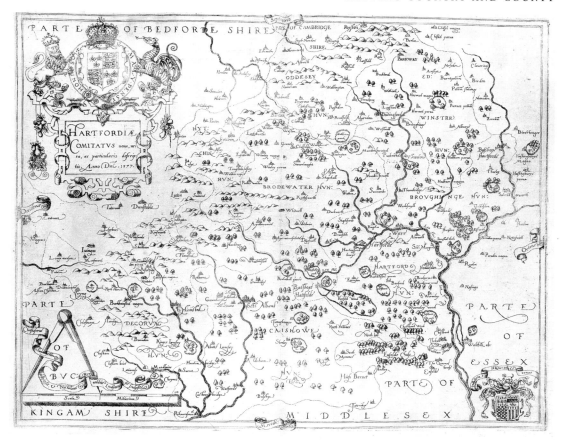

3.18 An atlas map: Christopher Saxton's map of the county of Hertfordshire. Final version, with the names of the map-maker and engraver (Nicholas Reynolds), the Royal arms, and the arms of Thomas Seckford. The cardinal points are indicated but there is no latitude or longitude. 1577. 40 × 50 cm. British Library, G.3604. Reproduced by permission of the British Library Board.

certain whether Francis Scatter (two maps) and Nicholas Reynolds (Hertfordshire) were. Remigius Hogenberg (nine maps), Lenaert Terwoort (five maps) and Cornelis de Hooghe (Norfolk only) were either Dutch or Flemish in origin.[65] Burghley added the proof copies of Saxton's maps to his reference collection as they were sent to him and must have inspected them – or even used them – instantly, for it would have been he who suggested at least some of the alterations which were carried out before further printing, with the result that Burghley's copies are different (State 1) from all others known.[66] By 1577, all the county maps were ready and Saxton was granted a licence protecting his rights to exclusive publication for ten years. So far, no mention had been made of any atlas, yet two years later all 34 county maps, a map of England and Wales as a whole, and preliminaries such as an engraved title-page, list of contents, and a list of towns and cities, were gathered and bound into a single volume and published in 1579.

In 1583, Saxton used his survey material to compile a wall map of England and Wales. The reduced scale of this map meant less information could be shown than on the county maps, even though the wall map was huge. Each sheet measured 28.5 × 44 cm and, when the 20 sheets were assembled, the whole map occupied some 1.7 × 1.5 metres. It must also have been very expensive. The copy left on Andrew Perne's death in the library at Peterhouse, Cambridge – valued for probate at 20 shillings – was by far the single most expensive item in Perne's collection of nearly 3,000 books and it would not be surprising if the map was comparatively rare even in its day.[67]

Looking at the maps in Saxton's atlas (Fig. 3.18), the initial impression is of homogeneity.[68] The map sheets vary little in size, measuring on average some 40 × 51 cm (only Yorkshire needs an especially large format, 54.5 × 74 cm, folded to fit the bound volume). They are similar in orientation (mostly more or less to north), in the provision of cardinal points, a Latin title, decorative cartouches, coats of arms, the acknowledgement to Saxton, the depiction of open dividers, usually over a scale bar, and, some-times, the engraver's name. There is also a certain degree of homogeneity as regards geographical content. Physical features like coastal sandbanks and the occasional rock, rivers, lakes and estu-aries, hills and woods are depicted. Features of human geography include towns, villages and country houses (or, rather, their named parks), county boundaries (the boundaries of hundreds more rarely), some bridges, and occasional features of interest like the coal pits in Somerset, minerals in Cumberland, antiquities (such as Hadrian's Wall and various standing stones), and the occasional windmill.

Analysis, however, reveals an underlying lack of uniformity from map to map. Map signs are not standardised. Hill signs, for example, not only come in three sizes but are styled in at least seven different ways and shaded in five different ways, and there are five different designs for the park sign. Not all of this variety can be attributed to the range of engravers employed, for individ-ual engravers were themselves inconsistent. Augustine Ryther used two different styles of park sign over two years and Remigius Hogenberg two styles over three years; Lenaert Terwoort used three styles of hill signs in two years. Distances were calculated in customary miles (equal to 1.3 statute miles). When it came to draft-ing each map, individual maps (or group of maps) were made to fit the page.[69] Saxton may, or may not, have been consistent in his own drafts but he clearly made no attempt at standardisation and the engravers worked largely as they pleased.

Students of English literature, intrigued by the post-modernist question of authorship in general, have considered the issue in rela-tion to Saxton's atlas. Richard Helgerson questioned the name which should be associated with a work like Saxton's atlas, point-ing out that Lord Burghley had supported it, Thomas Seckford had financed it (and had been rewarded for so doing by Queen Elizabeth), the Queen herself had approved the project, and that Saxton was (as stated on his pass of 1576) merely 'servant to Master Seckford, Master of the Requests'.[70] While Saxton travelled and toiled, measured on the ground, observed and drew, the pro-

ject would have been associated (the argument runs) primarily with
Thomas Seckford, Queen Elizabeth, and her Privy Council. At first
only Seckford's coat of arms appeared on the proof copy of each
map. Even those of Queen Elizabeth are missing from what may
have been the first of the maps to have been printed (Norfolk,
1574), although they do appear, in what Helgerson sees as 'a tardy
entry, crowded into a narrow margin as an apparent afterthought'
on the second sheet (Oxfordshire, Buckinghamshire, Berkshire,
1574).[71] Saxton's name does not appear until later maps (Denbigh
and Flint, 1577; Gloucestershire, 1577). Only in the final published
format of the entire set (1579) does the now-familiar threefold
acknowledgement (Seckford's arms, the royal coat-of-arms and the
inscription 'Christophorus Saxtonus descripsit') appear on each
map, setting forth, as Helgerson puts it, 'the whole system [of pro-
duction] – from royal authority, through gentry patronage, to
commoner craftsmanship'.[72] In fact, the authorship of the atlas of
England and Wales published in 1579 was not questioned at the
time, and the atlas was always acknowledged as the work of
Saxton; by William Camden in 1584, Abraham Ortelius in 1595,
John Norden in 1596, and John Speed in 1611.[73] After the death
of Queen Elizabeth (1603), the royal arms were changed to those
of James I and were dropped altogether after the Civil War.[74]
However, deconstructive exercises such as Helgerson's do more
than merely spotlight the importance of carto-bibliography; they
can serve to remind us of the principle of deep context, the need to
identify all possible inputs in the map-making process.[75]

Saxton is often cited as the embodiment of English map-making
in late sixteenth-century England. His was not merely the first systematic survey of the kingdom but a huge undertaking,
unparalleled in Europe (apart from Philip Apian's Bavarian survey)
and dependent on a high degree of organisation and financial support. At the same time, there was nothing peculiarly English in
either Saxton's methods of mapping or the appearance of his maps.
His county maps reflect cartographical practices that had long been
common across the Channel. In fact, a number of features commonplace on printed topographical maps made on the continent,
such as coordinates of latitude and longitude, an explanation or
key, the occasional road or route, are all missing and would not be
regularly found on printed English topographical maps for another
hundred years or so.[76] Specific features on Saxton's maps, such as
the pre-eminence of the sign for parks, reflect the peculiarities of
English social structure in a post-feudal age rather than cartographical innovation. Overall, Saxton's survey and atlas were, as Brian
Harley expressed it, 'first of all an English expression of developments pioneered in Renaissance Europe'.[77]

John Norden, William Smith and John Speed

Saxton's maps were not received without critical appraisal. There
were complaints about the way he had crowded several counties
together on one sheet for no very obvious reason. George Owen

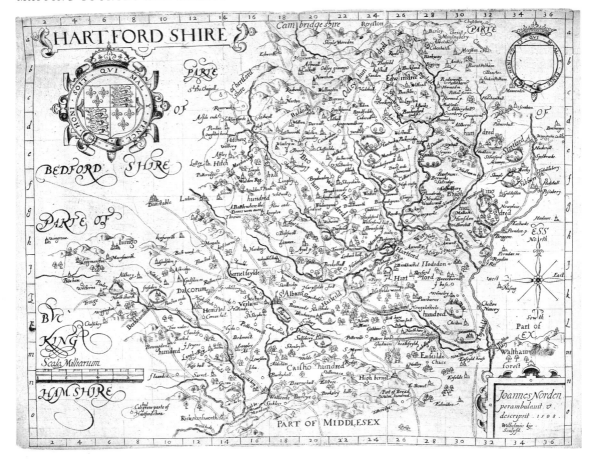

grumbled that 'soe neere together' were such counties that one town was thrust upon another.[78] Much of John Norden's work in the 1590s and William Smith's at the start of the 1600s was prompted by the need to remedy some of the shortcomings of Saxton's work. John Speed also updated the maps, at the same time turning them into decorative – and thus highly saleable – artefacts. For the rest of the seventeenth century and more, map printers and publishers lived largely from the achievements of Saxton and his immediate successors.

Like Saxton, John Norden was an accomplished and at times busily employed estate surveyor.[79] In 1590, he embarked on his *Speculum Britanniae* which was to be a pocket-sized county-by-county historical description of the country in which text (in imitation of William Camden's *Britannia* of 1586) would be combined with maps. Norden's project was never completed, possibly because the death of Lord Burghley in 1598 removed a key supporter. Only two volumes of the *Speculum* were published during his lifetime, Middlesex in 1593 and Hertfordshire in 1598

3.19   A map for a county topography: the map of Hertfordshire from John Norden's *Speculum* (1598). Only half the size of Saxton's map of the same county, Norden included more villages and parks than did Saxton and more roads than on his other county maps. Three battle sites are marked. Cardinal directions are shown and a reference grid is given in the margins. 20 × 25 cm. British Library, Maps C.7.b.23, p. 1.
Reproduced by permission of the British Library Board.

(Fig. 3.19). The ten maps (nine hundreds and one general) and text for the Description of Cornwall were printed only in 1728 and for Essex in 1840. Norden's larger maps (Hampshire, Surrey, Sussex) were re-issued for William Camden, after some alterations to the plates, early in the seventeenth century. Both map and text for Northamptonshire remain in manuscript.

Norden's maps include many of the features omitted by Saxton, notably a key to map signs, a network of selected roads (on some maps), boundaries of hundreds as well as counties, a plan of the county town (as an inset), and a reference grid (but no latitude or longitude) – cartographical apparatus which Norden had either seen for himself on printed topographical maps imported from the continent or about which he had been informed by his friend William Smith on the latter's return from his years in Nuremberg. Thus, either influenced by Smith or in response to his own originality, Norden's maps are interesting for their cartographical novelty in the context of English map-making. They are also interesting for some of the more arcane aspects of their content. It would be nice

3.20 The printed version of William Smith's map of Hertfordshire, showing that the changes made on the manuscript draft (see Plate 5) were adhered to. For the basic geography of the county, Smith followed his friend Norden rather than Saxton. c.1602. 39.5 × 49 cm. British Library, Maps C.2.cc.2 (3).
Reproduced by permission of the British Library Board.

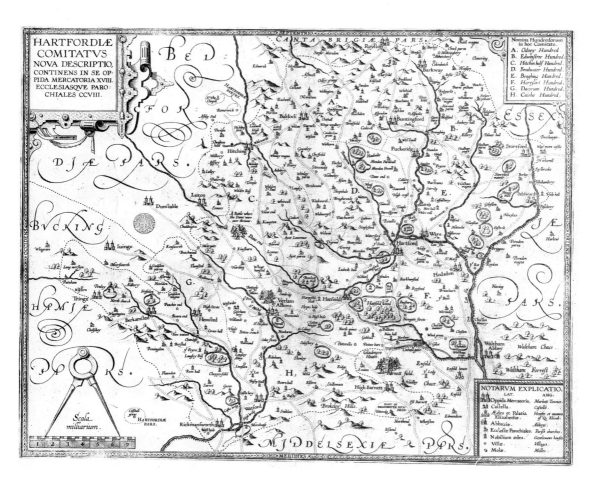

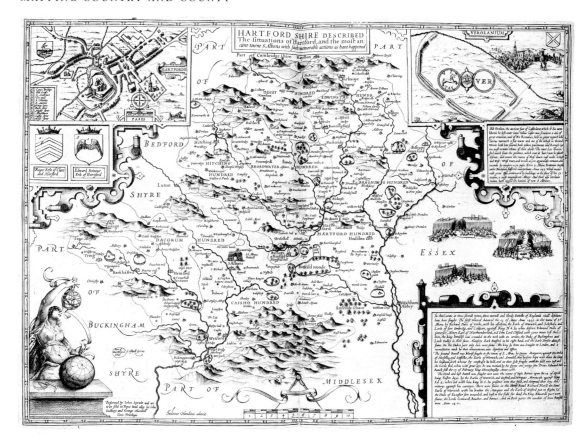

to discover, for instance, what prompted Norden to elaborate on the idiosyncrasies of the English settlement hierarchy and to differentiate so clearly 'houses of the Queen's' and 'houses of nobility' on the one hand and 'decayed places' (deserted villages) on the other. Why was he so painstaking over the inclusion of chapels-of-ease, an ecclesiastical category found on no other English, or European, printed topographical map? Easier to understand is Norden's indication of mineral deposits on the hundred maps of Cornwall and his urging, in the text, that King James should promote their exploitation.[80] Copper – in which Cornwall was rich – was by now in great demand from the printing industry and a valuable, and exportable, commodity.

Norden never went abroad nor does he seem to have had much to do with foreign cartography. In contrast, his friend the herald William Smith spent four or five full years in the cartographically important centre of Nuremberg.[81] One wonders how much Smith encouraged Norden to add to his maps the novelties which he, Smith, brought home from his experiences abroad. Or had Norden rather unfairly pre-empted his colleague? Of Smith's regional

3.21 John Speed's map of Hertfordshire may have been based on Saxton's but the final cartographical image is very much his own creation. Insets contain plans of towns (often surveyed by Speed himself). Views of antiquities, representations of local dignatories, historical or topographical notes and coats of arms of the county's leading families decorate the map. Orientation, reference grid, and coordinates are lacking. 1610, engraved in Amsterdam by Jodocus Hondius but printed in London. 32 × 39 cm. British Library, G.7884. fol. 39r.
Reproduced by permission of the British Library Board.

mapping (as opposed to town plans), a dozen printed county maps survive, together with – a rare combination – the manuscript drafts for four of them (Cheshire, Hertfordshire, Warwickshire and Worcestershire) (Fig. 3.20 and Plate 5).[82] All Smith's county maps date from about 1602 or 1603 (most carry no date) but almost nothing is known about his cartographical work; a major study is needed. It has been surmised that the map of Essex (dated 1602) was the first of Smith's county maps. It follows Saxton's very closely but with the addition of roads (mainly as shown by Norden). Not surprisingly, given their alleged friendship, Smith's other county maps follow Norden rather than Saxton. Smith's keys (in English and Latin) are always neatly set out and are less detailed than Norden's. The one on his map of Surrey, for instance, contains eight items as opposed to eleven on Norden's. In some of Smith's maps (Northamptonshire, Staffordshire), west is placed at the top of the map, rather than north.

John Speed took good note of his predecessor's and contemporaries' work. His map of Sussex acknowledges authorship as 'Described by John Norden, augmented by John Speed', that of Cheshire as 'Performed by John Speed, assisted by William Smith'. It is difficult to disentangle responsibility for every novelty when personal relations were so close. Speed, for instance, had his eye on the commercial value of his maps. Not having the official financial support to conduct ground surveys from scratch, as had Saxton, Speed was free to concentrate on corrections in place-name spelling or location, some new information, and on surveying a number of towns to supplement the plans he copied from Norden. Driven by his need to recoup his expenses, Speed's outstanding contribution was to render each map as visually attractive as possible, with vignettes of features of particular interest to potential buyers, a plan of the county town, an array of the coats of arms of the leading county officers and families, and portraits of costumed figures (Fig. 3.21). His comment was self-deprecating: 'I have put my sickle into other men's corn' but Brian Harley's appreciative sketch of Speed as 'not riding briskly through [the countryside] as Saxton had done – but poring over maps, manuscripts and printed authorities; keenly waiting the replies of his learned correspondents; copying, adapting and editing the work of others' is probably a fairer assessment.[83] Speed's maps were engraved, mostly by Jodocus Hondius in the Netherlands, and published as an atlas, the *Theatre of the Empire of Great Britaine* (1611).

A CENTURY OF RE-PUBLISHING

By the end of the first decade of the seventeenth century, all the English counties had been systematically surveyed (by Saxton) and represented on maps at least twice (by Saxton and by Speed), even, in a few cases, up to four times (by Norden and/or Smith). It would be more than a century before much new surveying for county maps would be again undertaken, although the second part of the seventeenth century did see yet more county maps coming on to an increasingly buoyant market. The vast majority of these, however, were reissues, copies or derivatives of the Elizabethan and early Stuart maps. The seventeenth century brought little or no advancement to county mapping. Instead, what we see in a little-studied period is a broadening of the cartographical front and a deeper entrenchment of the map in society.[84] Maps were printed on silk for use on screens or as handkerchiefs, were embroidered, were customised to incorporate the owner's geneaology, were displayed on study and library walls as a status symbol, and were embossed on the reverse of the Great Seal of the Commonwealth of England and on medals in general. Expensive atlases filled with county maps and maps of other parts of the world graced the shelves of domestic libraries, and the county maps themselves were used in plotting parliamentary election strategy. But little was done to improve the cartographical depiction of individual English counties, either through re-surveying or the re-design of the face of the map.[85] Closely linked to this late seventeenth-century diversification and intensification of map use (indeed, underpinning it) was an increasingly energetic map publishing industry.

The disparate pattern of map production in England over the seventeenth century reflected a number of factors, not the least of which were the political uncertainties of the period. There was the interruption of the Civil War (1642–1651), the creation of a Commonwealth, and the Restoration. Curiously, the Civil War brought major infrastructural enterprises like the draining of the southern Fenland to a halt but failed to create a significant demand for new cartography. Whatever need there was for maps generated by the hostilities, it was evidently met by the reissue of old material. Some of Speed's maps were pressed into service, for instance, and Saxton's great wall map was reduced and printed on a number of variously-shaped sheets to provide army quartermasters with a reference map which could be rolled up to fit into a saddle-bag and used in the field to locate billeted troops.[86] Even after the Restoration, Parliament tended to dampen rather than encourage enterprise, its 'general suspicion . . . of any lavish official expenditure' leading to a reluctance to sanction any cartographical project.[87] Nor did the seventeenth century bring wealth to England. While individuals might be monied, the Crown on the whole suffered a shortage of funds, especially at the beginning of the century. After 1660, the Exchequer's fortunes had improved sufficiently to permit a limited revival of royal patronage, although this would never again be on a scale equal to that which had enabled Saxton to survey the whole country in the 1570s.

Other factors discouraging further initiative in county mapping were external. From the 1650s, the primacy of the Dutch printing dynasties and engravers, and their dominance of the European map trade, was successfully challenged by new, scientific and crown-sponsored cartography in France.[88] But, as London print and map sellers began to establish their own spheres of influence, they found themselves responding to demands generated by England's diplo-

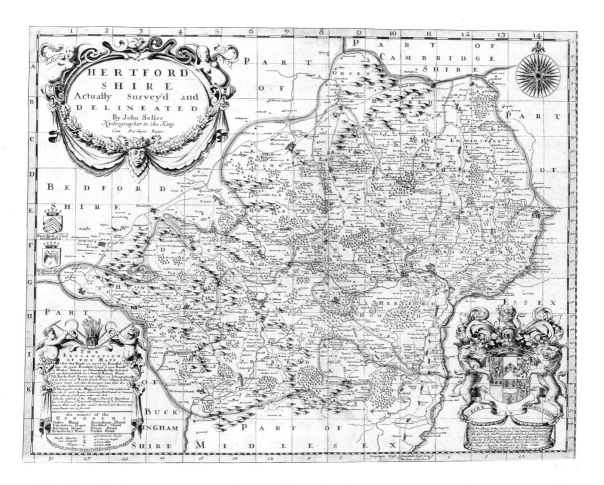

3.22   John Seller's map of Hertfordshire was the first new survey (carried out by John Oliver) since Saxton's. It was also the first to use the prime meridian of St Paul's, London, for the coordinates of longitude instead of the Azores. The coordinates are given in the bottom and right-hand margin, the other margins contain a reference grid. A compass rose gives cardinal directions. Towns are represented in plan and related to a road; parish churches, villages, 'Gentlemens Houses' and 'Ordinary houses' are indicated by pictorial signs. 44.5 × 49 cm. 1676. British Library, Maps K. Top. 15.43. Reproduced by permission of the British Library Board.

matic and military involvment in Europe and elsewhere, by her American colonies, and by the growing importance of English overseas trade. While hostilities in Ireland, France and Flanders meant that charts of coastal waters and maps of the seats of war were perceived by the government as useful propaganda as well as operational aids, private individuals seeking patronage and finance for their nation-wide county map-making projects (like John Ogilby in 1671, John Adams in 1681, and John Seller in c.1694) found none forthcoming.[89] There were some positive factors. The century saw two foreign monarchs coming to rule England, bringing with them new outlooks. As individuals, though, the Stuarts too had shown enthusiasm for maps. James I had wanted a full-scale survey of the royal forests until his project foundered for lack of finance and Charles II is said to have been 'genuinely excited by maps and charts'.[90] It was probably Charles who was responsible (in 1683) for the order to the Board of Ordnance to make a collection of military maps to be housed in the Tower of London. It was certainly Charles who gave support and encouragement to John Ogilby for his map-making and to Greenville Collins for his chart-making.

As far as the printed county map goes, then, the emphasis had shifted in the seventeenth-century to the repackaging of existing surveys in one form or another. There were attempts at something new. Philip Lea, who added roads to many of the maps in his edition of Saxton's atlas (*c.*1693), had previously published *A New Map Of England and Wales with the Direct and Cross Roads* (1687), information on the latter being taken from John Ogilby's survey of the postroads of England and Wales. Lea's map was printed in four strips with the intention of making it into a folded booklet 'so contrived to carry in the pocket . . .'.[91] There were a few new county surveys, all at small scales (that is, half-an-inch, or less, to one mile), like John Seller's *Hertfordshire Actually Survey'd and Delineated* in 1676 by Seller, John Oliver and Richard Palmer, one of the first, if not the first, county map to be based on the meridian of St Paul's Cathedral rather than the Azores (Fig. 3.22).[92] Together with new surveys of all the counties of England and Wales, the Hertfordshire map was to form part of a projected folio atlas, the *Atlas Anglicanus et Cambria* planned by Seller and his partners.[93]

Rather than new surveys and single-sheet maps, the last decades of the seventeenth century saw the take-off of the great tradition, as it was to become, of the English county atlas. Other forms of atlas were also being published, often by the same publisher, sometimes with the same maps in one form or another. There were idiosyncratic atlases like John Ogilby's atlas of postroads, *Britannia* (1675), maritime atlases like John Seller's *The English Pilot* (1671) and Greenville Collins' *Great Britain's Coasting Pilot* (1693), and general or world atlases like Seller's *A New System of Geography*, the latter advertised in the *London Gazette* towards the end of 1684 as 'accommodated with new Maps of all Countries in the whole World'.[94] Despite a number of projects like Seller's, and others, the general rule with both county atlases (discussed below) and their maritime counterparts (see Chapter 5), was to re-cycle existing – and sometimes, for sailors, dangerously old – material. When, about the turn of the century, Robert Morden produced several series of county maps, including those for Edmund Gibson's edition of Camden, his description of the maps as 'all new engraved either according to surveys before published or according to such as have been made and printed since Saxton and Speed' was true only to the extent that he did try to incorporate new topographical and antiquarian information (including roads) supplied by correspondents.

The system of up-dating provided no guarantee of accuracy. Morden, for example, unwittingly gave a new lease of life to the mythical Wiltshire village of 'Quare' through his uncritical copying of Speed's 'query' as to a missing name on the original he in turn was copying – Saxton's map of 1576.[95] Far more seriously, Collins' carelessness on a chart of the Isles of Scilly in the *Coasting Pilot* led to 'one of the greatest natural disasters to a British fleet recorded', the wrecking of the *Association*, Sir Cloudesley Shovell's flagship, and two other ships, on a group of islands misplaced on the chart by nine minutes of latitude (nine nautical miles). The lives of 2,000 men were lost on the outlying Western Rocks that October night in 1707.[96] It is tempting (if not always fair) to apply Gough's sardonic comment a century later to maps as well as to books, which Gough dismissed contemptuously as 'meagre compilations, booksellers catch-pennies, or, to speak most favourably of them, perfectly uninteresting'.[97]

New county mapping may have lagged throughout the seventeenth century – so that in the middle of the eighteenth century, Cheshire still had had no newly-surveyed map since Saxton's (1577) – and the accuracy of the printed maps so confidently marketed may be questioned, but this was not matched by stagnation in other contexts. The seventeenth century heralded the 'age of the improver' in towns, and above all in the countryside.[98] In the transformation of the economy and landscape, maps were vital agents. In the private domain, especially, increased personal wealth as well as the need to intensify output encouraged investment in land drainage, enclosure, forestry, and mining. Economic success also laid the foundations for the post-Restoration wave of interest in the landscaping of parks and gardens. Local surveyors were needed, and the results of their work were recorded on maps. In the case of mining, only surface features were represented.[99] Forest maps, too, tended to be based on the most elementary surveying procedures. By this time, the decline in the density and extent of the nation's woodland had become an issue. Exploited by metal and glass industries and required by the navy and the mercantile marine for ship-building, English woodland was a fast-vanishing asset. By 1617, Norden had surveyed and mapped well over 50 coppices and manorial woods, and Sherwood Forest – one of the most important of the royal forests represented in detail on maps since medieval times – was mapped again in the seventeenth century, at first for James I (1609), Charles I (1636) and again under Cromwell (1657).[100] Needwood Forest in Staffordshire was mapped in 1636 and 1657, in connection with enclosure and dis-afforestation proposals.[101] Sketchy as these mappings may be, they would have been rather more informative than the representation of the Forest of Dean on the map by Speed that Samuel Pepys was being shown in 1662 when, as Secretary to the Admiralty, he was negotiating a contract for timber from one of the navy's main sources.[102] But no single economic proposal matched the 'great designe' for the drainage of the Fens set in train by Parliament's passing of an 'Act for the recovering of many hundred thousand Acres of Marshes . . .' (1600), a proposal which resulted in topographical maps of a region as opposed to a county.

A regional map: the Fenland

The distinctive landscape and physical geography of the Fenland is not contained within a single county but spreads into (mainly) four surrounding counties: Lincolnshire, Huntingdonshire, Northamptonshire and Cambridgeshire. Although this has long been one of the wealthiest agricultural regions of England, capable of enriching medieval monastic foundations and Tudor estates, its former wetland environment ensured that it remained sparsely inhabited.[103] The absence of fixed boundaries in areas of fen vegetation, the wide-spread practice of inter-commoning, interference (or alleged interference) with the free flow of water in rivers and ditches ('sewers'), and failure to maintain the dykes all contributed to disputes for which, from the Middle Ages onwards, maps were frequently drawn. A new situation was reached in the seventeenth century, for the General Drainage Act of 1600 encouraged the alienation of land to any prepared to undertake, or 'venture', the capital to drain it. Speculators were attracted into this traditionally self-contained region from outside and printed maps were found useful for one reason or another, either in the process of planning and effecting each new drainage act or to help advertise the work outside the region and beyond those involved.

Although Ralph (Radulph) Agas was writing to Lord Burghley in 1597 urging him to have the region properly surveyed and mapped to show ownership (and thus identify those responsible for the upkeep of drainage), and a map of the region is said to have been made for Humphrey Bradley, a drainage engineer, by Agas and John Hexham, the first known general map of the Fenland is William Hayward's manuscript map of 1604.[104] The original of Hayward's map, once owned by by Sir Robert Cotton, is now lost but two different copies survive.[105] One of these was used in the Netherlands by Henricus Hondius for the first printed topographical map of the region, *A generall Plott and description of the Fennes* (1632), later included in Henry Hexham's English edition of Mercator's atlas (1636).[106] The fact that only the Dutch seem to have been printing maps of this particular English region is no coincidence. While London printers might have hesitated over publishing a map of a sparsely-inhabited and somewhat remote English region, the Dutch printing firms were sufficiently financially secure to take the risk of a map of possibly limited commercial viability. Indeed, by including the map in their atlases, Hondius, Jansson and Blaeu were assuring themselves of a wider market for it. For the Dutch were actively involved in the Fenland in an 'uneasy partnership' which the English only gradually came to dominate.[107] There were Dutch engineers, notably Cornelius Vermuyden who had been working widely in eastern England and who finally, in 1630, took over responsibility as director for the most ambitious project of all, that led by the Earl of Bedford and his Adventurers' Company for the drainage of the Great Level (1630–1651); and there were Dutch prisoners-of-war (as well as Scots) amongst the ten thousand or so men toiling to build the dykes and cut the ditches.[108]

In 1650, after the interruption of the Civil War and as work was nearing completion (land was under cultivation on the Great, or Bedford, Level in 1656), Sir Jonas Moore was appointed Surveyor. The first version of his sixteen-sheet map of the new Fenland landscape was ready not long after, and almost certainly by 1658.

Despite its size (1.8 × 1.2 m), its large scale (two inches to one mile), and its remarkable detail, Moore's map covers only the southern part of the Fenland and another map, on a smaller scale and printed on two sheets to show the entire region, seems to have been issued about this time, perhaps as an index map to the larger one.[109] In fact, Moore's cartographical masterpiece is known only from a single surviving example of the original printed version, a single example of Moses Pitt's issue of 1684 and a number of reimpressions taken from Pitt's plates around 1706 by Christopher Browne.[110]

The rarity of printed examples of Moore's great map points in more than one direction. On the whole, one would not expect a map associated with a highly technical project of this nature to have been printed at all, or, if a few were printed, for more than a few examples of such a cumbersome specialist map to have survived the wear and tear of a major development project. But there were other factors. On the one hand, the Adventurers' Company wanted maps not only as practical tools but also as publicity, as is evident from their early resolve 'when there is a perfect Mappe made ... to have it printed'.[111] On the other hand, Moore, too, evidently anticipated some advantage to himself accruing from sales of a printed version of his great map. In the first place, as a means of recouping some of the costs of the initial surveying, the first purchasers should have been the 87 members of the Adventurers' Company whose arms Moore portrayed in the margins of the map. In the second place, he could use the map as an advertisement of his skill as a surveyor and as a way of demonstrating his 'association with a prestigious enterprise led by influential political figures'.[112] In the event, Moore's hopes backfired. The coats of arms he had included during the Commonwealth gave his map 'a political complexion [that] was, to say the least, inappropriate' after the Restoration in 1660.[113] However, over time and despite its rarity, Moore's map has stood as a giant amongst early English printed large-scale topographical maps; for the part of Cambridgeshire covered, it remained the only large-scale map until the nineteenth century.

RE-MAPPING ENGLAND: LARGE-SCALE COUNTY MAPS

The half-century after 1750 has been diagnosed as 'a crucial and distinctive period', a veritable 'cartographic revolution' for English topographical mapping.[114] Not only were new surveys carried out but they were conducted according to the scientific principles of the Enlightenment and with a concern for the most precise measurement of distances and angles, the calculation of latitude and longitude, and from astronomical observation made with increasingly accurate instruments. Triangulation replaced the relatively easy but less accurate method of road traverse. To be useful in the new context, the maps had to be on scales larger than those used by Saxton, Norden, Smith and Speed. The county continued to be the unit of mapping, although by the end of the century dissatisfaction with its uncritical use as the basic administrative unit was being expressed. For instance, by the last decade of the eighteenth century, William Marshall was urging the adoption of 'Natural, not fortuitous lines' and 'Agricultural, not political, distinctions' for the Board of Agriculture's influential county reports.[115] County mapping continued, though, until eventually the Ordnance Survey started to map the country according to the straight lines of rectangular sheets rather than county boundaries.

Mid eighteenth-century England was already modern in its values and judgements. By the 1730s, Enlightenment thinking had put paid to Renaissance doubts as to whether the moderns should be allowed to override the authority of the ancients. Modern supremacy in fields of inquiry such as natural history, geology, chemistry, mathematics, astronomy, geography and history was insisted upon.[116] Maps took their place amongst the many 'advances which mankind are daily making in useful knowledge' and map-making instruments fascinated artists, scientists and philosophers alike.[117] Notions of 'progress' underlay advances in agriculture as well as industry. At the same time, a number of factors tied early eighteenth-century England to the past. Society was still rigidly ordered, although now the power base depended less directly on the size of an individual's personal estate or extent of landed property and more on his overall standing in London and in Parliament. Moreover, the power base now included those whose success came through the occupational hierarchies of industry and the professions as well as landownership. Henry VIII's Church of England remained the established church, although now joined by non-conforming Methodists, Baptists and Quakers. Consumer spending, literacy and education were within the reach of provincial as well as metropolitan middle classes. But the social order was still sufficiently important to ensure that each map-maker needed patronage to be commercially successful.

At the start of the eighteenth century, England was economically a relatively rich country with average incomes second only to those in the Netherlands.[118] It was also a pre-industrial economy, primarily dependent on agriculture and rural industry, although agriculture was already regionally specialised and the traditional metal working and textile industries had long supported a flourishing overseas trade. London was the largest city in Europe, housing a tenth of the nation's population.[119] As the century advanced, entrepreneurial drive, expanding consumer demand, and the buying power of a growing mass market fuelled by rising standards of living contributed to the underpinning of a seemingly insatiable map-market. Maps were needed, too, to keep up with widespread and sometimes radical landscape changes which were rendering existing maps obsolete. People were leaving the country for the towns. Industrial centres, powered at first by water and later by the coal of the midlands and the north, were expanding. Raw materials, finished products, and foodstuffs needed transportation from region to region and from the regions to the metropolis.

Traditional forms of transportation – coastal traffic, navigable rivers, and roads for goods, postal and carrier services – were continuously subject to 'improvement'. By 1730, two-thirds of the rivers in England eventually made navigable had already been attended to, and the 'canal mania' of the last decade of the century saw to it that not only lowland England but also much of northern, upland, England had access to a waterway, the cheapest means of inland bulk transportation.[120] Heavy four-wheeled waggons were in greater use than ever and responsibility for the road surface of many miles of main roads passed from the parish to the turnpike trust. So many turnpikes were created in the last three decades of the eighteenth century that by 1836 at least a fifth of all main roads was administered by turnpike trusts. Railways developed from the old horse-drawn waggon ways or tramways connecting mining districts with a waterway, as on Tyneside from as early as 1723. The replacement of wood by iron for the rails, and the exclusive use of the steam-powered locomotive, marked the real start, in 1830, of the railway era. By the close of 1835, 269 miles (430 km) of railways authorised by Parliament were open.[121]

All these developments involved maps in one way or another: as working plans, as a means for helping raise capital, or as a record of the resultant landscape changes. Working plans tended to remain in manuscript but versions were often printed in the regular and widely-circulating monthly journals. For example, nearly a quarter (53) of the maps published in the highly reputable *Gentleman's Magazine* between 1736 and about 1780 concerned road improvement and projects for canals and waterways in England.[122] The involvement of the Society of Arts (see below) may have been one of the key factors in the eighteenth-century cartographical revolution but the tide had already turned against poor quality small-scale county maps by the end of the seventeenth century. When John Strachey started field work on a map of Somerset in 1708 he complained: 'I have a list of above 150 Errors I have found myself in Speeds mapps & could not trust . . . it any where'.[123] When Strachey's own map was published in 1736, after much delay, similarly scornful charges were laid against it. The grosser of Strachey's inadequacies included the absence of any indication of scale and the omission of the postroad from Taunton to Minehead. The refrain first voiced in the 1670s by Ogilby and Adams was taken up in the mid-eighteenth century: 'It is true', claimed Peter Burdett in 1768, 'that *new* Maps of England are daily published; but it is equally notorious that they only serve to transmit to us the Errors of those from which they were copied, and generally with *new* ones'.[124] Even in the late 1770s, when Richard Gough was collecting material for his review of maps, he found plenty to be critical about, commenting that 'map-making must be at a low ebb with us when our neighbours consider [Henry Carington] *Bowles* as the Delisle or Robert [de Vaugondy] of England'.[125]

As in the mid sixteenth century, there was no shortage of skilled surveyors and county map-makers in eighteenth-century England.

Again as in the first period of county map-making, it could take time before a specific project was put into action. The training of surveyors through apprenticeship meant that a surveyor acquired his skills in one county but had to wait for that map's completion before he could work on his own account elsewhere. As much as the system engendered loyalties, there were also rivalries between different teams of surveyors. And there was always a shortage of funds. Projects were drawn out through lack of money, many were aborted, and there were bankruptcies. Thomas Jefferys's hitherto successful map business failed either by pure coincidence or as a result of over-stretching his financial commitments when he turned to county maps.[126] Less dramatically, William Yates took nearly a dozen years to launch and complete his map of Lancashire. Even so, his experiences reveal something of what must have been the day-to-day problems of many an apparently successful county map-maker.[127] There was a great need for a new map of Lancashire, for the county was one of the most rapidly developing industrial regions, but Yates had trained in Derbyshire under Peter Burdett and could not begin work in Lancashire until Burdett had renounced his own intention to survey the county. Meanwhile, Yates had other problems. His business partner, John Chapman, died and it took Yates time to unravel their joint affairs. He also had a full-time job as a customs officer. Money for the project was perhaps especially hard to come by in the late 1770s, when the map trade was depressed by the American War of Independence and the loss of lucrative trans-Atlantic markets. Exactly when Yates had formulated his idea of mapping the county is not clear, although it would have been logical for him to start on it once his work for George Perry's *Map of the Environs of Liverpool* was completed in 1769. In the event, Yates's map did not reach the engraver until 1786, eleven years after the announcement of his project in the Liverpool newspapers and six years after completion of field work.

New demands, new maps

Paul Hindle's published list of large-scale county maps suggests that twenty such maps appeared between 1699 and 1764 and another 29 between 1765 and 1783.[128] However, the list is not all it seems. To start with, many of these maps were not the result of a comprehensive new survey on the ground. When map-sellers such as Philip Lea, Robert Morden, or John Sellers in the last decades of the seventeenth century, or Philip Overton in 1715, say in the title of their maps that the map was 'actually surveyed', we have to be cautious. What such words usually meant, at best, was that the map had been *compiled* from a patchwork of existing maps and surveys. For Robern Morden's reissue of Camden's *Britannia* in 1695, re-translated and edited by Edmund Gibson, the 'new engrav'd' maps of the Proposal were not to come from any new fieldwork but from existing base maps, printed or manuscript, which were to be sent out to the counties to be corrected and updated from local knowledge.[129] Map sellers and publishers

3.23 Emanuel Bowen map's of Lancashire was made for the *Large English Atlas* (1752). The map contains little new topographical information compared with a late-sixteenth century county map, except that mileages are indicated for major routes and market days are listed below the place-name. The extra space gained from the use of an unusually large scale (one inch to $3\frac{1}{2}$ miles), and from a stylistic neatness, is filled with lengthy notes. Reproduced here from the second printing (1753). Area shown about 74×58 km (46×36 miles). British Library, Maps C.10.d.18. Reproduced by permission of the British Library Board.

continued to put out such secondary works throughout the eighteenth century. In 1752, the title of Emanuel Bowen's map of Lancashire, published at an unfashionably large three-and-a-half inches to one mile, contains the boast that the map was 'based on the best authorities' but this was true only insamuch as these authorities were none other than Christopher Saxton and John Speed (at several removes).[130] For all the space afforded by the larger scale, Bowen's map was scant improvement on the older maps in its geographical detail, for he merely filled the gaps with copious historical notes (Fig. 3.23). Apart from Jonas Moore's map of the southern Fenland, published at two inches to one mile about 1658 and reissued in 1684, only a dozen English counties had been completely surveyed anew and mapped at, or larger than, a scale of one inch to one mile prior to the announcement of the Society of Arts's premiums. Little is known about the way these maps were

prepared. It is not clear, for instance, whether the results of the field survey were transferred on to the paper at the scale of the published map or whether this had been significantly reduced. The basic science of eighteenth-century map-making remains out of sight, awaiting investigation map by map. Meanwhile, the quality of many of the 'new' large-scale county maps appearing on the market before 1761 has to be treated with circumspection.

An outstanding exception is Joel Gascoyne's nine-sheet *Map of the County of Cornwall newly Surveyed*, published in 1699 on a scale of about one inch to one mile (Fig. 3.24).[131] Gascoyne was a member of the Drapers' Company and had acquired, unusually, both chart-making and land surveying skills. His advice on charts was sought by many in London, not least by Samuel Pepys, Secretary to the Admiralty. Gascoyne left chart-making and London, though, for south-west England and worked there for the rest of his life as a land surveyor. His map of Cornwall anticipated the standards set by the Society of Arts over half a century later. It was based on a field survey which took six years to complete. Besides his estate surveying, Gascoyne must have had access to official records, such as lists of the 201 parishes and 2,475 places shown on the map (the latter keyed into an accompanying gazetteer), and the active support of the Lord Lieutenant of the county, for whom the completed map would have been an indispensable administrative aid. Apart from Gascoyne's map and those

3.24   Detail from Joel Gascoyne's one inch to one mile map of Cornwall. The county's major resources are represented in the decoration around the cartouche. Parish boundaries are marked as well as those of the hundreds. 1699. Area shown about 19 × 29 km (12 × 18 miles; 1:63,360).
Reproduced by permission from a private collection.

of Thomas Martyn (Cornwall, 1748) and John Rocque (Middlesex and Berkshire, 1754 and 1761 respectively), parish boundaries would not be shown regularly on one-inch topographical maps until the last quarter of the nineteenth century.[132] Towns are represented 'ichnographically' (in plan) on Gascoyne's Cornwall map, with buildings differentiated from the adjacent yards and gardens (Fig. 3.25). Churches with spires or with towers are distinguished (not always accurately), as are open and enclosed roads. Hamlets, villages, 'Gentlemen's Seats', wind and water-mills are shown, although curiously Gascoyne makes no mention of the tin and copper mines, recorded much earlier by Norden and later faithfully depicted by Martyn. But overall, the detail and clarity of Gascoyne's map was both highly original and far ahead of its day.

The half-century which passed between Gascoyne and John Rocque was not devoid of new county maps on a relatively large scale, but few of these approached either Gascoyne or Rocque in quality of original survey, amount of information given, or standard of presentation. Most, too, were rather less than one inch to one mile (1:63,360) in scale. Some were not even the work of qualified surveyors, like the map of Shropshire by Bas[il] Wood (1710) and the map of Norfolk by George Forster (1739). Philip Overton's two-sheet map of Oxfordshire (1725) was printed from plates acquired by him.[133] Richard Budgen's map of Sussex (1723) was perhaps unfairly condemned by contemporaries despite some original fieldwork and the inclusion of much hitherto unpublished material.[134] Other maps were based on original surveys, like the six-sheet map of Middlesex, Essex and Hertfordshire produced by John Warburton, Joseph Bland and Payler Smyth (c.1724) and Henry Beighton's excellent map of Warwickshire (c.1727, one of the few at a full one inch to one mile).[135] For his four county maps alone, though, Rocque – who between 1734 and 1762 published over 100 maps, plans, road books and indexes while running a successful map business – stands as the outstanding cartographer of the period.[136] Rocque came to county map-making relatively late in an already impressive career as a cartographer. His 24-sheet plan of London, published in 1746 but planned in 1738, is one of the milestones of urban mapping and he had mapped three provincial towns before he turned to a county map.[137] His very first maps, though, were neither urban nor county but of parks and gardens, for Rocque was also a *dessinateur des jardins* and the distinctiveness of his contribution to county mapping owes much to his application of the conventions of estate and garden plans to topographical maps. Having published in 1752 a one inch to one mile map of Shropshire on four sheets, with its innovative (but usually notional) distinctions between arable, pasture, heath and other land uses, Rocque started surveying in Berkshire. The Berkshire map was published in 1761, on eighteen sheets, and at a generous scale of two inches to one mile, accompanied, like many of the county maps of the period, by an index map and by a thirteen-page gazetteer.[138] The quality of the surveying conformed to the standards the Society of Arts would later stipulate, and

3.25   Joel Gascoyne represented towns 'ichnographically' (in plan) on his one inch to one mile map of Cornwall (see Fig. 3.24) and distinguished between buildings or messuages and yards or closes.
Reproduced by permission from a private collection.

Rocque had evidently taken care to use the best instruments and to carry out more accurate triangulation. He also followed the still relatively new practice of using the longitude of St Paul's for his prime meridian.

It is in their visual impact that Rocque's maps are strikingly original. While Society of Arts award winners like Benjamin Donn, Peter Burdett, and William Yates were all traditionalists as far as the content of their maps was concerned, Rocque's maps are uniquely detailed. For example, his map of Berkshire shows five categories of land use (arable, common, meadow, parkland, woodland) and what appears to be every field in the county (Fig. 3.26). However, the extent to which Rocque or his assistants did or did not actually survey every field boundary has been the subject of debate.[139] To survey all the fields (an exercise which would have anticipated by nearly 100 years the best efforts of the Ordnance Survey) would have involved a phenomenal amount of work for which there is no unambiguous evidence. Some modern historians would like to see absolute veracity in every line on the map, but Paul Laxton points out that 'the [enclosed] fields portrayed are too large for a two-inch survey' and suggests that while these are consistently under-represented (by an approximate factor of seven to ten) the extent of the open fields does seem to have been 'with few exceptions. . . . faithfully portrayed'.[140] Consensus rests with the conclusion that, while the boundaries of the enclosed fields were certainly not individually surveyed, they are by no means entirely fictitious, having been for the most part 'probably sketched in by eye from vantage points along the surveyed roads'.[141] Similar qualifications have to be made about Rocque's reliability as regards the specific details of land use. Some laxity, either in the field or in overseeing the final preparation of the different sheets, is revealed in the way land use along the inner edges of the Berkshire map fails to match from sheet to sheet.[142] When Rocque's map of Surrey (on nine sheets at the same scale of two inches to one mile, engraved by Peter Andrews) was put on the market in 1768, six years after Rocque's death, the problem had disappeared. In their different ways, Rocque's maps can stand with some of the best cartographical achievements of the eighteenth century.

The Society of Arts premiums

In the dynamic economic and intellectual climate of the English Enlightenment, the insufficiencies of county maps, especially those in traditional small-scale mould, gave rise to an outcry. In September 1755, William Borlase, the Cornish antiquarian and natural historian, had complained in a letter about the way 'Our Maps of England and its counties are extremely defective'.[143] A year later, in December 1756, he was complaining again how the shops 'are full of boasted surveys [which] when it is examined accurately will be found . . . excessive low, oppressed . . . with errors' due to 'the ill-capacity of common Map-makers' who had to work 'from hasty observations without a variety of good

3.26   The remarkable detailing of fields on John Rocque's 18-sheet map of Berkshire gives the impression that all field boundaries had been individually surveyed but this was not the case. Even so, the map accurately reflects the distribution of open field land. The town of Newbury is shown in plan. 1761. Scale: two inches to one mile; area shown about 27 × 5 km (17 × 3 miles; 1:31,680). British Library, Maps C.11.c.13.
Reproduced by permission of the British Library Board.

instruments'.[144] Borlase's letters were addressed to his friend Henry Baker, a fellow founder-member of the Society for the Encouragement of Arts, Manufactures and Commerce (created in March 1754 and usually referred to as the Society of Arts). In a sharply-worded memorandum, Baker insisted the Society do something about the problem of inadequate maps. At the same time, the lead in establishing a national trigonometrical survey had been taken in France by Cassini de Thury. On 10 March 1759, the Society at last announced it would offer a premium for the best county maps, resolving 'to give proper surveyors such encouragement as may induce them to make accurate surveys of two or three counties towards completing the whole' and to award an extra prize for 'an exact and accurate Level of the Rivers in any County Surveyed that are capable of being made navigable'. The basic cartographical standards were listed: the scale was to be one inch to one mile, coastlines were to be 'correctly laid down together with the Latitudes and Longitudes', triangulation was to be employed, the instruments used were be the most up-to-date and map-makers were to use 'the Theodolite or plain Table' for horizontal distances and the perambulator for road measurement (see Fig. 3.11, p. 61).

The direct impact of the Society's money prizes and medals was limited to the eleven English and two Welsh county surveys completed between 1765 and 1809 and deemed to merit an award.[145] Indirectly, though, the Society's prescription for the highest cartographical standards undoubtedly reached out much more widely, either through unsuccessful applicants or through the work winners went on to do in other counties. Local enterprise was also a major factor in the sponsoring and promotion of new maps. High standards were expected from all, and although the military officers of the Ordnance Survey later found aspects of the new county maps wanting, their judgements involved different criteria.[146]

The new cartographical face of England is best seen in the maps made by men whose efforts had been rewarded by the Society of Arts or whose work after 1761 leads one to suppose they had been influenced one way or another by the Society's criteria for a 'good' county map. In detail, the quality of the maps is uneven, although all depends on what is being judged. Most can be faulted for inaccuracy on this or that point but all were well-presented and most were skilfully engraved, like the maps of the early award winners: Benjamin Donn (Devon, 1765), Peter Burdett (Derbyshire, 1767), Andrew Armstrong (Durham, 1768), and also William Yates, who eventually won an award for his map of Lancashire (1786), having earlier completed a map of Staffordshire (1775). The impact of a few otherwise good maps, notably those by Isaac Taylor (Hampshire, 1759; Dorset, 1765), was weakened by poor engraving.

Failure to obtain an award is not in itself a guide to the intrinsic quality of the maps. Thomas Jefferys had started out as a map-engraver and publisher, re-working original copper plates (including those for Saxton's atlas) and providing maps as illustrations for books and periodicals like the *Gentleman's Magazine* (for which he engraved 21 maps between 1746 and 1757), and making a name for himself as a maker of maps of the Americas, which he made up from whatever materials could be obtained, foreign or English.[147] About 1768, however, he turned his attention to the English counties, perhaps partly attracted – as Gough suggested – by the hope of an award from the Society of Arts.[148] Concentrating on two parts of England, the southern and eastern midlands and northern England, Jefferys launched no fewer than eight county surveys: Bedfordshire (1765), Oxfordshire (1767–9), Huntingdonshire (1768), Buckinghamshire (1770), Westmorland (1770), Yorkshire (1771–2), Cumberland (1774) and Northamptonshire (1779). The survey of Nottinghamshire was commissioned but not completed, and only two of the twenty sheets of the Yorkshire map were published in his lifetime.[149] All Jefferys's maps are at a scale of one inch to one mile (except for the maps of the smaller counties of Bedfordshire and Huntingdonshire, which are at two inches to one mile), the scale which had gained favour since John Ogilby's time as small enough to allow most counties to be represented on a single sheet while also leaving room for a clear delineation of individual landscape features. Jefferys applied twice to the Society of Arts, in 1765 and 1769, but it seems to have been professional rivalry and simple bureaucracy that blocked recognition, not any fault in the maps.

Most map-makers who had an eye on the Society of Arts's premium made sure their large-scale county maps carried details of their methods of survey and, usually, a diagam of their triangulation. Perceptive map-makers and users had long complained about the way shape and position were incorrectly represented on the old or poorly-surveyed county maps. Bowen, for example, had got the latitude wrong for Lancashire with the result that his outline for the county is distorted along the east-west axis. One of the significant contributions made by the new insistence on accuracy in measurement and calculation was a major improvement in the depiction of the shape of each county which meant that, as good new surveys began to accumulate, the boundaries of neighbouring counties fitted together in a way they had previously failed to do. By 1793, a whole series of triangulations (based where possible on common survey stations) by Burdett, Yates, and John Prior effectively made a composite map covering much of the west and north-west England, and laid a conceptual foundation for the Ordnance Survey (Fig. 3.27). Increasingly, too, map-makers in the vanguard of surveying, like Jefferys, were adopting the Greenwich meridian instead of positioning each map on a local prime meridian, as John Rocque had done in Shropshire (selecting Shrewsbury) in 1752 and Isaac Taylor in Hampshire (Winchester) in 1759, or on the St Paul's meridian. Where the eighteenth-century maps failed was in the representation of the third dimension of the landscape and in their measurement of altitude. Jefferys quoted the heights of peaks in Yorkshire in yards where he should have said

⊙ Stations common to
    two counties

○ Other stations

Lancashire

Cheshire

Derbyshire

Staffordshire

Leicestershire

Warwickshire

0   20 miles

20   40 kilometres

3.27   By the end of the
eighteenth century, the county-
by-county triangulation of six
adjacent midland and north-
western counties effectively
formed a single interlinked
group. Cheshire and Derbyshire
had been mapped by Peter
Burdett, Leicestershire by John
Prior, and Lancashire,
Staffordshire and Warwickshire
by William Yates. Redrawn
from J. B. Harley, 'The Re-
Mapping of England' *Imago
Mundi* 19 (1965), pp. 56–67,
Fig. 2.

feet, making Pendle Hill, for instance, '1,568 yards', a foot higher
than Ben Nevis, Britain's highest mountain.[150]

A better idea of the context in which the new maps were produced may be gained by focussing on some of the individual maps
and their makers. Four have been selected to be discussed briefly
below. Three were winners of the Society of Arts awards: Benjamin
Donn (the first to be so honoured) for his map of Devon; Peter
Burdett for his map of Derbyshire; and William Yates, for his map
of Hampshire. The fourth, Isaac Taylor, only narrowly, and in the
view of some, unjustifiably, failed to get for his map of Dorset the
award which went to Donn. While their maps stand alone on their
own merits, they are also fully representative of the large-scale
maps of the counties of England produced in the second half of the
eighteenth century.

Benjamin Donn: In announcing the new prize early in 1759, the
Society of Arts asked map-makers to declare their intentions by 2
November of the same year. On 16 October, the 30-year old
Benjamin Donn wrote from Devon with details about his plans for
mapping that county.[151] Despite a number of mathematical publications to his credit, he had to persuade the Society that he was
qualified to produce the standard of survey required but he was
eventually allowed until March 1762 to complete his project.[152] In
the end, the map arrived early in February 1765 and in November
the Society adjudicated in favour of Donn's 'actual and accurate
survey of the County of Devon', setting aside Isaac Taylor's map of
Dorset for later consideration (Fig. 3.28).[153] Donn had committed
himself wholeheartedly to his map, raising the necessary finance
through subscription and by addressing potential sponsors personally, not only in his native county (especially in Exeter) but also
further afield in London, Oxford, Bristol and Bath. Altogether, 750
copies were subscribed for prior to publication, the money helping
to cover the expenses of field work, drafting, and Thomas Jefferys's
engraving – a total of some £2,000.

In compliance with the Society's regulations, the survey was
based on trigonometry, and Donn's surveying, as William
Ravenhill has found through checking some of his latitudinal and
longitudinal calculations, was impressively accurate: an error of
three minutes in the position of Start Point was one of the larger
mistakes, but at Lyme Regis Donn 'is almost correct'. The accolade
would have delighted the map-maker, who had recorded his sur-
prise that his astronomical observations and calculations were in
'harmony . . . to a greater degree than I myself expected, in a work
of so great extent . . .'.[154] Again as urged by the Society, Donn may
have taken especial care over the coastlines, marking rocks
(naturalistically when close to the shore; with triangles when further away) and noting good and bad anchorages about the island
of Lundy, for example. Inland, he seems to have concentrated on
indicating the most abrupt changes of gradient, by vertical shading,
rather than to have attempted to depict every hill or all upland. He

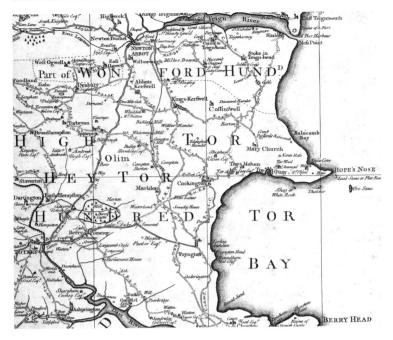

3.28 The first map to win the Society of Arts award was Benjamin Donn's twelve-sheet map of the County of Devon, completed in 1765. Compared with sixteenth- and most seventeenth-century county maps, on which few if any roads were shown, Donn's detailing of the road network is one of the outstanding features of his map. Five categories of road (all measured in the field) are identified in the key: actual and projected turnpikes, enclosed roads, open roads, and roads open on one side only. Donn also claims to show isolated farms and cottages as well as great houses and nucleated settlement. Scale: one inch to one mile; area shown about 17.5 × 19 km (11 × 12 miles; 1:63,360). British Library, Maps C.11.c.7.
Reproduced by permission of the British Library Board.

showed a full range of settlement types, from county towns to villages and from the houses of the great to isolated farms and cottages. Copper and tin mines are shown and moor, common and down are distinguished. Two town plans occupy spaces which would otherwise be empty sea, those of Exeter and Plymouth. Roads received careful attention, being verbally described as *Inclosed* or *Open*, as over commons or downs, or if hedged on one side and open on the other. The potential usefulness of the map in planning travel must have been one of Donn's selling points in raising subscriptions, for the plan of Exeter includes a key to the city's inns, and distances between places are given in figures along the roads on the main map. A further convenience was that the entire map, which had to be printed as twelve sheets (measuring when assembled some 1.8 × 1.8 metres) could be made available as a bound book, in which a general index map was included.

Peter Burdett: While Donn came from respectable, professional stock (his father was a schoolmaster), Peter Burdett was classed as a gentleman.[155] He belonged to a small entourage of cultured friends who met regularly in the city of Derby and who kept each other informed of the latest developments in the arts and sciences (Fig. 3.29). One of these friends was Earl Ferrers, who had been made a Fellow of the Royal Society for his astronomical observations, and another (with whom Burdett regularly performed music) was the painter Joseph Wright, a figure well known not only for his skill as a portrait painter but also for his interest in the sciences. Burdett would have received much informal advice and encourage-

ment from this well-informed circle for his project, the mapping of Derbyshire.

Unlike Donn, Burdett described his method of survey on his map with a diagram of the triangulation network (Fig. 3.30). He sought a suitable extent of flat land (probably the Derwent flood-plain immediately north of Derby) for his base line. A theodolite was used for bearings. Some of Burdett's sightings involved considerable distances, such as 48 miles (77 km) to the Wrekin, which suggests the use of telescopic sights (and an atmosphere much clearer than that of today). A magnetic compass adjusted to allow for the discrepancy between true north and magnetic north was used. The outcome was a map engraved on six copper plates by Thomas Kitchin in London (Fig. 3.31). Viewed by modern historians and historical geographers, Burdett's large-scale county map affords a unique view of Derbyshire in the 1760s, a time when no more than fifteen per cent of the county's parliamentary enclosures had been carried out and when the contrast between narrow, curving, hedged fields and great expanses of openfield cultivation and the unenclosed moorland grazings and heathlands was even more

3.29 Portrait of Peter Burdett. From a mezzotint after Joseph Wright's *Three Persons Viewing the Gladiator*. 1765. Derby City Museum, Acc. no. 859–15–39. Reproduced by permission of Derby City Council.

3.30 Triangulation diagram on Peter Burdett's one inch to one mile map of Derbyshire (1767). The display of such diagrams on maps aspiring to a Society of Arts award advertised the map's scientific basis. British Library, Maps 2050 (7). Reproduced by permission of the British Library Board.

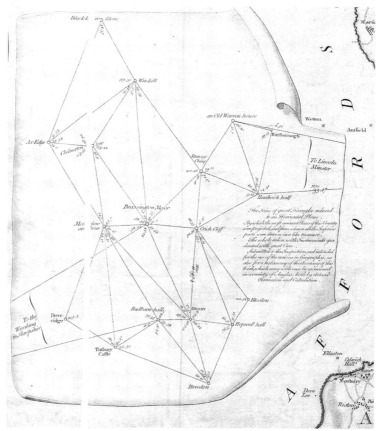

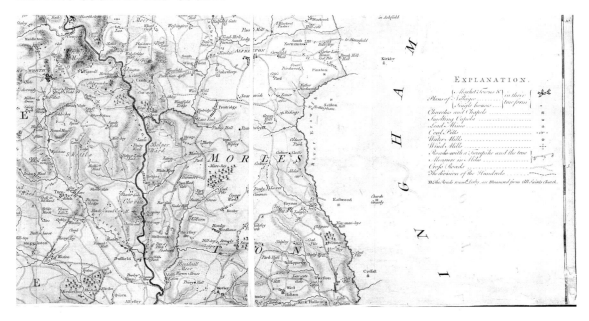

striking than it is today. The map documents the distribution of common pastures and other open land and the location of coal pits and lead mines. The distribution of manufactures is marked on the plan of Derby. Burdett moved to Liverpool after the publication of the Derbyshire map, together with his assistant William Yates, but his plan to survey Lancashire foundered for lack of sufficient subscribers and he later started his map of Cheshire before moving to Germany.

William Yates: William Yates worked at first in Liverpool as an assistant to George Perry, delaying his own attempt to survey Lancashire until Burdett had left (and, as already noted, other problems had been resolved). Yates's map of the County Palatine of Lancashire, again at one inch to one mile, was published in 1786 (Fig. 3.32). Like his master's Derbyshire map of nearly twenty years earlier, Yates's map also boldly asserted the new cartographical standards and can be held up as an example of the 'culmination of pre-Ordnance Survey cartography'.[156] Attitudes to measurement had greatly changed since Saxton's day and Yates, like Burdett, worked hard at the minutiae of mathematical precision. Certainly he would have referred to existing sources wherever appropriate but his map was primarily the result of a painstaking field survey. For angles he used a new theodolite, 'an instrument graduated with the greatest exactness'.[157] For his base line, he sought the flattest available land, in his case the broad, sandy beaches to the north of Liverpool which offered the additional advantage that no further calculations were needed to reduce measurements to sea level. Linear measurements were taken with a chain for maximum

3.31   Detail from Peter Burdett's one inch to one mile map of Derbyshire, the second county map to win an award from the Society of Arts (1767). Unlike Donn, Burdett showed only two classes of road (turnpiked roads, and secondary or 'cross' roads), but paid more attention to industrial resources, showing smelting cupolas, lead mines, coal pits, and wind and water mills. Settlements are in plan 'in their true form'. Land area shown about 19×17.5 km (12×11 miles; 1:63,360).
British Library, Maps 2050 (7).
Reproduced by permission of the British Library Board.

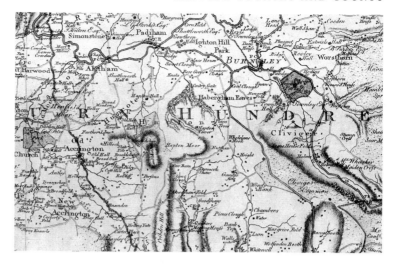

3.32 Until the use of contours, the accurate portrayal of relief presented the map-maker with many difficulties. For his award-winning map of Lancashire (1786), William Yates (like Donn in Devon) chose to emphasise the outline of major areas of upland rather than to attempt to indicate each hill and valley. Scale: one inch to one mile; area shown about 8 × 17.5 km (5 × 11 miles; 1:63,360). British Library, Maps 3155 (28) sheet 4.

Reproduced by permission of the British Library Board.

accuracy. To fix the county into the national space, Yates calculated the longitude of Liverpool at St Nicholas's Church. His observation (2° 54" W) left little room for improvement by the Ordnance Survey 60 years later, whose figure was 2° 59" W.[158]

Once the map's structure had been established by means of a system of triangles, and correctly positioned according to the prime meridian of London (probably St Paul's Cathedral), a similar degree of care was taken over the topographical infilling. Everything that could be measured was measured. Roads, rivers and canals were plotted from bearings obtained from the theodolite, and linear distances were measured by using a measuring wheel or perambulator (see Fig. 3.11, p. 61) or, perhaps, for the least accessible stretches of streams, a pedometer, a device attached to the waist and designed to register each pace taken. Measurements and observations were transferred to paper on a plane-table in the field but further instrumental work would have been completed in the drawing office by reference to sketches and notes made in the field. The final draft seems to have been Yates's own work and once this had been completed, he selected a local engraver (Thomas Billinge of Liverpool) rather than a Londoner for the task of transferring the draft on to copper plates in preparation for printing. On balance, the map was Yates's in every important respect. What we see today is the result of his selection of the map's projection and scale, his calculations of the adjustment needed to compensate for magnetic variation, and his decisions as to an acceptable degree of generalisation and as to details such as the style of map signs. Hachures replaced the old profile (mole-hill, sugar-loaf) signs for the representation of relief, but for other map-signs Yates was no innovator. For the way settlements and buildings, industries and vegetation are shown, Yates's map of Lancashire, like all the eighteenth-century county maps, remained at heart traditional.

Isaac Taylor: Unlike Donn and Burdett, Isaac Taylor was not a man of science but an estate surveyor with a keen interest in antiquities. Two of Taylor's large-scale county maps, as well as all three town plans, predate the Society of Arts's regulations.[159] They reaffirm the observation already made that, whatever the influence of those regulations, good map-makers were treading the pre-scribed path well before the Society's instructions. Yet, none of Taylor's three other county maps was granted an award, although there were those who judged his map of Dorset (1765) superior to Donn's map of Devon (Fig. 3.33).[160] Taylor's problem was his style of drawing, an untidy freehand which came unaltered from

3.33 Isaac Taylor failed to gain a Society of Arts award for his map of Dorset despite the high quality, in the opinion of many contemporaries, of its cartography, allegedly because of Taylor's poor engraving. Taylor's eye for landscape detail is reflected in many details, such as the plans of Iron Age forts and the note on the grading of pebble size on Chesil Beach. 1765. Scale: one inch to one mile; area shown about 19 × 19 km (12 × 12 miles; 1:63,360). British Library, Maps 2153 (3).
Reproduced by permission of the British Library Board.

his manuscript estate maps and which even professional engravers, like those who worked on his map of Hampshire, found impossible to formalise into something more appropriate for a printed county map. Dissatisfied with the unreliablity of those working on his map of Hampshire (1759), Taylor apparently decided to learn in order to engrave his map of Dorset himself – with no better results.[161]

It has to be admitted, though, that Taylor's portrayal of the rural landscape has an appealing evocativeness, due in large measure to the liveliness of his map language. Antiquities and architectural monuments, especially those in the formally laid out parks with their tree-lined alleys, are drawn as if from life. Hills with prehistoric earthworks rise from the map almost three-dimensionally. The key on the Hampshire map displays four different styles of sign for the 'seats and houses' of the great, two for farmhouses, three for water mills and two for wind mills, two each for ruined castles, bishops' palaces and ancient monasteries or churches and no less than five for ancient barrows. In some cases, the different signs distinguish different types of the feature in question (churches with spire, with tower, without either; post windmills, tower windmills, for instance), but in others it is difficult to follow Taylor's graphic nuancing. There are certainly long barrows and round barrows in Hampshire, but did Taylor really need three signs for the latter?

Other details would have been regarded at the time as useful. Like Gascoyne in Cornwall, Taylor in Hampshire marks the upper limit of navigation of each river. In contrast, his depiction of roads would have frustrated any traveller sufficiently rash to rely on his map for way-finding. While open and closed roads are clearly differentiated in the key, Taylor's freehand style blurs that distinction on the map, as well as any which might have been intended to identify principal and cross roads. Moreover, a number of tracks peter out in mid-course, failing to link with another road or with their supposed destination. Despite such criticisms, the contemporary judgement on Taylor's Hampshire map was that it was 'an accurate survey' and even the effect of the 'clumsy, gloomy and unequal Manner of the Engraving' of the Dorset map did not prevent the discerning from recognising an essentially good map.[162]

THE LAST OF THE PRIVATE COUNTY MAP-MAKERS

Between them, the eighteenth-century county surveyors produced the large-scale surveys which provided most English counties with modern maps for much of the nineteenth century. By 1800, Cambridgeshire and Rutland were the only English counties lacking a modern large-scale map. Elsewhere, eighteenth-century copper plates were passed from map-publisher to map-publisher, to be corrected, updated and re-used. Jefferys's plates passed to William Faden and it was Faden who re-issued Yates's map of Staffordshire in 1799, making the necessary alterations but extensively re-engraving only one sheet, to which he added the plan of

Lichfield.[163] Not all map-publishers were as scrupulous as Faden. When Henry Teesdale reissued Christopher Greenwood's map of Yorkshire (1818) in 1828, no mention was made of Greenwood. Indeed, after Greenwood, much as the original work of Saxton, Norden, Smith and Speed had been used and reused throughout the seventeeenth century, so too was much of the work of eighteenth-century county map surveyors plagiarised, copied, and in one way or another reused to meet all sort of purposes throughout the nineteenth century.

The turn into the nineteenth century saw the final phase of English county map-making by private individuals. It was not a question of decline but of 'the climax of an essentially eighteenth-century tradition', epitomised by Christopher Greenwood, arguably the most distinguished figure in the last years of the private surveyor's heyday.[164] With his younger brother John, Christopher Greenwood started a county map-making business in his native Yorkshire with his first survey and the publication in Wakefield in May 1817 and Leeds in August 1818 of his nine-sheet map of Yorkshire, at a scale of just under three miles to one inch. By 1818, when Greenwood moved to London, he had already determined 'upon making a complete Atlas of the English and Welsh Counties, the first ever attempted from Actual Survey', and to map the whole of England at one inch to one mile and Wales at three-quarters of an inch.[165] Together, the two Greenwood brothers ensured the survey of 33 English and four Welsh counties at a scale of one inch to one mile or thereabouts, all of a high standard of engraving.[166] In addition, there was the re-publication of the maps as an *Atlas of the Counties of England*. Issued in parts, between 1829 and 1834, the atlas contained 42 maps of the English counties and four of the Welsh counties, each decorated with vignettes of town views and the county's major buildings, at a scale of three miles to one inch.

It was not plain sailing, though. Like most of his predecessors, Greenwood found that 'long-term security eluded him'.[167] Surveying was a slow and costly exercise and the atlas far too expensive to make him money. Inadequate capital and falling sales in a highly competitive market ultimately led to failure, despite the high quality of the maps which he did achieve, his skills as a cartographer, the preparations with which he tried to ensure the success of each venture, and the support of a family firm. Bad luck must also be counted a contributory factor. For example, within a few days of issuing his proposals for the map of Lancashire, 4 July 1816, the two surveyors with whom Greenwood had worked on the Yorkshire map, the brothers Francis and Netlam Giles, published their proposals for an identical map.[168] In the event, their project was eventually abandoned, or bought out, and Greenwood's map of Lancashire was published in 1818.

The struggle against competitors was always present. In 1820, John Tapperell twice published proposals for a map of Worcestershire, only to be successfully challenged by Greenwood.[169] The dispute with Thomas Hodgson over the mapping of Westmorland was more acrimonious, and revealing; apart from anything else, professional reputations were at stake, and implications that Hodgson was short of finance for his map or that he lacked a sufficient number of subscribers were resented.[170] In the end, Greenwood's map of Westmorland was published in January 1824, Hodgson's in October 1828. More serious than controversy over a single map was competition from Andrew Bryant, whose scale of operations matched Greenwood's. Within the four years between 1822 and 1826, Bryant put out twelve county maps, including one of the East Riding of Yorkshire, Greenwood's home territory. There were occasions when the two map-makers raced each other to publication, Bryant beating Greenwood over the map of Surrey (June and September 1823 respectively), and Greenwood beating Bryant over the map of Gloucestershire (November and December 1824 respectively). But Greenwood in the end lost. His map of Yorkshire was reissued as early as 1828 by Henry Teesdale without any reference to the Greenwoods on it. By the 1840s, the Greenwoods' business had foundered, a number of plates had been sold and, in 1855, Christopher Greenwood died, the last of the outstanding private county surveyors.

Initially, at least, Greenwood, and his competitors benefitted from the concurrent surveying of the Ordnance Survey, founded in 1791 (see Chapter 7).[171] Private map-makers had ready access to the Board of Ordnance's trigonometrical data, published in William Mudge and Isaac Dalby's *An Account of the Operations carried on For Accomplishing A Trigonometrical Survey of England and Wales* (three volumes published between 1799 and 1811) as part of a policy aimed at the production of 'more correct maps of the counties over which the triangles have been carried'.[172] The stations were marked in the field with small stakes which had been placed over the stones sunk in the ground and were open to all. Indeed, J. B. Harley has suggested that a sub-period in the history of English county mapping could be defined as an era 'in which the trigonometrical framework was founded on official data, but much of the topographical data was compiled by traditional methods'.[173]

The growing influence of the Ordnance Survey is reflected in the style, content and standards of Greenwood's maps. From the start, and his first map, the nine-sheet map of Yorkshire, Greenwood recognised the 'unquestionable accuracy' of the Ordnance Survey's data.[174] Like the Survey, he based his maps on the Greenwich meridian. Unfortunately, by the time he was working on his maps of Worcestershire (1822) and Gloucestershire (1824) – counties for which only the principal triangulation had been published – the Ordnance Survey had stopped issuing detailed updates of their work. Private county map-makers like Greenwood had to construct their own networks of minor triangles within the official framework, incurring extra expense and leading to the sometimes variable quality of their surveys. Greenwood worked fast on such surveys, averaging 2,900 sq. miles (7,500 sq. km) a year. Since the distribution of Ordnance Survey triangulation stations was

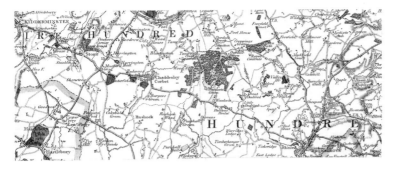

3.34 Christopher Greenwood's map of Worcestershire (1822). The Greenwoods represent the last of the private county map-makers. Although they worked independently, they were able to use geodetic data provided by the Ordnance Survey. Scale: one inch to one mile; area shown 8.5 × 16 km (5 × 10 miles; 1:63,360). British Library, Maps 23.c.19. Reproduced by permission of the British Library Board.

unrelated to county boundaries, he was able to work on the triangulation of several counties together, an investment of time and money which goes a long way to explain his anxiety to continue with his nation-wide project without competition from other map-makers. Thus the triangulation for Worcestershire and Gloucestershire was probably achieved as a single exercise. The speed at which Greenwood worked, though, may point to less than thorough methods, such as simple pacing instead of measurement with chain or perambulator.[175] Even so, Paul Laxton can point to 'Greenwood's impressive achievement' in the geodetic accuracy of the map of Lancashire (1818), seen in the remarkably close match of the parish, township and hamlet boundaries shown by Greenwood with those on the Ordnance Survey's first edition six-inch map surveyed over 25 years later.[176]

In content, though, Greenwood's maps tended to be both more traditional and more open to criticism. To some degree, this reflected his dependency on not only his own observations but also those of others – agents sent out to supply him with the desired information – and on the quality of existing cartographical sources.[177] For instance, by no means all the estate plans Greenwood used would have been particularly up-to-date or reliable, although the range of official sources he could have consulted (such as census data, Board of Agriculture reports, and parliamentary commission reports), as well as directories and gazeteers, guide-books and county histories, was way beyond anything available to earlier county map-makers. At first glance, Greenwood's map of Worcestershire (Fig. 3.34) gives the impression of a comprehensive portrayal of the rural landscape. Woodland, heath and common are distinguished and the layout of parks is represented in detail. But arable and pasture are not differentiated, nor has any attempt been made to record the importance of orchards to the regional economy, even in the Vale of Evesham. Industry, too, has received scant attention, despite its rapidly growing importance towards the Staffordshire border and the fringes of what was already well on the way to becoming the Black Country. The scattered coal mines of the Forest of Wyre are represented by a single group of pits, and the iron industry, for which Worcestershire had long been known, is scarcely acknowledged.[178] Greater attention seems to have been paid to the network of 'main and cross roads' which would have been of immediate concern to the majority of subscribers to his map. In particular, the county turnpike system, organised by 23 Trusts, was relatively accurately described, and the tollgates – which might have been seen as potential commercial sites – are marked.[179] The number of members sent to parliament by each borough or hundred is also given, as on many maps of the period.

COMMERCE AND CARTOGRAPHY

The county map which dominated English topographical map production for nearly two and a half centuries, from the 1570s to the mid-nineteenth century, had few if any parallels elsewhere in the world. Prior to the mid-ninetenth century, the county or shire, as opposed to its constituent hundreds, was of little administrative significance. As the focus of regional and political loyalty, however, it has been an inextinguishable force in the history of England, defying centralisation. When mapped, the county also had a commercial value. From John Speed's time onwards, there was always a map-publisher or map-seller ready to try to turn the county map into a profit-making venture. In some cases, as in Speed's, the driving force was the need to recover the expense of making the map in the first place. But increasingly, the commercial aspects of map production became self-interested and self-satisfying. Whereas in the later sixteenth and early seventeenth century, the novelty of the printed regional map might almost have been sufficient to justify its production on grounds of patriotism (quite apart from national need), by the end of the eighteenth century the country was overstocked with county maps, good and bad, and crowded with publishers and sellers of maps in pursuit of profit. Subscription lists, map sales, and the success of rival atlases of the English counties are but three strands of the commercialisation of English cartography in the modern period.

The cost of surveying, drafting, engraving and publishing a large-scale county map was all but prohibitive. The process of subscription as a means of financing the making and publication of maps was not new in the mid eighteenth century. John Ogilby had used every device possible – lotteries, advertisements, subscriptions, royal patronage – to raise cash for his various printing projects, but it was his rival Richard Blome who appears to have been more successful in raising money, for three years after publishing his Proposal, Blome's *Britannia* was actually published (1673).[180] A lengthy list of the right kind of supporters was essential in determining the success or failure of a map or atlas and Blome's atlas of county maps includes lists of the names and honours of the nobility and gentry who had subscribed. In the 1680s, though, as we have already seen, John Adams was less fortunate with his scheme for mapping all the counties of England, which had to be dropped for lack of response.

In the next century, almost every map-maker's first concern was the raising of finance. Patronage, if it could be obtained, was still one key to successful underwriting of the costs of surveying and printing. It is generally assumed that some of Peter Burdett's costs were defrayed by loans from friends (Joseph Wright certainly, perhaps also Earl Ferrers) and subscription.[181] The problem of obtaining finance was liable to delay production, especially if support from the local gentry was as slow as it had been in the 1730s for the 'outsider', Londoner John Strachey in Somerset. Burdett was at least living in Derbyshire, and had social and professional connections throughout the county. His friend Joseph Wright painted portraits for the descendants of Derbyshire's ancient families and for the county's gentlemen, clergymen, sportsmen, merchants and manufacturers, precisely the socio-economic classes which were as likely to patronise a local cartographer as an artist. In return, Burdett would have been expected to make sure the parks and houses of landowners who supported his venture were clearly marked on his map. In the event, one has to suppose Burdett's efforts at raising funds were successful, for his map appeared only months after his advertisement in the *Derby Mercury* (24 April 1767) announced that the forthcoming map of Derbyshire would cost 'one Guinea, half to be paid at the Time of subscribing, and the Remainder on the Delivery of the Map'.[182]

Maps were made to generate money in every way possible. One map could be published in a variety of formats. It could be sold separately or it could be re-used (as was often the case) in an atlas of English counties. As a much-reduced copy it might appear in a monthly journal like the *Gentleman's Magazine*. Popularity was courted, as in the early 1720s, when John Warburton embarked on the production of wall maps which were to be the 'most usefull and ornamental for libraries, staircases, galleries, etc that were ever published', and designed for modest subscribers.[183] Serious map-makers added town plans or views of famous landmarks around the edges of their county maps to broaden their appeal. Taylor included John Stair's plan of the site of Silchester (made in 1741), the first to show the streets of a Roman town in England, on his Hampshire map. Less serious map-publishers added almost anything they could think of which might catch a purchaser's eye as different. The margins of the county maps in Herman Moll's small atlas, *A Set of Fifty New and Correct Maps* (1724), are filled with representations of antiquities or natural features; two rows of fossils in the case of the map of Shropshire. The trend continued in the nineteenth century. Between 1830 and 1837, Thomas Moule, a bookseller turned writer, created a series of popular maps to illustrate his topographical survey *The English Counties Delineated*, sold also separately, in which he made striking use of architectural and heraldic decoration in a Gothic Revival style, targeting a fashion-conscious market with romanticised and nostalgic images of a long-lost English countryside (Plate 6).[184]

Another commercial development, from the late eighteenth century onwards, was the use of maps in unusual and inventive ways.

Many were for the education of children, like jigsaws and other map games and inflatable globes, but one large board game showing the English counties and over 100 places of interest and called *Wallis's Tour through England and Wales, A New Geographical Pastime*, published by John Wallis in London in 1794, was aimed at an adult market.[185] The depiction of the counties of England and Wales on playing cards dates back to 1590; three hundred years later the theme was no less profitable, although given a new twist in John Jaques & Son's pack of county caricatures for a game called 'Skits-A Game of the Shires'(1890).[186] Maps were printed on fans (1780s) and even on a glove, like the map of London produced for visitors to the 1851 Great Exhibition and maps also decorated domestic or library screens.[187] Curiously, none of the known extant screens bears a county map, possibly because the relatively small size of most English county maps before the middle of the eighteenth century rendered them unsuitable for the large surfaces of a screen. Instead the screens tend to show maps of the world or of towns and cities. Hooker's map of Exeter, for example, was copied onto a leather screen, either late in the sixteenth century or (opinions differ) around 1700.[188] In 1747 John Rocque suggested his map of London might serve for a screen. Another screen contains fourteen small views of Venice, nine smaller maps and George Willdey's four-sheet map of the world.[189] The maps would have been selected by the purchaser of the screen from existing stock. Particularly interesting, therefore, is the screen made about 1750 with its specially-made map, advertised by George Willdey as early as 1721 as a map of the world 'around which is added . . . a set of twenty different new sheet maps of the principal kingdoms and states of Europe with particular historical explanations'.[190] From the global dimension of the two-hemispheres world in the centre, the viewer's perspective is drawn down first to the countries of Europe (twelve sheets), then to the British Isles (three sheets), and finally to focus on the three areas assumed to be of greatest personal interest, those of twenty miles around London, Oxford and Cambridge respectively. Here was a screen designed for an educated and intellectually sophisticated English market.

Unlike his sixteenth-century predecessors, the eighteenth-century map-maker operated within a well-defined professional sphere. As John Strachey had discovered to his disadvantage in Somerset, subscriptions depended on the reputation of the map-maker. The activities of map-publishing houses must also have been powerful factors in enhancing, or otherwise, support for an individual map-maker. These businesses certainly marked a shift from reliance on the patronage of a Lord Burghley, for example, to dealing with a largely anonymous buying public and on something approaching a mass market through advertising. The late seventeenth century had already witnessed a significant expansion of prominent map- (or map and print-) selling houses, whose numbers in London rose from half-a-dozen in the 1660s to as many as sixteen by 1690.[191] The increase was a function of the growth in prosperity of post-Restoration years. Although none of the early

eighteenth-century map houses was a large concern, their professional and social heterogeneity gave them a broad marketing
outlook. For instance, the first Thomas Bowles (whose firm was so
despised by Richard Gough) was a member of the Joiners'
Company who had taken to selling maps and prints only in 1702;
George Willdey was a spectacle maker; William Berry was the son
of a baker; and John Seller senior was the son of a cordwainer.
Other map sellers had come to London from the provinces, like
Peter Stent (son of a Hampshire yeoman farmer), Robert Greene
(son of a Dorset stocking maker), Richard Mount (son of a Kentish
yeoman farmer).[192] Some, like John Senex, were sons of gentlemen.

Map selling in the eighteenth century tended to be a highly integrated form of publishing. The Bowles family were running two
businesses by 1728, as were the Overton brothers. In 1715, George
Willdey took over the business that Christopher Browne had
acquired from Robert Walton. The output of these family firms
might appear prolific – Sarah Tyacke lists over 400 advertisements,
most of them for more than one map, for the period 1660–1720
alone – but in fact well-known map sellers such as Emanuel
Bowen, John Cary, Samuel Dunn, Andrew Dury, William Faden,
John Gibson, Thomas Jefferys, Thomas Kitchin, Herman Moll,
Robert Morden, Robert Sayer and John Senex relied on a great
deal of 'recycling': the copying and reissuing of each others' maps
and atlases, the holding of shares in each others' projects, even the
trading of each others' copper plates. A few, like William Faden,
who commissioned new county maps and who printed the first
published Ordnance Survey map (Kent, 1801), were makers and
sellers of serious maps for serious purposes.[193] There was little
specialisation within any one printing house and the aim was to
produce a map that sold. John Cary, for instance, produced large
numbers of canal-plans, road-maps, maps, road-books and itineraries, geological maps and sections, as well as terrestrial and
celestial globes, and astronomical books.[194]

The majority of county maps produced between the end of the
sixteenth and the middle of the nineteenth century were either
published or re-published in atlases, and the English county atlas
(or, more accurately, the atlas of English counties) has come to be
recognised as 'a distinctive cartographic form, not exactly paralleled in any other country'.[195] When, though, in 1579, Christopher
Saxton gathered his 34 maps of the counties of England and Wales
into the single volume that has come to be known as his atlas, it is
unlikely that he had any more profound motive than expediency
and convenience. Nor was John Speed likely to have been seeking
anything deeply subliminal from the publishing of his maps as the
Theatre of the Empire of Great Britain (1611); his primary concern, in the absence of patronage, was financial. It was the early
continental world atlases of Abraham Ortelius, Maurice
Bouguereau and Gerard Mercator, not the English county atlases,
which were presented by their compilers as philosophical and
moral statements about the 'theatre' of the world or of France.[196]
In England, appeal to national patriotism, regional loyalty or (as in

the later case of Thomas Moule's *English Counties* of 1836)
sentiment and nostalgia, would not have been much more than the
gloss on the fundamental and practical concerns of successfully
marketing a nation-wide reference work for government and
adminstration, a library necessity, and a pedagogic tool.

From the publishers' point of view, an atlas was nothing if not
durable. Few map printers would have been disturbed by notions
of plagiarism or of out-of-date topography; their main concern was
to ensure that the dedications were appropriate and that the content of the maps would not give offence to any subscriber. A Latin
title was seen as a marketing aid, designed to attract (or to give the
impression of being worthy of) foreign purchasers by advertising a
product which was both 'universal' and erudite (see Plate 5).
Saxton's atlas may have lost some of its popularity when Speed's
Theatre appeared but it was reissued in 1645 from the original
plates (with appropriate changes, such as the substitution of
Charles II's coat of arms for Elizabeth's) by William Web; another
edition was projected in 1665; three editions were put out by Philip
Lea in the seventeenth century, two in English (*c.*1689, *c.*1693);
one in French (also *c.*1693); and another three appeared in the
eighteenth century, from George Willdey (*c.*1730), by Thomas
Jefferys (*c.*1749) and C. Dicey & Co. (*c.*1770).[197] Saxton's maps
were used in other atlases too. When, in 1607, the fifth edition of
William Camden's topographical and antiquarian survey of the
English and Welsh counties (*Britannia*) was for the first time provided with maps, the majority (43 of the total of 50) were reduced
copies of Saxton's. The same maps were republished in the English
translation of 1610, but a new set was engraved for Edmund
Gibson's 1695 edition by Robert Morden – one of the few publishers who can be said to have made a positive contribution to the
content of these essentially sixteenth-century maps, in this case, by
adding roads.[198]

Speed's *Theatre* was reissued in 14 editions up to 1770 (the last
was again Dicey's). The 1627 edition, published by George
Humble and his nephew Robert Sudbury, was issued together with
an entirely new atlas, composed of maps of other parts of the
world, and a map Speed had produced as a broadside in 1600 as 'A
Briefe Description of the Civill Wares, and Battails fought in
England, Wales, and Ireland', here called 'The Invasions of
England and Ireland with al their Civill Warre since the
Conquest'.[199] The 49 maps in Richard Blome's *Britannia* (1673) –
a work condemned by Gough as 'a most notorious piece of plagia-
risation' – were based mainly on Speed's.[200] All these atlases were
folio editions. Cheaper, small format editions widened the market.
About 1605 there had been the Dutchman Pieter van den Keere's
collection of 44 maps of the British Isles in an oblong octavo format. At the end of the century, Blome went on to issue a reduced
set of his maps in a work entitled *Cosmography and Geography*
(1693).

Finally, in the seventeeth century, there were John Overton's
and John Seller's atlases. Overton had acquired Peter Stent's stock

of copper plates on the latter's death in 1665. Stent's stock included the plates for William Smith's county maps which, together with other Elizabethan maps from Stent's stock, comprised the core of Overton's atlas editions (first issued about 1670, last edition, by the second Henry Overton, in 1755). Seller – whose main interests as Hydrographer to the King were the making of nautical and surveying instruments, sea atlases, and the teaching of navigation – had announced in 1679 an ambitious folio county atlas which was to embody the fruits of a completely new county-by-county survey and which was to be called the *Atlas Anglicanus*. Despite some support from the King, and the completion, with John Oliver and the engraver Richard Palmer, of the maps of Hertfordshire (see Fig. 3.22, p. 77), Middlesex and Surrey, the project for a new survey came to nothing and the mock-up of the atlas presented to King Charles contains mostly impressions of Speed's maps and some of William Smith's. What Seller eventually published, in several editions, was the modest, small format, *Atlas Contracta* (*c*.1694), aimed at the popular market.

In the carto-bibliographies of English county atlases, the first half of the eighteenth century is dominated by names familiar from the previous century: notably Morden, Blome and Overton (not to mention Speed). Of these, only Morden, with his partners, offered a new work, the *New Description and State of England*. This was printed in a handy octavo edition in 1701 and again in 1704, when a quarto edition, with maps engraved by Herman Moll, was also put on the market.[201] Morden's maps show a number of post-roads, the information for which was taken from Ogilby's road surveys of the early 1670s. On some maps (for example, Hertfordshire, Middlesex and, most strikingly, Rutland), major towns are represented in plan instead of by stylised pictorial place-signs. Morden reissued his maps in 1708 in his *Fifty Six New and Accurate Maps of Great Britain, Ireland and Wales*, adding further notes about the postroads, and again in a much augmented edition of Camden called *Magna Britannia et Hibernia, Antiqua et Nova* (1714 etc., and 1739). The fact that the usually curmudgeonly Gough approved of the *Magna Britannia* probably owes much to the way in which Morden had sent a questionnaire to correspondents throughout the counties asking them to send in 'what Accounts they think proper'.[202] In fact, *Magna Britannia* was far more than a mere reissue of Camden. Morden intended it to be the Great Britain and Ireland section of the *Atlas geographicus*, a comprehensive world geography covering both ancient and modern periods which was to contain 100 maps engraved by Moll. This was distributed in monthly parts in 1708, one of the earliest books to be issued by serial publication. In the end the *Magna Britannia* section contained only the 92 parts covering England. The maps in Moll's own small atlas, *A New Description of England and Wales* (1724), were based on those he had engraved for Morden. Although new cartographical material would have been available to him, such as John Warburton's maps of Northumberland (1716) and Yorkshire (1720), Moll does not appear to have made use of this to make any useful changes. Instead, he created his own 'improvements', notably by adjusting the county boundaries of each map so as to make them fit each other neatly but wholly unrealistically.

Emanuel Bowen's activities as a publisher started with his pocket editions (1720 onwards) of John Ogilby's road-book (*Britannia*, 1675, see Chapter 5). Indeed one hallmark of atlas publishing from the turn of the century and in the early eighteenth century was the proliferation of small atlases as well as, or instead of, the expensive folios and library editions. Another hallmark was the way atlases were often issued in weekly or monthly parts in order to spread costs for both publisher and purchaser alike, although even this provided no guarantee against failure. Thus, a work of 52 engraved plates initially launched by M. Payne, Marshall Sheepey and James Brindley at the end of 1748, was finally gathered into a single volume and published by Thomas Jefferys and Thomas Kitchin as *The Small English Atlas* at the end of 1749, with editions continuing to 1787 (the last under a new title, *An English Atlas, or Concise View of England and Wales*).

Publishing partnerships changed too, sometimes with bewildering rapidity. In the same year as Jeffreys and Kitchin produced the *Small English Atlas*, Kitchin was working with Emanuel Bowen in engraving maps for the *Large English Atlas*, 'by far the most important eighteenth-century English atlas to be published before John Cary's *New and Correct English Atlas* (1787)'.[203] At last, here was a publisher, John Hinton, organising the production of maps to a standard specification, using the best available base maps (mostly from Morden's edition of Camden) to which as much information from the latest available county surveys had been added (notably more roads) and place-name spelling revised.[204] Until then, even in the eighteenth century, the market for folio atlases still depended primarily on Saxton (whose elderly and altered plates were re-run for Willdey in 1732 and Jeffreys in about 1749), Speed and William Smith. At least nine of Bowen's and Kitchin's maps appeared by 1753, at intervals which became more irregular as, presumably, financial difficulties intervened. By the time all 45 maps were ready for binding and sale as a complete volume in May 1760, Hinton had sold out to John Tinney, and Tinney in turn had entered into partnership with three of the most successful map and print sellers of the period, Thomas Bowles, John Bowles, and Robert Sayer.[205]

Perhaps surprisingly, in view of its size and cumbersomeness, the *Large English Atlas* was a commercial success, becoming the standard upon which new county atlases were based, and later encouraging Bowen and Kitchin to embark on another, but less successful, atlas. Over the next thirty years or so, the *Large English Atlas* ran to seven editions, the last edition being published in 1787. This contained 50 maps. In addition to the county maps, there was, as announced in the title, 'a Map of the Country 35 Miles around London, a Plan of London and Westminster, and general Maps of Scotland and Ireland'. The title goes on to note the content of the county maps: *all the Cities, Towns, Villages and Churches, whether Rectories or Vicarages, Chapels, many Noblemen and Gentlemen's Seats etc.etc'* and gives details of '*the Cities, Borough and Market Towns,* [and] *the Number of Members returned to Parliament, of Parishes, Houses, Acres of Land etc. And Historical Extracts relative to the Trade, Manufactures and Government of the Cities and Principal Towns* . . . .

The successor to the *Large English Atlas* was supposed to be the slightly reduced *Royal English Atlas*, published by Bowen and Kitchin about 1764 and measuring about 47 × 31 cm compared with the 61 × 43 cm of the *Large* atlas. But presumably, as Donald Hodson suggests, the difference between the two atlases in either size or price (two guineas for the smaller version, three for the larger) was insufficient to attract somebody buying a copy for his library to the smaller work, and the *Royal English Atlas* went to only two further editions.[206] Better marketing judgement seems to have been employed in the case of John Ellis's *New English Atlas* (1765) which really was much smaller (approximately 25 × 20 cm) and cheap enough for use in schools and for purchase by the average middle class family. Ellis's atlas was a distinct commercial success, running to fourteen editions and remaining on sale over the next 60 years until well into the 1820s.[207]

In both Bowen and Kitchin atlases, the *Large* and the *Royal*, written text was given as much space as the maps. The second hugely influential eighteenth-century county atlas, John Cary's *New and Correct English Atlas* (1787), concentrated on maps alone. Cary's large-quarto atlas, a highly popular production which ran to a dozen editions before the end of the nineteenth century, was a model of visual economy, cartographical accuracy and, above all, high-quality engraving, much of it Cary's own work. Cary, whose working life spanned nearly three-quarters of a century, was the first to adopt the meridian of Greenwich on his *New Map of England and Wales* (1794).[208] As Surveyor of the Roads to the General Post Office, Cary must have travelled many of the 9,000 miles of the roads on which his *Itinerary* (1798) was said to be based. For all the variety of his work, however, it was the county atlas which, in an age of county atlases, ranked Cary with Saxton.[209]

By the middle of the nineteenth century, the county atlas had reached the end of its life. There was no shortage of new titles, and the rapidly expanding railway network lent a new urgency to keeping the maps up-to-date, but the majority of these saw no further editions beyond 1850. Those in the smallest formats (octavo or duodecimo) tended to do best, like Samuel Leigh's *New Atlas of England and Wales* (1820, eleven further editions to 1843) and Sidney Hall's *Travelling County Atlas* (1842, thirteen further editions to 1885). Of the larger ones, a few, like Cary's *New and Correct English Atlas* and the boxed 'atlas' published by John and Charles Walker, went on into the second half of the century. The final edition of Cary's *New and Correct English Atlas* was published by George Cruchley in 1862. The Walkers' *British Atlas* went to seventeen editions before the last in 1879. The only large – and relatively expensive – atlas of quality in the nineteenth century was the Greenwoods' *Atlas of the Counties of England* (1834), composed of reduced versions of, in the main, their own county surveys.

In the fifteen years between 1765 and 1780, some 25 large-scale county maps were published, covering some 65 per cent of England.[210] This was an impressive achievement on all counts but perhaps especially so given that the entire responsibility rested on private individuals – their finances, their organisational, surveying and cartographical skills, and their business acumen. By 1840, notwithstanding a slackening in activity during the Napoleonic Wars, another 80 or so maps had appeared, many of them by Christopher Greenwood and representing a second survey of counties already mapped in the eighteenth century. The eighteenth-century re-mapping of England had been carried out by a handful of map-makers working on their own initiative, within the constraints of what was possible for an individual surveyor to undertake, and to remarkably high technical standards. The project was not at all systematic, and it was certainly not national in the way that Christopher Saxton's had been, or that the Ordnance Survey's would be. As in the seventeenth century, when Ogilby, Adams and Seller had each offered to undertake a countrywide survey, there were those who had in mind to survey the entire country, John Rocque among them.[211] There was no question of sponsorship, though, and a mid or late eighteeenth-century national survey would have had to pay for itself.

In sum, the eighteenth and early nineteenth century re-mapping of England, carried out at a time of critical change in the human geography of the country, was undoubtedly of practical use in its day. Individual maps also remain useful, especially for scholars quarrying for historical information. Those large-scale county maps provide a comprehensive cartographical source for the rural and industrial changes of the period. Some thought, though, should be addressed as to how these maps should be judged. They are at once so close to those of the Ordnance Survey and yet are still deeply rooted in cartographical traditions which go back to the late sixteenth century, and earlier still. To what degree can modern historians count on them for a reliable portrayal of contemporary landscape, county by county?

As regards their content, the maps have been found to be tantalizing. On the one hand, the increased scale meant that a map-maker like Joel Gascoyne had room to plot names of small places which until then had never been recorded on any map. On the other, probably not one of the maps can be regarded as infallible as regards content. A historian seeking a definitive distribution of any category (let alone all categories) of information shown is doomed to disappointment.[212] So much can vary from map-maker to map-maker, and even from map to map, according to exactly who, within the map-making firm, was primarily responsible for drafting each sheet. Moreover, consistency in map content or appearance was not yet the criterion it would come to be with the work of the Ordnance Survey. Contemporaries remarked on the 'inaccuracies' they had noted, meaning they had identified places or features that were missing or wrongly described. They might occasionally have complained, like Edward Cave, about being led '40 miles to a fair on the wrong day'.[213] Modern scholars, used to the uniformity of *conventional* (in the strict sense) map signs, remark on different matters, such as the way the same name can be spelt differently on map and index, on lapses in the coverage of each category of information, and on the 'wrong' shape of the coastline or position of other physical features. Burdett's award-winning map of Derbyshire, for example, seems to have shown only those forges and iron works the surveyors would have passed in the course of a river traverse, and his depiction of woodland has been deemed 'erratic'.[214] But given the shortage of money and pressures of time all these private surveyors were working under, it is understandable that even top-ranking county surveyors felt obliged to ignore certain details. As Harley remarks, for Greenwood to have 'mapped the coal-pits of the Forest of Wyre, or the orchards of the Vale of Evesham, would have taken the surveyors into every field, increasing the cost of the survey beyond Greenwood's resources'.[215] County map by county map, the same conclusions are arrived at: the maps are not wholly reliable.

It is sometimes difficult for the modern map-user to judge early maps in context, and to comprehend just what achievements even these modern county maps – let alone those of Saxton – represent. It is hard for us to grasp the degree to which places in England persistently escaped the cartographer's net, to remain 'unknown' – meaning unmapped and unindexed – until modern times. Witness the problems John Adams had in 1681 when, depressed by the inadequacies of his already remarkable *Index*, he was exchanging letters with Sir Daniel Flemming about 'the two Deaneries of Carli[s]le and Alndale mentioned in your [letter], there being only Cumberland and Coupland in the Kings books . . .' in a county on which (and in which) he had been working since 1676, and for which he had already sought information from Flemming on the precise location of Longtown and Wigton, 'two Market Townes in Cumberland . . . not projected in any Map now extant'.[216] Likewise, we may be tempted to reproach Gascoyne for the inconsistency in the spelling of place-names on his map of Cornwall and in the accompanying Index until we remember that many of these may have never been written down at all. For a map made, probably, at the Lord Lieutenant's suggestion for county administration, it was sufficient to see the place represented. Spelling did matter, of course, and mis-spelling could lead to confusion, as on the occasion Adams had to reassure Flemming that 'Chorley is not omitted but written Charley'.

A private map-maker had to be pragmatic over the selection of scale, too. For an atlas, Yorkshire, the largest English county by far, was usually printed at a smaller scale than other counties simply to fit it onto a double page or, at most, a manageably-folded double page. So, while the geodetic errors of the eighteenth-century county maps undoubtedly offer many, and some risible, inaccuracies, these were not concerns of the first order of importance.[217] The indication of triangulation stations on a map would have informed, as Burdett put it, 'the curious in Geography' but would have made little practical difference to those who purchased it for display or reference at home. When in 1801, the first national census was conducted and an attempt was made to calculate the total area of parishes, existing large-scale county maps were usefully studied for what they revealed about the county boundaries.[218] The Ordnance Survey, with its small army of surveyors and Parliament to answer to, could afford assiduously to plot the full outline of every patch of carr, moor or timber omitted by Burdett in Derbyshire and Greenwood's missing coal pits, orchards and ironworks in Worcestershire. But while doing so, the military surveyors must surely have admitted that they had much to live up to in the best work of the best county surveyors of the later eighteenth and early nineteenth centuries.

Mapping Property:
Private Land and the State

Probably a majority of maps concern ownership in some way. Even a county boundary on a topographical map implies an administrative responsibility for that portion of the nation state. In this chapter, the focus is directed to maps of landed property belonging to private individuals, to institutions, or to communities. The first category examined, that of private estate maps, is heterogeneous and includes maps of varying degrees of complexity and function. The second category, that of cadastral maps, is in contrast more homogeneous as regards format and visual appearance. The reasons for the mapping also differ, being quite specific in the case of cadastral maps (enclosure and tithe maps) but diverse and often elusive in the case of estate maps.[1] Yet all types of property maps share similar identifying characteristics; they can be thought of as a 'family of maps'. In the first place, all property maps have as their main function the portrayal of lines demarcating land owned or occupied by one person from that owned or occupied by another. These property boundaries may be coterminous with the physical boundaries of land parcels (and in enclosed countryside and in towns they often are), but in open-field, communally farmed areas the boundaries of the tenurial *parcelles* portrayed on the map may have no physical expression in the landscape. Secondly, a property map commonly is associated with a written document (a terrier, a book of survey, a cadastral register) in which details of the constituent parts of each property, such as the area and land use of individual land parcels, and the names of owners and occupiers are recorded.[2] Thirdly, maps of landed property are nearly always large-scale (for example, 1:10,000 or larger) and are usually manuscript, although some cadastral maps were printed. Finally, English and Welsh property maps, with a few nineteenth-century exceptions, are usually the work of private land surveyors who were working on the basis of individual commissions. Consequently, a group of property maps will tend to share similar technical and stylistic cartographical characteristics which can be traced back to a particular surveyor.

While an estate map records the boundaries of land owned by a particular person (or group of individuals or an institution), cadastral maps record property falling within a particular administrative unit, for example a parish or township.[3] English cadastral maps thus concern not a single individual but either the entire local community (enclosure maps) or some combination of community and national interests (tithe maps). They usually identify property owners within a parish by linking the properties represented on the map with entries in a register in which details of the property are recorded. Thus, while the private estate map can be (and has often been) classed as cadastral, we have preferred to narrow the definition to that described. At the same time, our view is broader than that of those who would restrict the appellation 'cadastral' to taxation mapping alone.[4]

ESTATE MAPS

Estate maps are a sub-group, or genre, of property maps. They are maps of individual domains, properties in which a single land owner has an interest. Why estate maps were produced is a much-debated matter. In Frederick Emmison's terms, estate maps were produced 'for the information and pleasure' of land owners or they may have been produced for 'inventorying, management or improvement' of the property.[5] However, such pragmatic roles cannot be ascribed to all estate maps.

Estate maps vary considerably as regards the extent of territory covered. A straightforward case would be a map which portrays all the land owned by a particular individual or institution. Such an estate might comprise a handful of scattered strips in open fields, or a small group of land parcels worked as a single farm unit, or a much larger territory composed of a number of separate farms. Thus an estate may cover part or the whole of a parish or even extend across parish boundaries. However, not all the land comprising an estate need be contiguous. Large estates may comprise spatially fragmented parts, each separated by several miles, in different parishes or even dispersed among several counties over the length and breadth of the country. In the case of such properties, a single estate map tends to represent only a small proportion of the total land holding. These smaller maps have survived in vast numbers, although it is clear that probably at least as many have perished. They may show a few plots or a whole farm, pieces of woodland, land on which roads or canals are to be built, or the distribution of entrances to mines or the extent of ore or coal deposits on an estate. Occasionally, all constituent parts of a large, dispersed estate were mapped separately and the individual maps bound together to form a comprehensive estate atlas (Plate 7).

It is not always easy to draw the line between a map of an estate and a map of just a few plots of land, particularly for medieval maps. For example, on the fifteenth-century map from Shouldham, Norfolk, the openfield *parcelles* depicted belonged not to one but to two owners (manors) and the Shouldham map is thus by our definition not an estate map (Fig. 4.1).[6] The many property maps made in connection with a boundary dispute are also ambiguous – a map which simply illustrates features along the boundary in question cannot really be considered an estate map proper.[7] That said, there is no doubt that the legal process was a major stimulus to property mapping at least in the sixteeenth century.[8] Finally, we may ask how much of an owner's property needs to be shown on a

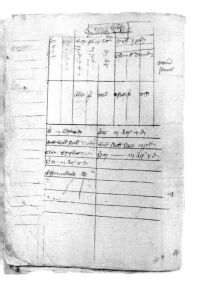

4.1 Map of open fields at Shouldham, Norfolk, *c.*1441. One of two maps in a draft terrier, it shows the layout of land parcels in a small part of the parish. The text records the names of tenants and parcel acreages. 29 × 22 cm. Norfolk Record Office, Hare MS 2826, fol. 16v.
Reproduced by permission of Norfolk Record Office.

map for it to qualify as an estate map? What are we to make of a map which shows a few square yards of urban property, on which only a single house or building is sited, for example? Provided that the ground is owned by a single person (or institution), the answer is technically yes, it is an estate map, although in practice we suspect that such maps have usually been ignored in searches for estate maps.

The Roman-medieval discontinuity in property mapping

Whatever the arguments for an underlying Romano-medieval cartographical continuity (see Chapter 2), they do not apply to the mapping of properties. A discontinuity in property mapping is found not only in England but across the whole of Europe. Roman Empire administrators had a well-developed sense of map awareness, and maps of properties were used as a means of exerting and maintaining social and political control, particularly in the context of colonial settlement, in areas of centuriated land grants, and in accounting for state revenue. The *Corpus agrimensorum*, a collection of illustrated manuscripts of varying dates assembled in the fourth century AD, served as a textbook for training the *agrimensores* (land surveyors). The *Corpus* provides evidence of the way Roman surveyors worked, at least in the context of expropriated colonial lands or wherever state rights to land revenues had to be re-established.[9] In view of the Roman use of the property map as a powerful instrument of social and political control, we might be forgiven for assuming that, once invented, the idea would have been taken up avidly by successive rulers. Yet with the fall of Rome, the use of maps to describe and record landed property was effectively discontinued. In Anglo-Saxon England, description of property boundaries in land charters was wholly verbal.[10] Medieval authorities clearly preferred to identify the extent of a property through written descriptions of, literally, the 'extent' of each parcel within the property and of the topological relationships involved which were set down in terriers or manorial extents. The result was a virtually complete abandonment of the use of property maps in feudal England.

A number of reasons may help explain, at least in general terms, the rarity of medieval property maps and, conversely, the increasing frequency with which they were compiled during the Renaissance. It has been suggested that a community would have been concerned with the boundaries of its land only under pressure from neighbouring communities, or from intruders, and that clashes of interest over the all-essential 'waste' – peripheral woods and grazing – and over rights to water and to the seasonal exploitation of marsh, heath, and moor only surfaced when the reservoir of unappropriated land was perceived to be nearing exhaustion.[11] A more persuasive factor is the way that medieval income from land was calculated not by reference to area-based quotients but to yield. Income was thus ultimately derived from the possession of rights to land use, not from rights to the land itself, as defined by manorial custom. The manorial extent was concerned first and foremost with the productive contents of a manor. It was an enumeration and valuation of the assets and rentals, not a description of the location of the buildings, orchards, pastures, woods, and ploughlands; it was an economic account, not a geography.[12] Nevertheless, some descriptions are highly elaborate. When the land of a particular manor is detailed, parcel by parcel, according to the layout on the ground and described in terms of bounds and abuttals, we have a verbal map.[13] Their compilers had yet to find a reason to rediscover the critical threshold across which lay the more succinct method of graphic description. Another factor in traditional medieval property description is that boundaries of estates and manors were marked primarily by topographical features, easy to describe adequately in words for the requirements of the local community. Where special trees, hills or streams were insufficient, man-made features such as mere-stones and balks were placed specifically for the purpose. The continued existence of all boundary markers was regularly checked by perambulation.[14]

How far medieval land surveyors were technically qualified to locate precisely a piece of land on a map and exactly to measure its area before the rediscovery of Euclidean geometry is a questionable point.[15] There is no doubt that land parcels could be represented graphically in their correct topological relationships, like the arable strips at Shouldham.[16] To us it may seem axiomatic that a map is the medium *par excellence* for the description of an arrangement of land division as complex as the open-field system, but the fact remains that the Shouldham map is one of the earliest surviving examples of an attempt to map open fields. Early property maps are not drawn to scale, but they show in diagrammatic form the kind of information which would have been set down verbally in a written extent: matters such as farm boundaries, names of tenants and freeholders; arable, meadow, pasture and commons; and the ownership of various tracts of land.

The process by which any one landowner might have moved towards the graphic representation of his land as he became conscious of the potential of property maps has been demonstrated from the documentary record of two small West Midland manors.[17] John Archer, who owned the land between 1472 and 1519, was closely involved in their management. At some time about 1500, when the usual written survey of his manor was being compiled, Archer constructed a schematic sketch map to serve as an *aide-mémoire* of the spatial relationships of part of his land. From an analysis of family papers, Brian Roberts has been able to show the way Archer 'was clearly feeling the need for a map and seems to have been taking the critical mental step between a written survey describing the locality of each piece of land and doing the same more accurately and simply by means of a map'.[18] Other sketches pre-dating the appearance of the estate map proper have been cited as the end product of an estate survey.[19] The process could also work the other way. In the sixteenth century, a map of some fields at Dedham, Essex, made in 1573, is thought to have

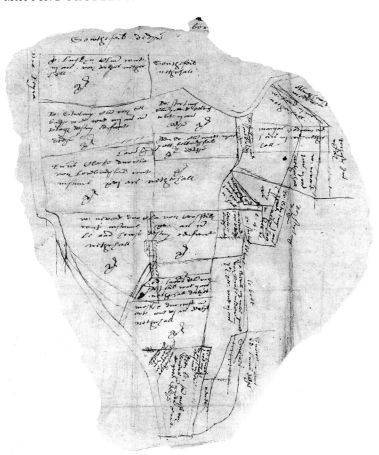

4.2   Map of fields at Dedham in Essex, 1573. One of a number of sketch maps thought by P. D. A. Harvey (*Maps in Tudor England*, 1993, p. 85) to have been used to compile a written survey. 22 × 30 cm. Public Record Office, MPC 1/77, fol. 4r.
Reproduced by permission of the Keeper of Public Records.

been one of a group of maps used to help compile a written survey of the manor to which the Dedham holding belonged (Fig. 4.2).[20]

Feudalism to capitalism: the new land surveyors

The complex process known as the transition from feudalism to capitalism which occurred in England in the sixteenth century changed both the social and economic valuation of land. The land itself, rather than its produce, became a commodity. Power relationships were expressed through control of the means of production which, of course, were usually land-based. As feudalism was overtaken by capitalism, so specific parcels of land acquired monetary, and symbolical, value. Land was now worth mapping, and the new maps could be used as a tool in the development of new systems of exclusive (private) rights to land.

Professional land surveyors were themselves crucially important in the emergent capitalist economies in bringing about changes in attitude towards graphic representation of topographical information within society in general, and its governing institutions in particular. The economic imperatives of the new capitalism encouraged developments in the science of surveying. At first, mid-century authors of treatises describing methods of land surveying, like Richard Benese (1537), Leonard Digges (1556) and Valentine Leigh (1577), gave instructions only for compiling the final terrier. It was not until the end of the sixteenth century and the first years of the seventeenth century and the publication of two treatises, one by Ralph Agas (1596) and one by John Norden (1607), that any mention is made of a map as an end-product of the survey together with the book.[21] The need for a landowner to 'know his own' was felt strongly in the seventeenth century. The phrase, which recurs like a motif from surveying manual to surveying manual, 'draws attention to an emergent desire to apportion objective rights of ownership over goods and land'.[22] Land was changing hands in a way it never could have done in feudal England and each new prospective owner needed to know what it was he might be purchasing. At the same time, both landlords and tenants were encouraged to alter their appreciation of 'one's own', and to learn to maximise production 'as a socio-economic outlook dominated by moral standards and interpersonal relations gave way to a discourse which facilitated economic individualism and competition'. In other words, the spur to the surveyors' activities was not the joy of intellectual rediscovery, but the gain to be made out of a changing economic and social situation.

Both Ralph Agas and John Norden were enthusiastic about the utility of private estate maps. Estate maps, Agas pointed out, could define the situation of a manor and locate the manor house, the various tenements, curtilages, barns, fields, stables, and cottages 'in their full number measure and forme'. A map, he continues, may display chases, warrens, parks, woods, fields, closes, pastures, and 'every parcel of land lying within the boundes thereof, in their exact measure: fashion and quantitie. Boundaries and abuttals are so wholly put downe, as no booke may bee comparable with the same'. He notes that the type of boundary can be specified on a map, whether wall, pale, hedge, ditch, river, lane, or path, and that the 'true placing whereof, bringeth perfection to the woorke, and may in time to come bee many waeis most necessarie and profitable'. He recognised that the map could be a management tool: 'heere have you also every parcel ready measured, to all purposes: you may also see upon the same, how conveniently this or that ground may be layd to this or that messuage'. The map was also a record of change. Agas had perhaps in mind the sometimes scandalous replacement of arable by pasture when he remarked that when 'bounders and meeres were moved, or when names were changed, notwithstanding their auncient and faire bookes, for the abuttals thereof: the surveigh by plat, suffereth no such inconvenience, but shall be for continuall evidence'.[23]

Agas was clearly aware that a map had one great advantage over a written description or terrier. Property boundaries could be identified for all time, even when the landscape of balks and hedges had changed, as it had been changing since the late fifteenth century, particularly as a result of the process of enclosure. Once the topographical markers used in a written description of boundaries no longer existed, it was impossible to reconstruct their former position on the ground. With a map as record, the boundaries could be repositioned by carrying measurements from the map to the ground. Agas advanced three main advantages for survey by plan rather than written cadastre alone: precision of location, efficiency of land management, and permanence of record. John Norden, who, like Agas, had also made other types of maps, was also a committed advocate of property survey by maps. To the question posed in his *Surveyor's Dialogue*, 'Is not the fielde itselfe a goodly map for the lord to looke upon, better than a painted paper?', his reply was 'A plot rightly drawne by true information, discribeth so the lively image of a mannor, and every branch and member of the same, as the lord sitting in his chayre, may see what he hath, and where and how he lyeth, and in whole use and occupation of every particular is upon suddaine view'.[24]

Notwithstanding the output of such instructive treatises at the end of the sixteenth and early in the seventeenth century, it is clear that most landed properties were in fact being managed as in the Middle Ages without the aid of maps. Land was bought and sold, farmed and exploited, surveyed for private individuals and on behalf of crowns and governments for valuations and tax assessments by written description alone. The post-medieval history of estate mapping in England is not characterised by a uniform and progressive adoption of this 'new' medium of communicating local cadastral detail. David Fletcher's study of the mapping of estates owned by Christ Church, Oxford, from the seventeenth century onwards, concludes that map consciousness amongst land owners was not something which developed either uniformly or consistently.[25]

Estate maps in land management

There is some circumstantial evidence of a relationship between surveying and mapping on the one hand, and agricultural improvement on the other. Figure 4.3 shows the intensity of surveying and mapping activity in the counties of England and Wales between 1470 and 1640. It is a pattern which mirrors the contrast between the grass-growing regions of the west and north-west of the country and the mixed farming regions of the south and east.[26] Other patterns can be identified. For example, there is the contrast between the capitalist, corn-growing villages of East Anglia and Kent, 'deeply involved in commercial dealings in food', and the population of the peripheral counties, which was 'dispersed in lonely farmsteads, some still preserving vestiges of the clan spirit, still almost completely isolated from the commercial world'.[27] It was market opportunity which encouraged agricultural improvement in early modern England. It can be argued that the demand for agricultural improvement also generated commissions for estate

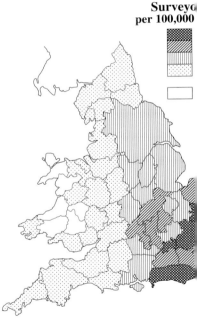

**Surveyo
per 100,000**

4.3 Surveying and mapping in England and Wales, 1470–1640. Compiled from entries in Peter Eden's *Dictionary of Land Surveyors*, 1975, 1976, and 1979.

maps. Where there was a general absence of improvement, as in Scotland, estate surveys are also lacking, for 'there was little agricultural improvement to require plans, for the traditional infield-outfield system remained wholly unchanged until the eighteenth century'.[28]

While it is possible to demonstrate a spatial correlation between those parts of the country characterised by early agrarian improvement and the intensity of surveying activity, simply plotting available evidence provides no support for a causal relationship between mapping activity and agricultural improvement. It is more difficult to document instances of map use in the everyday running of an agricultural estate. The wear and tear of some surviving maps suggests frequent use in an estate office; one map of manors in north Dorset (c.1569–74) seems to have been drawn for use at Sherborne Castle, 'for the map is rubbed at this point, as though from frequent handling'.[29] Other maps have been annotated by users; Lord Burghley commented on land use and tenure on a map of Cliffe Park, Northamptonshire (1593). The replacement of an outdated, or worn, map by a new one is evidence of a continuing need. A new copy of Christopher Saxton's map of a manor owned by St Thomas's Hospital at Aveley, Essex, heavily folded as if much used in the field, was made in 1782, indicating that the original map was still in use some two centuries later.[30] Similarly, Thomas Langdon's map of Salford, Bedfordshire (1596), was still being used in the All Souls College estate office in 1769, while an anonymous map of Radbourne, Warwickshire, made in 1634, has annotations which suggest that it was in active use for at least 150 years thereafter.[31] Other maps were used in the planning of later changes. A draft of a map of the manor of Wotton Underwood, Buckinghamshire (1649), shows, by the superimposition of a new layout, how the estate was reconstructed to accommodate a new house and its park and avenues leading to a new house.[32] Perhaps one of the clearest instances of the use of a map in land management comes from the Earl of Northumberland's Spofforth estates in Yorkshire, which were mapped by Saxton in 1608. He identified the tenant and the land use of most of the fields, by name or initial letter. Additional symbols seem to indicate the relative soil quality field-by-field.[33] Christopher Saxton's map, taken with maps of the same properties made by Robert Norton in the early seventeenth century, show that at Spofforth maps were not only a considerable item in annual estate expenditure but also an integral part of the programme of estate management.[34] Revealing his reliance on printed treatises like those by Agas and Norden, the ninth Earl of Northumberland wrote to his son in 1609 laying down his first principle of estate management: 'understand your estate generally better than any of your officers . . . I have so explained and laboured by books of surveys, plots of manors, and records that the fault will be your own, if you understand them not in a very short time better than any servant that you have'.[35]

By the eighteenth century, when the emphasis was on greater mathematical accuracy, carefully measured surveys had become almost the norm. William Sampson's large map (145 × 127 cm) made in 1792 of properties owned by the Dean and Chapter of Christ Church, Oxford, at Midsomer Norton, Somerset – a map protected by heavy rollers – suggests 'a huge and expensive undertaking, but clearly one justified by the resources [coal deposits] at stake'.[36] Some landowners, like Darcy Burnell of Winkburn, Nottinghamshire, in 1766, included a plan of each farm on the estate in the estate's field book.[37] An insight into the day-to-day process of surveying, about which we hear little from any period, comes from Epperstone, Nottinghamshire, where in 1734 the sur-

England

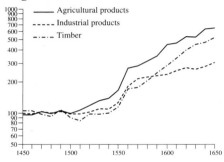

Bowden, 1967 p.595

Grain prices in western and central Europe in grams silver per 100 kg

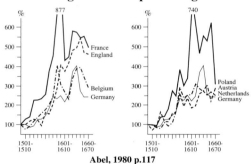

Abel, 1980 p.117

English rents 1510-19 to 1650-9

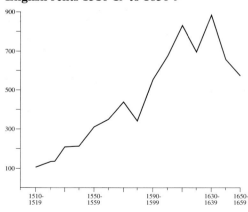

Abel, 1980 p.125

Netherlands

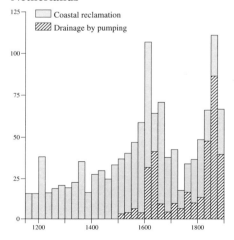

Pounds, 1979 p.196

veyor, Thomas Badeslade, checked on the accuracy of his map as a record of the estate's tenurial arrangements with a jury of tenant farmers who went with him 'all over the Fields . . . and saw their names wrote upon their pieces of land in the mapp'.[38] Private surveyors continued to be employed by land owners to make maps of their properties through to about 1840 when, as we shall see below, large-scale mapping of property was undertaken on behalf of the government for a quite different purpose, the reformation of the way the established Church of England was financed from tithes. From 1840, these tithe maps were commonly adopted as the basis of estate maps. Then, towards the end of the nineteenth century, Ordnance Survey mapping on a scale of 1:2500 provided a base map on to which property details could be plotted. The day of privately-commissioned property mapping was over.

Estate maps as fiscal and social symbols

Estate maps served a number of purposes. It has to be said, however, that for none of these was the employment of a map absolutely indispensable. The medieval estate steward had organised the buying and selling of land, its drainage and improvement, enclosure, valuation and day-to-day management without maps, and many of his successors worked without maps throughout the early modern period.[39] If maps are to be related to the process of agricultural improvement in early modern England, and to their usefulness as records of property characteristics, what was it that encouraged their proliferation to such an extent that by the later seventeenth century they had become commonplace? The post-feudal economic and social order may be part of the explanation but there are other aspects of the transition to capitalism which merit attention, notably those related to the rising value of land itself in the sixteenth and seventeenth centuries.

It has been suggested that demand for land as a factor of production and as symbolical space was the motor which drove the early modern property-mapping revolution throughout Europe.[40] Agricultural expansion characterised the sixteenth century but in most places this had halted by about the middle of the seventeenth century, from which point some sectors of the economy experienced crisis, recession, and an actual contraction.[41] Price inflation also affected the period from 1500 to 1650 (Fig. 4.4). The existence of an active land market only exacerbated the effect of the general inflationary pressure on rents. England had her own problems. The dissolution of monasteries brought land speculation in its wake. Wealthy merchants saw land-purchase and reclamation schemes as outlets for their growing supply of investment capital. In our terms, land had become a 'blue-chip' investment, with rents tending to rise faster than either production or prices. The picture in England was spectacular from about 1580; 'with land values continuing to rise, rent rolls on estate after estate doubled, trebled and quadrupled in a matter of decades'.[42] It was these rising rents which made imperative the clearest and most accurate delineation

4.4 Sixteenth- and seventeenth-century price indices. Full references to the abbreviated sources for each graph are listed in the bibliography. See also Fischer, *The Great Wave* (1996).

possible of estate and farm boundaries. Better than the written terrier, a map might reveal a tenant's 'concealments' and infringements, matters which the owner could no longer afford to ignore. Nor, as land became relatively more scarce and more valuable in money terms, could inaccuracy of survey or plotting of maps be tolerated.[43]

The quest for accuracy, attributable to the rise in land values, was also a spur to the improvement of instruments and the technical execution of surveys. The cost of an estate map could be quickly amortised by the higher rents which should follow from the landowner's full or better knowledge of his property. Significantly, estate maps were frequently drawn just before leases were due to be renewed or entry fines or rents were to be raised, as Sarah Bendall has shown by a cost-benefit analysis of properties owned by Cambridge colleges.[44] William Fowler's maps of the Bridgewater estates in Shropshire, made in 1650–1, were also used in this way as part of a concerted effort to obtain the maximum rental income from the estate to pay off debts.[45] In south-west England, George Withiell's 'true plott' of Luttrell property at Kilton Park, Somerset, led to the recovery of more than seven statute acres, together worth £98, which had been concealed by the old 'false plott being drawne with dark lines' (Fig. 4.5).[46] In the last analysis, then, estate maps were sometimes made expressly for the profit they might bring in the new capitalist world. But while there can be no doubt that one element of that profit was reckoned in money, an estate map, like possession of the land itself, came to be invested with a symbolical value. A landed estate, with its fields, woods, mansion, farms, cottages and tenants, was the means of entrance to landed society. It was 'a little commonwealth' in its own right.[47]

Thus, investment in land meant far more than just fiscal gain. It was the key to social advancement, the way by which an urban merchant or manufacturer could approach, if not always attain, the noble status to which he aspired. In the Tudor and Stuart age of mansion-building, as the possession of land became the prime indicator of such status, the non-pecuniary aspects of land owning became more highly valued amongst an ever-widening group of aspirants.[48] We see this clearly at Laxton, Nottinghamshire, where in 1625 the Duke of Buckingham sold the estate he had recently acquired to Sir William Courten, the recently knighted son of a Flemish refugee and trader in silk and linen, and himself a highly successful merchant trading with Africa and the West Indies.[49] Like Buckingham, Courten had no intention of settling in Laxton and had bought the estate purely as an investment, selling it less than twenty years later. The magnificent map which the surveyor Mark Pierce produced for Courten in 1635, an update of a now-lost map drawn at the time of Buckingham's sale, together with the usual written terrier, reflects in some of its details Courten's *arriviste* interest in the estate (Plate 8).[50] Any estate map pictured the miniature cosmology of the estate, and served as a touchstone to the rights and privileges which its possession entrained, but decorated,

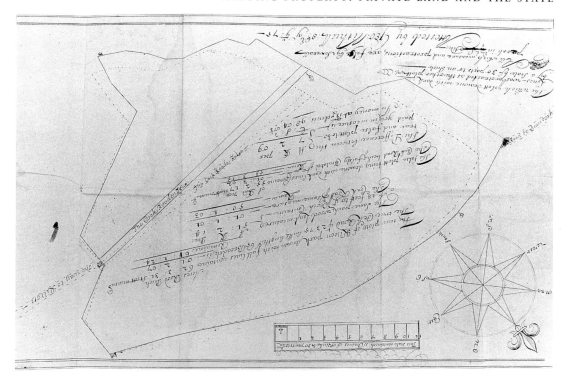

4.5   George Withiell's map of Kilton Park, Somerset, late seventeenth century. An example of a map used to maximise rental income from an estate, Withiell's 'true plott' led to the recovery of more than seven acres, worth £398 in annual rent, which had been concealed by the old 'false plott with dark lines'. 27 × 40 cm. Somerset Record Office, DD/L II 6/33).
Reproduced by permission of Somerset Record Office.

as many were, the map became 'a seigneurial emblem, asserting the lord of the manor's legal power within the rural society'.[51] As Harley continues, the map was one badge of the owner's local authority and the 'family coats of arms added within the margins were certainly more than mere decoration, for the right to these heraldic emblems also incorporated an individual's right, rooted in the past, to the possession of land'.

So decorative are some estate maps that they can be regarded as minor works of art (Plate 9). Moreover, the decoration was seen at the time as an integral element of the map, even if it did not entirely justify the production of a map as well as a terrier.[52] 'Your plot', observed William Leybourne in his instructive treatise (1653) 'will be a neat ornament for the lord of the mannor to hang in his study, or other private place, so that at pleasure he may see his land before him'.[53] Leybourne's use of the word 'land' lies at the heart of the Renaissance property-mapping revolution, for not one of the maps reviewed here would really have been needed had 'land' meant in England in the late fifteenth and sixteenth centuries precisely the same as it had meant in the Middle Ages. Once land began to be counted in specific pieces and specific amounts, rather than being regarded as the source of a 'bundle of assorted rights over different bits of territory', it is possible to comprehend why men like John Archer in late fifteenth-century Warwickshire had felt the need of a map of their property.[54] We begin to understand too, how, as society became more commercially- and cash-orientated, socially ambitious and litigious, the surveyor was required to expand his activities and improve his techniques in order to meet the new demands for maps for multifarious practical and symbolical motives. At the beginning of the sixteenth century, such maps were not unknown but they were a rarity, and surveying was a self-conscious, nascent profession. By 1678, though, the English surveyor John Holwell was sufficiently confident of the standing of his profession that he could begin his *A Sure Guide to the Practical Surveyor* with the declaration: 'I shall not trouble my self to write any thing in commendation of the art, its use being sufficiently known'.[55]

ENCLOSURE MAPS

Like the estate map, an enclosure map is of no administrative use without the accompanying register, in which each plot of land delineated on the map (usually identified by number) has a corresponding entry. Unlike the maps just discussed, an enclosure map represents a survey of property ownership within a parish and thus concerns a multiplicity of land owners as well as tenants. Within the clearly-marked boundaries of a parish, each individual *parcelle* of arable, each piece of meadow and all moors and pastures of the old open-field farming system have been carefully surveyed. The proposed allotments of the new system of farming *in severalty* (privately, each man for himself), together with the names of their future owners are set out.[56] Although the process of enclosure itself could take place at any time after the Statute of Merton (1235), maps like that of Crewkerne illustrated in Figure 4.6 are characteristic of the period of wholesale enclosure authorised by act of

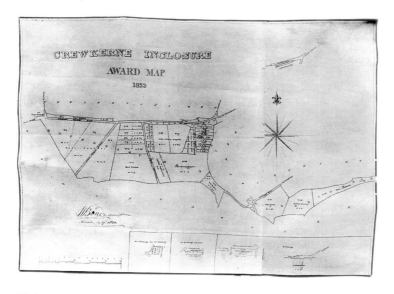

4.6 Enclosure map of Crewkerne, Somerset, 1823, by Thomas Tilbrook of Horningsham, Wiltshire. A good example of the type, showing how only the newly enclosed area was mapped. 62 × 73 cm, 1 inch to 4 chains (1:3168); the area enclosed covers about 70 hectares, or approximately 2.3 × 0.7 km. Somerset Record Office, Q/RDe 26.
Reproduced by permission of Somerset Record Office.

parliament (parliamentary enclosure) which gathered pace in the 1750s and continued to the 1860s, by which time the nation-wide process was largely complete. Maps are not normally associated with the earlier periods of medieval and early-modern piecemeal enclosure. Parliamentary enclosure was a process in which the central government was much involved, particularly after 1801 when a succession of general enclosure acts was passed in order to simplify, cheapen, and thereby encourage, enclosure. The actual mapping and apportionment was not conducted by public agencies but was entrusted to private surveyors, that is to the same men who were hired by private landowners to make private estate maps.

Maps and the enclosure process

The enclosure and redistribution of communally-held land, and of land divided into small strips or *parcelles*, seems to have been instituted for a variety of reasons. One was to introduce what was deemed to be a more efficient farming structure, one which would allow individual farmers to crop as they liked on a compact block of land (with larger fields) to a degree which could not easily be accommodated in the small, intermingled strips of communally-farmed, open-field systems. Another was to expand the acreage under the plough, or to create pastures of a better quality, by 'taking in' moorland or heath.[57] The precise reasons for enclosure in any one place at a particular time, and the often profound social and economic effects of the process, are complex and far from clear to modern agrarian and social historians. The authors of seventeenth-century English surveying treatises, however, had no doubts that enclosure was a distinct agricultural improvement. It was, they argued, an improvement that could be done even better with accurate measurement and maps. Perhaps not entirely disinterestedly, surveyors were both advocates of capitalist farming on enclosed land and promoters of the cadastral map as a means to attain this state.[58]

To a twentieth-century observer, the utility of a map to record a new cadastre resulting from the exchanges and consolidations of land necessitated by enclosure seems obvious. We know, however, that much enclosure was carried out in the sixteenth, seventeenth and early eighteenth centuries without maps, although the absence of maps in many cases may be simply a case of non-survival. In contrast, by the end of the eighteenth century, enclosure maps were being made in increasing numbers. The utility of the map for enclosure purposes was certainly well-established in the public mind by the 1790s, as can be seen from the testimony of witnesses called before select committees of the House of Commons and the House of Lords; from the observations and opinions of agricultural commentators; and in the wording of general enclosure bills put to parliament. For instance, Sir John Sinclair's enclosure bill of 1797 required that 'the Surveyor or Surveyors shall, with all convenient Speed, make an exact Survey of all such Common Arable Fields, Common Meadows, Common Pastures, Wastes, or Commons'.

William Marshall, a leading and outspoken commentator on many agrarian topics at the turn of the eighteenth into the nineteenth century, was a convinced advocate of the value of cadastral maps in estate management. In his major work, *On the Landed Property of England* (1804), he discusses the utility of maps for reorganising inefficiently disposed farm holdings on an estate and remarks on the value of maps for 'promptly exhibiting the several farms and fields, as they lie . . . it is to him [the improver] what the map of a country is to a traveller, or a sea chart to a navigator'.[59] Even more tellingly, maps headed his list of the tools a modern estate office required.

Marshall was questioned on the narrower issue of enclosure by House of Commons committees in relation to two separate enclosure bills, and published his views in a pamphlet entitled *On the Appropriation and Inclosure of Commonable and Intermixed Lands* (1801). His proposals for a general act of enclosure included the preparation of 'a map or maps, of such lands, with a number of other distinguishing marks, – the quantity in acres, and the estimated rental value, by the acre, marked upon each of the several parcels'.[60] Maps were needed, Marshall argued, so that the commissioners 'may be able to conduct the business of appropriation with facility and due effect'.[61] In his view, two maps ought to be produced, one of the existing state of the land and a second of the land divided into enclosed parcels. Cadastral maps were also advocated by the successful 1801 General Inclosure Act (*An Act for Consolidating in One Act Certain Provisions Usually Inserted in Acts of Inclosure*), which required that 'a Survey, Admeasurement, Plan and Valuation of the Lands etc. to be inclosed shall be made, and kept by the Commissioners, which shall be verified by the Persons making them. . . . Maps made at the Time of passing Acts may be used, without making new ones, if the Commissioners shall think fit.'[62]

It is clear that, by the turn of the nineteenth century, there was a general conviction of the essential role of maps in English land enclosure. That this had already been the case at least a generation earlier may be discerned from H. S. Homer's *An Essay on the Nature and Method of Ascertaining the Specifick Shares of Proprietors upon the Inclosure of Common Fields* (1766). Homer's essay was published at a time when the activity of parliamentary enclosure was accelerating markedly. In it, he discussed in detail the surveying procedures to be followed upon enclosure but remained silent as regards the value of recording the surveyed enclosures on maps, an omission which may be taken to indicate that it was a routine matter, not worth mentioning. Homer's advice to surveyors is precise. First, the yearly value and the names of furlongs and plots comprising the common land of a district were to be ascertained and then, 'when this Point is settled, the next Thing to be procured is a Plan and Survey of the Number of Acres in each Division; which, when compleated, is called the General Survey: Afterwards the known Property of every owner in each Division is separately to be measured, and this when finished, is called the

Particular Survey'.[63] As to the method of survey, 'The most approved Method of surveying and planning any Tracts of Land, is to measure the Outside thereof from station to station with a chain, taking Offsets from the straight line at convenient Distances, where there is any Irregularity in the Figure; and at each Station also to mark the Angles or Bearings between the Lines; which is done either by means of a graduated Instrument, as a Theodolite, Circumferentor, or Semicircle; or otherwise it is done by means of a Telescope and Plain Table'.[64] All working lines, Homer said, should be left on the plan so that the maps could be tested.

For all its impact on the landscape, on peoples' lives, and on the quantity of surviving maps, parliamentary enclosure affected only about a quarter of England. This means that an overwhelming majority of land had already been enclosed by local, private, agreements between the sixteenth and the late eighteenth centuries, most of which seem to have been effected without maps.[65] Moreover, few maps are found accompanying the earliest parliamentary enclosure awards, those taking place before the last quarter of the eighteenth century.[66] Why, then, did enclosure by award and map become so quickly, and almost universally, a *desideratum* in the final quarter of the century? It is a question to which there seems to be no clear answer. What is certain is that the making of a map for enclosure was an already established practice under private acts by the end of the eighteenth century, so the inclusion of clauses requiring maps and specifying their nature in the 1801, and subsequent general enclosure acts, simply enshrined contemporary local practice into general acts of parliament. A possible explanation for the growth in the use of maps among enclosers in the eighteenth century is that the use of maps in private enclosures arose out of the then established tradition of private estate mapping. In other words, it might be contended that enclosure mapping was not a product of the process of enclosure itself but simply a reflection of changing awareness of the utility of land mapping. This would certainly help explain the written nature of enclosure agreements *vis-à-vis* the map-recorded parliamentary enclosures which became more frequent at precisely the time of increasing output of private estate maps.[67]

Parliamentary enclosure

English parliamentary enclosure falls into three distinct periods (Fig. 4.7). The earliest date at which the government became involved in the process of redistribution and enclosure of communally held land in England was in 1604, when an act of parliament was obtained for the enclosure of Radipole, Dorset. Thereafter some 5,250 parliamentary enclosures were effected, covering three million hectares or about a quarter of England (in comparison, the tithe surveys of the 1840s covered three-quarters of the country).[68] In a few regions, parliamentary enclosure involved more than 50 per cent of the parishes. Elsewhere, even within the great midland belt of open-field farming, it affected less than 30 per cent of

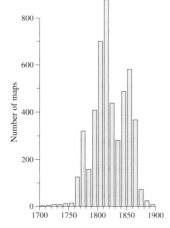

4.7   Progress of English and Welsh enclosure mapping from 1580 to 1930. Compiled by Roger Kain and Richard Oliver; the data used are available from the Economic and Social Research Council History Data Service, University of Essex (Study No. 3820).

parishes. In the west and north of England, and in much of the south, parliamentary enclosure was quantitatively insignificant.

The first surviving enclosure map comes from Haselbeech, Northamptonshire (1595).[69] It is one of a small number of surviving maps prepared for enclosure by agreement. For the next century and a half, the number of mapped enclosures remained a comparative rarity and only about 60 are known, almost exclusively from the midland counties of Leicestershire, Warwickshire, Rutland and Northamptonshire. The 1760s and 1770s experienced the first main wave of parliamentary enclosure, affecting about 600,000 hectares. Again, this is concentrated especially in the eastern part of midland England from the East Riding of Yorkshire in the north to Northamptonshire in the south and extending west to include the county of Warwickshire. Undoubtedly the greatest amount of enclosure dates from the two decades from about 1795 to 1815, a period when farm produce could command high prices as a result of the shortages of imports during the Napoleonic Wars. This second phase of enclosure was mostly concentrated on the eastern flank of the main midland belt from Cambridgeshire south to Middlesex and also in the west of England from Herefordshire to Somerset.

The first general enclosure act was passed in 1801, during the Napoleonic Wars, to facilitate enclosure and thus to promote increased agricultural production to help the war effort. From that time on, it was widely accepted that land enclosure brought economic advantage. A further general act was passed in 1836 for enclosing open arable fields, to be followed by yet another act in 1840 dealing with other categories of communal land. According to the provisions of the general acts, if two thirds of local proprietors (counted by their numbers and by the value of the land in question) agreed, enclosure commissioners could be appointed and enclosure could proceed without any further reference to parliament. Since the 1836 and 1840 acts had effectively disenfranchised small proprietors from the decision-making process, a further act was passed in 1845 establishing a central Enclosure Commission. This Commission, with its body of assistant enclosure commissioners and surveyors, was entrusted with the local administration and the provision of safeguards for all property interests.[70] Enclosure was all but complete by the middle of the nineteenth century. The last community to enclose its land was in Gloucestershire in 1918, and only Laxton, Nottinghamshire, remains as a unique relic of an open-field farming system which once covered by far the greater part of lowland England. It is still largely unclear why the eighteenth-century procedure of enclosure by act of parliament became the norm when in preceding centuries communities had found local agreements sufficient. At the most pragmatic of levels, an act of parliament enabled a majority of proprietors in a village community who advocated enclosure to overrule landowners who were opposed to it. From the point of view of the history of English maps, parliamentary enclosure provides a rich cartographical vein.

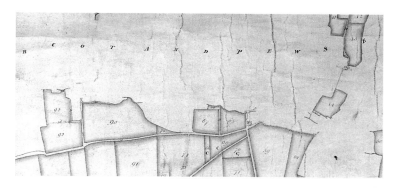

4.8   Enclosure map of Pewsey, Wiltshire, 1775, by Walter Dutton and John Hand. The illustration shows clearly the use of colour banding to indicate field boundary ownership, a device often employed on enclosure and estate maps. 102 × 81 cm; 1 inch to 8 chains (1:6336). An area of about 1.2 × 2 km (0.8 × 1.2 miles) is illustrated here. Wiltshire Record Office, 1634/34.
Reproduced by permission of Wiltshire Record Office.

Parliamentary enclosure maps

Each enclosure award comprised three elements: a written record of the allotments to the new owners; and usually two copies of the map, one to be deposited or 'enrolled' with the county clerks of the peace, the other for local use. Not all enrolled copies of enclosure awards were accompanied by their maps. By law, the nineteenth-century commissioners were required to enroll only the written award and, no doubt for the sake of economy, some adhered strictly to this limited requirement. Nor have all deposited maps survived. For parliamentary enclosures effected before 1770, the map survival rate is less than 33 per cent; for those enacted after 1810 it is more than 90 per cent; and for the period as a whole it is just over 70 per cent, figures which may be modified when current work on English and Welsh enclosure and other large-scale maps is completed.[71] Furthermore, only half of all extant enclosure maps show the whole of a parish; the rest deal exclusively with land to be enclosed, which often appears on the maps as 'islands' among earlier enclosed properties (Fig. 4.8).

The provisions of each enclosure act were normally put into effect by enclosure commissioners, usually three in number in order to represent the interests of the lord of the manor, the ecclesiastical or lay tithe owner, and the other proprietors respectively. After their appointment, these men in turn recruited a clerk and a surveyor (Fig. 4.9). As the commissioners made their decisions about each of the new allotments which they judged a fair equivalent of pre-existing open lands and common rights, these were set out on

4.9   Cartouche from the enclosure map of Henlow, Bedfordshire, c.1795 showing the surveyor and his assistants at work in the field.
Bedfordshire Record Office, MA 5/1.
Reproduced by permission of Bedfordshire Record Office.

the map in substitution of the pre-enclosure cadastre. It was important that all existing tenurial and land use arrangements were
well-known to the commissioners and some maps of the pre-enclosure cadastre were made for this purpose in association with
enclosure, as at Cotgrave, Nottinghamshire, where the maps are
dated 1789 and 1790 respectively, and at Kilnsea, East Yorkshire,
(Figs. 4.10 and 4.11).[72] At Headon-cum-Upton, Nottinghamshire,
the surveyor, anxious that there would be no confusion over anybody's rights, explained in a note on his map that 'The letter B
placed in each furlong, in the Open Fields signifyes which side of
the furland to begin to number the lands see the book'.[73] Pre-enclosure plans prepared for enclosure commissioners tend to be rare
before about 1830. Where they are found, it is usually with other
relevant plans and documents, since the commissioners were duty-
bound to investigate the evidence of old surveys when framing their
award.[74]

Each enclosure map had to be attested, on oath, by all the sur-

4.10 *Left:* Draft enclosure map for Kilnsea, East Riding of Yorkshire, 1818, with additions to 1840. Surveyor unknown. This draft map shows the proposed new enclosures superimposed on the open-field strips. 24 × 82 cm; 1 inch to 4 chains (1:3168). An area of about 2.0 × 1.2 km (1.2 × 0.7 miles) is illustrated here. East Riding of Yorkshire Council Archives and Record Services, DDX/92/4.
Reproduced by permission of East Riding of Yorkshire Council Archives and Record Services.

4.11 *Right:* Confirmed enclosure award map for Kilnsea, Yorkshire East Riding, 1840, by Richard Fowler of Salthaugh Grange, Hedon. Unlike the draft map (Fig. 4.10), this shows only the post-enclosure landscape. 83 × 67 cm; 1 inch to 8 chains (1:6336). An area of about 2.0 × 1.2 km (1.2 × 0.7 miles) is illustrated here. East Riding of Yorkshire Council Archive and Record Service, PC 15/21.
Reproduced by permission of East Riding of Yorkshire Council Archive and Record Service.

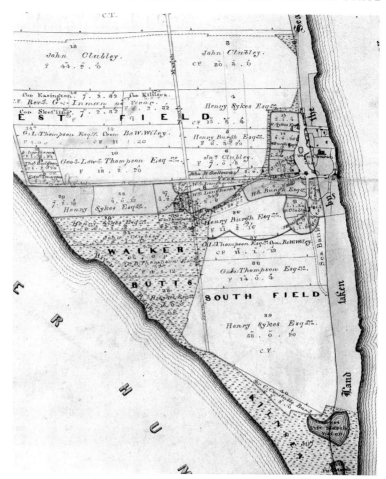

veyors concerned as true and accurate, although often only one name is recorded on the map. Parcels depicted on the plans are linked by reference numbers to the enclosure award. The features shown on enclosure maps are usually those found on estate maps; indeed, estate maps and enclosure maps were often made by the same local firm of surveyors. Typically, an enclosure map will show the new allotments and their boundaries and roads, often distinguishing public carriageways from private roads, bridleroads, and footpaths. Some are crude sketches in black and white, others are highly ornamented in colour. The map of Haddenham, Buckinghamshire (1834), for example, shows old enclosures tinted in green and distinguished from the new enclosures which are coloured yellow (Fig. 4.12). Few enclosure maps are as ornate as the one from Snaith, Yorkshire (Fig. 4.13). A relatively early map, dating from 1754, it has a distinct visual affinity with some of the highly decorated private estate maps of the period.[75] Elaborate enclosure maps become infrequent by the end of the century and

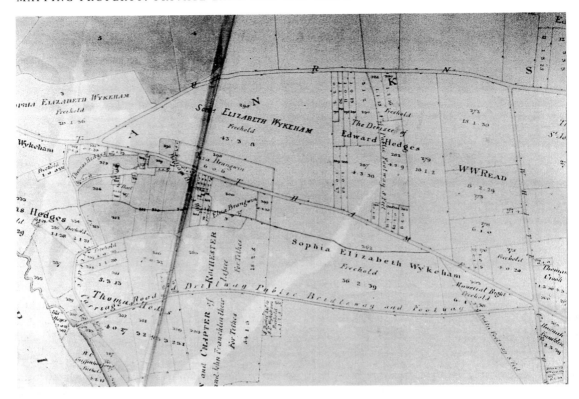

are very rare indeed in the nineteenth century. As such, enclosure maps parallel the broader cartographical trend in England for cadastral maps to become increasingly utilitarian in appearance as the large-scale map became accepted as an axiomatic adjunct to enlightened land management.

TITHE SURVEYS

The tithe surveys of the early Victorian age represent the most detailed and important national inventory of the land of England and Wales to be taken before our own times, with the possible exception of the Domesday survey of 1086.[76] In the wider context of the history of the cartography of the western world, the middle of the nineteenth century marked a high point in the making of cadastral surveys. Over immense areas of the world, detailed, large-scale plans of landed property were drawn with great precision. In France, the plans of the Napoleonic *ancien cadastre* recorded every field of more than 100 square metres in every *commune*; in Denmark, Sweden, Norway and Finland the land of almost every village was enclosed and the agrarian system was reorganised with maps used as a base; in India, the British government's Revenue Surveys mapped the farmland of the subcontinent in incredible

4.12 Enclosure map of Haddenham, Buckinghamshire, 1834, by Martin Nockolds, Stansted, Essex. The map shows enclosure allottees, acreages of allotments, and public and other roads and paths. In two sheets, 135 × 75 cm and 78 × 41 cm; 1 inch to 6 chains (1:4752). An area of about 1.1 × 1.9 km (0.6 × 1.2 miles) is illustrated here. Buckingham Record Office, IR/101.Q.

Reproduced by permission of Buckinghamshire Record Office.

detail; and in the United States the Federal Land Survey was engaged on the colossal task of measuring and mapping the whole of the public domain from the Ohio river to the Pacific coast.[77] In England and Wales there were the tithe surveys.

Tithes and the Tithe Commutation Act (1836)

All these cadastral surveys were in a sense also final reckonings, concluding statements on fast vanishing ways of life. Usually tithes represented a tenth of annual agricultural production and for about a thousand years were the heaviest direct tax on farming and, from their mode of collection, the most repugnant.[78] Tithe was gathered locally, each farmer paying his dues in kind or cash to his priest. Over the centuries these payments had become hedged about with a multiplicity of customary arrangements and local practices. Courts of law sanctioned both gross oppressions by the Church and flagrant evasions by tithe payers but problems remained, especially where tithe was still collected in kind. From the late eighteenth century, this increasingly anachronistic annual selection of a number of corn sheaves from the field, or of so many piglets from the litter, or lambs from the flock etc., was regularly caricatured by political satirists. The problem was by then compounded by the expansion of the non-conformist churches, whose members, particularly Quakers, found making payment to a Church they did not recognise doubly repugnant. Farmers resented the fact that it was only they who, as owners of farmland, were paying to support the established Church, while the increasing proportion of the population who obtained their incomes from manufacturing and other non-agricultural pursuits were exempt. Moreover, the Church itself invested nothing in the risky business of farming, while receiving a tenth of all profits. The injustice rankled: while it was the landowner who put capital into costly land improvements such as drainage or the enlargement of fields, the tithe owner took part of any resulting increase in income.

The situation was coming to a head after the Napoleonic wars when uncertain grain prices, a succession of poor harvests, and a decline in household-based industry on top of crippling taxes, rates and tithes, brought poverty and distress to many farmers and hunger to many labouring families. William Cobbett, the social commentator and advocate of political reform, repeatedly reminded farming audiences that the privileges enjoyed by the clergy could not last much longer: 'there must be a settlement of some sort; and that settlement never can leave that mass, that immense mass, of public property, called "church property", to be used as it now is'.[79] When protest and violence erupted in the Kent countryside and in other southern counties in the autumn of 1830, it was directed only in part against farmers' property and the workhouses.[80] In a great number of parishes, angry mobs attacked their parsons and many of the great tithe barns – potent landscape symbols of what was seen as the Church's extortion – were set on fire (Fig. 4.14). From being a source of local irritation, the tithe ques-

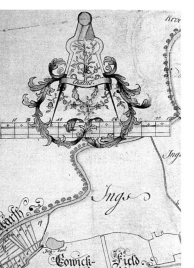

4.13 Enclosure map of Snaith, Cowick and Rawcliffe, East Riding of Yorkshire by John Read, *c*.1754, at a scale of 1 inch to 12 chains (1:9504). An unusually decorative and elaborate example. East Riding of Yorkshire Council Archives and Record Service, DDCL 3394.
Reproduced by permission of East Riding of Yorkshire Council Archives and Record Service.

4.14 Tithe barn at
Abbotsbury, Dorset.
Photograph by A. Teed.

tion had now become a concern for national government. Between
1833 and 1836 no fewer than four attempts were made to secure
the passage of bills to commute tithes; eventually, on 13 August
1836, Lord John Russell's Tithe Commutation Act received the
royal assent.

The settlement of 1836 produced an equitable arrangement
which lasted virtually unchallenged until the depressed years of the
1930s, when the tithe issue came again to the fore in another
period of hard times for farmers.[81] The Tithe Commutation Act
substituted a fluctuating money payment for all tithes payable in
kind and for all other customary payments. This money payment
(tithe rent-charge) was to vary from year to year with the price of
grain crops so that a parson's income would keep pace with inflation, while farmers would be relieved of their tithe burden in times
of low agricultural prices. The amount of tithe rent-charge due
from each parish was fixed according to the total value of the tithes
generated in the parish at the time of commutation. This global
figure was then apportioned among the landowners in the parish
according to the type and value of land that they farmed. A field
survey was thus essential in every parish in which tithes were
payable and involved field-by-field mapping. The resultant tithe
survey maps provide an exact depiction of the rural landscape of
England and Wales, and of many components of rural economic
and social structures, at the time of their compilation around 1840.

Tithe surveys are extant for about three-quarters of the country. Coverage is low for the midlands, where provision had often been made for voluntary commutation in earlier enclosure awards, but in eastern, southern and western England and Wales a majority of parishes was surveyed. In broad outline, the distribution of tithe surveys is a mirror image of that of parliamentary enclosure.

A Tithe Commission was set up in London to organise the implementation of the Act, forming a massive, paper-based central bureaucracy which, however, proved extremely efficient. The first task of the three Tithe Commissioners who headed the Commission was to find out where tithe was payable. Enquiries were sent to each of 14,829 tithe districts (usually parishes in southern England and townships in the north). Where tithes were found to be still payable, commutation proceeded in two stages: first the rent-charge was fixed for the parish and then this was apportioned amongst the landowners. The 1836 Act had been founded on the premise that voluntary commutation agreements should be encouraged wherever possible, so that if local land-owners and tithe owners could not agree on a sum to represent the initial rent-charge, the Tithe Commission was empowered to impose a compulsory award, sending an assistant tithe commis-sioner to hold meetings, collect evidence, and frame the award. Once the rent-charge was agreed or awarded, the tithe district was mapped by a surveyor, and a valuer was appointed to deal with the apportionment.

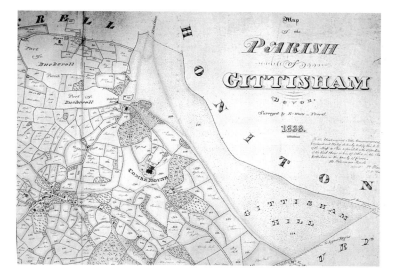

4.15   Tithe map of Gittisham, Devon, by E. Watts, Yeovil, 1838. On the original, land use and woodland are shown in colour; field names are also given. 75 × 116 cm; 1 inch to 6 chains (1:4752). An area of about 2.5 × 3.6 km (1.5 × 2.2 miles) is illustrated here. Public Record Office, PRO IR 30 9/189.
Reproduced by permission of the Keeper of Public Records.

Tithe maps

In content, a typical tithe map is not unlike an estate or enclosure map. Field boundaries, woodland, roads, streams and the area occupied by buildings are shown (Fig. 4.15). Each parcel of land (known as a tithe area) carries a reference number linking the map

LANDOWNERS.	OCCUPIERS.	Numbers referring to the Plan.	NAME AND DESCRIPTION of LANDS AND PREMISES.	STATE of CULTIVATION.	QUANTITIES IN STATUTE MEASURE. a. r. p.	Amount of Rent-Charge apportioned upon the several Lands, and to whom payable. PAYABLE TO VICAR.	PAYABLE TO Impropriator.	REMARKS.
Boggis Golding.	Golding Boggis.	284	Fish House Farm Wix Hill	Arable	7 : 34			
		309	Hop Ground Piece	do	5 1 21			
		310	Hop Ground	do	1 3 11			
		311	Eight Acres	do	9 . 30			
		312	Wood Field	do	10 3 3			
		313	Cross Path Field	do	6 1 25			
		359	Hunts Field	do	7 1 38			
		360	Parlour Inn Field	do	12 . .			
		363	Lower Spring Field	do	4 3 15			
		406	Miles Meadow	Pasture	1 2 38			

to the other element of a tithe survey: the apportionment (Fig. 4.16). This is a document in which are listed, under the reference number for each tithe area, the name of the owner and the tenant. Also usually given is the field name of a land parcel or the house name of an occupied dwelling, the land use if an agricultural plot, its area in statute measure acres, roods and perches, and the rent-charge apportioned to it. The third component of a tithe survey, the tithe file, contains a written record of the local process of tithe commutation, including such matters as the minutes of parish meetings.

The tithe apportionment is the legal instrument in which the commutation occasioned by the 1836 Act is recorded. The same question arises as already asked about estate survey maps and enclosure maps: why did government insist on a map of each tithe district? A field-by-field survey was certainly needed to value and apportion tithe rent-charge, but neither activity needed tithe areas to be set out on a map. In fact, tithes in about a quarter of English and Welsh parishes had been satisfactorily commuted prior to 1836 without a formal tithe map. The Tithe Commutation Act itself contained contradictory clauses as to the sort of map it demanded.[82] Clause 35 of the Act states that: 'the Valuer or Valuers or Umpire may, if they think fit, use for the Purposes of this Act any Admeasurement, Plan or Valuation previously made of the Lands or Tithes in question of the Accuracy of which they shall be satisfied; and that it be lawful for the Meeting at which such Valuer or Valuers shall be chosen to agree upon the Adoption for the Purpose aforesaid of any such Admeasurement, Plan or Valuation, and such Agreement shall be binding upon the Valuer or Valuers; provided always, that Three Fourths of the Land Owners in Number and Value shall concur therein.' So far, so good, but later clauses (63 and 64) required the Tithe Commissioners to sanction the accuracy of every map by affixing their official seal as a final duty when confirming the commutation of tithes in a district: '. . . if the Commissioners shall approve the Apportionment they confirm the Instrument of Apportionment under their Hands and Seal, . . . and every Recital or Statement in or Map or Plan annexed to such Confirmed Apportionment or Agreement for giving Land, or any sealed Copy thereof, shall be deemed satisfactory Evidence of the Matters therein recited or stated, or of the Accuracy of such Plan.' The situation was ambigu-

4.16 Part of a page of the tithe apportionment for Bures St Mary, Suffolk, 1838. For each tithe area (land parcel) separately identified on the tithe map, the schedule of tithe apportionment records the names of the landowner and the occupier, the map reference number, the name and description of the parcel, its state of cultivation (land use), area in statute measure, and tithe rent-charge. Usually, as here, the schedule is written in manuscript on printed pages measuring 48 × 54 cm. Public Record Office, PRO IR 29 33/81.
Reproduced by permission of the Keeper of Public Records.

ous. On the one hand, the Act entitled landowners to adopt any plan of their choice; on the other, the Tithe Commissioners were required to certify the accuracy of every map before confirming an apportionment.

A map was certainly required by the Act, but in the event it was left to the Tithe Commission and their map adviser, Lieutenant Robert Kearsley Dawson (on secondment from the Royal Engineers), to define the precise purpose and nature of tithe maps. On 29 November 1836, Dawson advised the Commissioners that in his opinion tithe maps should portray the boundaries of both the tithe district and the tithe area absolutely accurately, and that they should be drawn at a scale large enough to enable quantities to be computed from the map.[83] Dawson had an additional function in mind for the maps for which he was seeking the Tithe Commission's support and for which accurate maps would be a vital element. Dawson's idea was to extend the tithe mapping to produce what he termed a 'General Survey or Cadastre' of the whole country.[84] His proposal found immediate favour with the Tithe Commission, who saw in it not only a way of ensuring a set of conformable, accurate tithe maps but also a possible means of devolving some of the costs of tithe commutation to the government. The main argument in support of the case for accurate, large-scale cadastral maps for tithe-survey purposes rested on the permanence of the record that such maps would provide, a weighty consideration given that the fundamental objective of the Tithe Commutation Act was to produce a once-and-for-all settlement of the tithe question. To avoid future litigation, all field boundaries

4.17   Robert Kearsley Dawson's 'Instructions for the Preparation of Plans for the Purpose of the Tithe Commutation Act', 1837, contains a series of diagrams to help surveyors. This particular diagram advises surveyors to sight on major buildings, such as churches, when setting out the principal triangulation lines for a parish tithe map. From: 'Copy of Papers Respecting the Proposed Survey of Lands under the Tithe Act', British Parliamentary Papers (House of Commons), 1837, XLI.
Reproduced from a copy in a private collection.

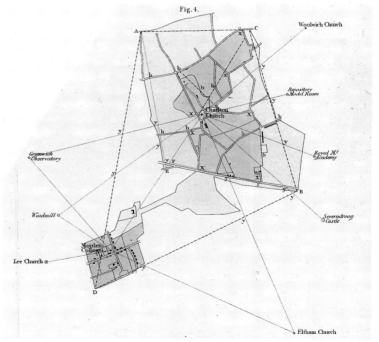

needed to be accurately portrayed on the maps, so that each tithe area and thus its rent-charge liability could always be identified.

On 4 January 1837 the Commission's secretary wrote to Dawson outlining its opinion on the role of tithe maps: 'Considering the permanent nature of the record which it is meant that the instrument of apportionment, with the Plan which is to be annexed to it, should constitute, it appears to the Commissioners to be their duty to require such Plans as will ensure a possibility, and, as far as may be, a facility for ascertaining precisely the extent and position of lands to be declared subject to or free from the rent-charges to be created under the Tithe Act, when the present boundaries of those lands have been altered or displaced. The Commissioners are of the opinion that the Scale recommended in your Report, [3 chains to an inch] and the lines of construction proposed to be left marked on the Plans, are essential for this purpose, and they authorize you to communicate this opinion to all parties with whom you are, or may hereafter be in correspondence or communication on the subject' (Fig. 4.17).[85]

The Commissioners were quite clear about the type of map necessary for tithe commutation. Tithe maps were to be first-class, accurate plans at a scale of 26.7 inches to a mile (1:2373) constructed according to a system of internal triangulation. Such specifications were also ideal as a basis for a full cadastral survey of the nation. The Commissioners were fully aware that cadastral maps had been used in, for example, the Netherlands, France, Denmark, Austria and some German states to record the tax liability of particular pieces of land. Dawson told the Tithe Commissioners: 'in many of the States of Continental Europe, Cadastral Surveys have long been in progress, at an annual expense commensurate with the importance attached to the possession of such documents'.[86] He also set out in the same letter an impressive list of advantages to be derived from a similar survey of England and Wales, which included the resolution of boundary disputes, easier transfer of real property, and identification of the best lines for new roads, railways and canals. The government, he said, would obtain an accurate statement of the 'real capabilities of the country' and would be able to decide where investment in improvements might be most beneficial. Dawson concluded by pointing out that: 'the necessity which now exists, for Surveys of nearly the whole country, for a specific purpose presents means for forming a General Survey or Cadastre, at such a cheap rate, that the opportunity cannot be lost without exposing those who ought to present its importance, to the certainty of future censure, if they fail to perform that imperative duty'.[87]

The Tithe Commissioners remained convinced that accurate maps were necessary for an effective and lasting commutation for tithes. By pressing for accuracy, however, they were increasing the already considerable cost of commutation for some landowners. In a letter to the Chancellor of the Exchequer, the Commissioners reiterated Dawson's arguments for combining tithe surveys with a general survey of the whole country and also outlined additional tax collection advantages which would accrue from a full cadastral survey. In particular, they referred to the maps needed in the assessment of poor rates under the New Poor Law.[88] The Poor Law Commissioners were in strong support of both the idea of a general survey and the method of producing parish maps described by Dawson. The Tithe Commissioners concluded their appeal to the government by asking whether 'such maps shall be attained at enormous expense at some future period, or whether the large sums of money which must now be expended on the maps, good or bad, supplied for the purposes of the Tithe Act, instead of being wasted for all other public purposes, shall be so expended as to be the means, as far as it goes, of supplying all the wants of the Nation as connected with surveys'.[89]

These arguments persuaded the Chancellor of the Exchequer to appoint a House of Commons select committee on 16 March 1837 'to consider the best mode of effecting the Surveys of Parishes for the purpose of carrying into effect the Act for the Commutation of Tithes in England and Wales'. The committee was to decide whether accurate maps were needed for tithe commutation; maps which might be combined into a general survey, and then be reduced, engraved, and published. However, the committee was not slow in appreciating the fact that it was not at all essential to have a map for the immediate purposes of commutation and apportionment, a point the chairman of the Tithe Commissioners, William Blamire, had to concede to the committee. The Tithe Commissioners' position was that an accurate written schedule would suffice for the immediate purpose of commutation; but it was the future confusion which might result from the adoption of such a system that concerned them.[90] After weighing the evidence, the select committee recommended in May 1837 that precise, accurate maps were not required for successful tithe commutation, a conclusion influenced by the question of cost to landowners and by the spirit of the 1836 Act, which was to encourage voluntary commutations wherever possible. The committee recommended some relaxation of clauses 63 and 64 in the Tithe Commutation Act to relieve the Commissioners of the duty of sanctioning map accuracy. In June 1837 a bill to amend the Tithe Commutation Act was brought before the House of Commons, becoming law on 15 July. Its passing meant the opportunity for a cadastral survey of England and Wales had been lost.

The result was that maps accepted by the Tithe Commissioners as satisfactory were made to various scales at a variety of dates and, as a body, were wholly unsuitable as the basis of a national cadastral survey. The country let slip 'its chance of a cadastral system on the continental pattern, with all that means in terms of cheap and simple property transfers'.[91] The first section of the amended Tithe Act relieved the Commissioners of the need to certify the accuracy of every map and its accompanying apportionment and permitted them to establish two classes of tithe map. 'First-class maps' were to be those which the Commissioners considered sufficiently accurate to serve as legal evidence of boundaries and areas. They are identified by the certificate of accuracy which they bear and by the presence of the Commission's official seal (Fig. 4.18). There are some 1,300 first-class maps (eleven per cent of all tithe maps), each having been checked, together with the accompanying field books, in London by Lieutenant Dawson's staff. The technical specifications which Dawson had prepared before the amendment to the 1836 act still applied to first-class maps, except that the Commissioners no longer considered a uniform system of conventional signs essential (Plate 10): 'The maps which will be most acceptable to the Tithe Commissioners [as first-class maps] are the plain working plans, with the lines of construction, names and reference figures shown upon them, and with no other ornament or colour whatever; and the most ready way of obtaining the seal of the Commission will be to send up the actual working plan'.[92]

Second-class maps include those which failed Dawson's tests for first-class quality and which were not subsequently corrected to the Commission's satisfaction. By far the larger number are those maps 'which three-fourths of the landowners are desirous to use, but which the parties do not mean to submit to the test of the Commission'.[93] Second-class tithe maps were not necessarily wildly inaccurate. Many were so classed simply because they were drawn at a scale smaller than one inch to four chains (1:3168), the smallest scale from which it was thought possible that the area of a land parcel could be computed directly from the map. At the same time, there are a large number of second-class maps which are little more than crude sketches. This is especially the case in upland parishes, with huge expanses of waste (moor and heath) or land of low agricultural (and thus tithable) value (Fig. 4.19).

4.18 The Tithe Commissioners' official seal, the presence of which identifies a first-class tithe map.
Reproduced by permission of the Archivist, Devon Record Office.

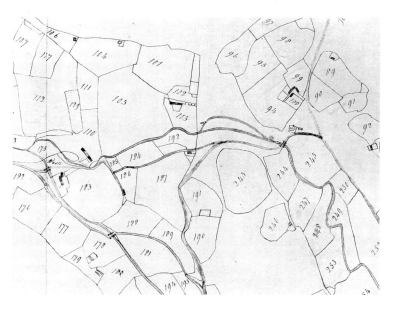

4.19 Second-class tithe map of Martindale, Westmorland, 1838. A few tithe maps, like this one, are little more than simple sketch maps produced at minimum expense and with no concessions to appearance or 'scientific accuracy'. 1 inch to 4 chains (1:3168); the area covered by this extract is about 0.7 × 1 km (0.4 × 0.6 miles). Public Record Office, IR 30 37/53.
Reproduced by permission of the Keeper of Public Records.

Many tithe maps are made at very large scales, commonly one inch to three chains (1:2376). In large parishes, or where the landowners were particularly anxious to reduce costs, the maps were made on a range of smaller scales, down to about 1:10,000. Like estate and enclosure maps, virtually all tithe maps were produced by local surveyors working close to home. Not surprisingly, they were often the same surveyors as had worked, or were working, on local estate and enclosure plans. Altogether, some 1,700 surveyors are known to have made tithe maps, but since many maps are unsigned, the total actually involved was doubtless even greater. The surveyor was appointed, employed by, and paid by the landowners of a parish. This fact, and the number of surveyors involved throughout the country, together with the lack of firm central control over the final outcome, explains why English tithe maps tend to vary a good deal in both style and content.

The history of property mapping in England shows how close the country came to producing a national cadastre. Inasmuch as tithe maps all derived from the national legislation of 1836, it might be tempting to see them as an at least partial national survey.[94] The 11,785 tithe maps cover about 75 per cent of the country. In addition, there are the enclosure maps which cover much of the residue of tithe-free parts of England and Wales. But Dawson's and the Tithe Commission's failure in the late 1830s to persuade central government to fund the mapping of both tithable and tithe-free land to a uniform standard was indeed a lost opportunity. Estate, tithe and enclosure maps, although in some senses a 'family' of privately-produced, large-scale maps, vary a great deal in date, scale, content, accuracy and style compared to the maps of the national mapping agency, the Ordnance Survey, whose products, unlike tithe, enclosure and estate maps, can be said to constitute a true national survey.

Maps and Travel

MAPS AND MOBILITY

One thing is clear. Whatever early maps were used for, it was not for finding the way in the manner in which most people today use topographical maps or road maps and atlases. Until the mid nineteenth century, the great majority of ordinary people confined their movements to a relatively restricted local area. They knew their territory from infancy and did not need maps to find their way about their fields or from village to market. At the same time, there has always been a minority who moved further afield, either on what we can call 'professional' business or, more rarely, for their own reasons (that is, as 'independent' travellers).[1] Planning the journey was very different for the two groups, but once on the road actual way-finding would have been much the same, a matter of constant enquiry and of looking out for sign-posts and land-marks, or of hiring local guides from town to town.

Travel for professionals was a consequence of social status or occupation.[2] The professional's journey was conducted on behalf of an institution of some sort – government, church, military, or commercial. Since Roman times, the business of government and administration has depended on letter writing and the delivery of messages at home and abroad. Indeed, throughout the Middle Ages, to rule was to travel; kings, queens and bishops spent much of their lives on official progresses about their territories. For all these people, knowing where to go and what route to take was a part of their duties. Way-finding at sea more obviously demands specialised travel knowledge (and skills), and navigation has tended to remain the jealously-guarded prerogative of ship masters and experienced or trained pilots. Land travel, though, also breeds its own specialists such as guides and couriers. In thirteenth-century England, there were 'brokers of carts' in London and full-time 'common carriers' worked across the country.[3] All these people knew the way from experience or through apprenticeship, or employed officials who did, or took local guides, or were issued with instructions. These instructions – the route – took the form of oral or written itineraries. They were almost never, until modern times, maps.

Independent travellers – those making a journey for its own sake or in the course of a non-essential activity – had to be responsible for their own travel plans. Except in the case of pilgrims, where social mixing was part of the experience, pre-modern independent travellers came usually from the upper ranks of society. They were setting out on a journey for some personal need, rarely more than once or twice in a lifetime. They planned their route and set out alone or with a few like-minded companions. On the road, most would have taken advantage of the protective company of experienced merchants going in the same direction. From the end of the sixteenth century the numbers of this class of travellers increased dramatically; it was this heightened activity that led to the adoption of maps as travel aids in addition to and, increasingly instead of, the written itinerary.

Despite allegations that roads in medieval England were 'impassable' or 'non-existent' and that there was 'neither desire nor opportunity' for travel, there is in fact no shortage of contemporary documentation for the close match between purpose of journey and method or route selected.[4] The conveyance of a verbal message or a letter on urgent government or military business involved single riders or, as during the wars between England and Scotland in the late fifteenth century, relays of riders.[5] The transport of bulky or heavy goods over long distances (such as the export of raw wool – the mainstay of English medieval international trade – and the import of the finished cloth from Flanders or Lombardy) involved trains of pack animals, waggons and carts under the guidance of a merchant or his representative. Another well-worn myth is that early travel was limited mainly to the summer months. Analysis of travel records both refutes this and shows that good roads and waterways made travel in medieval and early modern times both easy and reliable for most of the year in all but the remoter parts of the country.[6]

How did all these early travellers, professional or independent, find their way? To answer the question, we need first to distinguish route-planning from way-finding. A route can be defined as a set of abstract notions, incorporating a sequence of directions between intermediate points of arrival and departure on the way to a final destination. A record of a route tells us little other than that someone was once thinking about going that way, or went that way, or that a commodity was once transported to or through those places. Essential route knowledge, information which the traveller must have before setting out, need be only the name of the destination, although it is reassuring and probably prudent to know a bit more about what lies ahead, such as the names of at least some of the intermediate places. Before mechanised forms of transport, the distances covered each day were generally short; possibly up to 50 or 60 km for a horseman in easy terrain but at most perhaps 24 to 30 km for a merchant's train. Given the normal speed of land travel, the immediacy of contact with the surrounding countryside, and enforced intimacy with fellow-travellers, knowledge about the route ahead and about conditions on the road could be learnt in the course of travel.[7]

It is rare that one can be absolutely certain of the road actually taken by any medieval or early modern traveller. For one thing, the track between two places could be short-lived, even where the route was of considerable antiquity. For another, lists of places

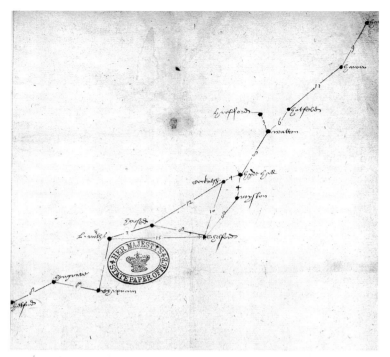

5.1 'A breeff shew of the scituation of the severall houses named in her Ma[jes]te with the nombre of myles between every of them'. A sketch of the proposed route from Thetford to Hampton and Richmond made in the course of planning Queen Elizabeth's Norfolk Progress of 1578. 20.5 × 19 cm. Public Record Office, SP12/125, fol. 98r (formerly 46r).
Reproduced by permission of the Keeper of the Public Records.

passed through by early travellers mention only major localities, usually those scheduled as overnight stops, and tell us next to nothing of what lay between them. Thus, we know the main articulations of Archbishop Sigeric's itinerary for his return to Canterbury from Rome in 990 but no more.[8] The same goes for William Worcestre, who has left us an account of his travels in southern England between 1478 and 1490 but not a word as to the ground he rode over.[9] In the second half of the sixteenth century, Queen Elizabeth's 'gestes' – itineraries normally in the form of lists of lodgings planned for one of her progresses about the country – were published at court in advance of the event. Neither from these nor from one of the court officials' planning maps (Fig. 5.1) do we learn about the road taken.[10]

WRITTEN DIRECTIONS

In fact, the normal manner of travel in pre-modern times involved setting out knowing the destination, and sometimes the names of intermediate places, but assuming that the way – the actual road itself – would be revealed stage by stage. Way-finding was expected to be an *ad hoc* and pragmatic affair. The whole process of travel was normally protracted, punctuated by the need to feed, water and rest horses and other beasts of burden and to find safe overnight lodgings. Like local guides, moreover, horses and carriages were hired only for use between one town and the next. John Dee's definition of the art of navigation – 'the shortest good way, by the aptest direction, and in the shortest time' – applied equally to travel by land.[11] The difficulty for sailors, though, was the knowledge that there would be no one *en route* from whom to enquire the way. The sailor has to be self-reliant in way-finding to a degree the land traveller need not be. Even so, traditional sailing directions (*portolani* or, in northern waters, rutters) comprised little more than lists of places (in geographical order), a note of the distance between them, and the sequence of wind directions a ship had to follow to keep on track and to enter port safely.[12]

Rutters

For those sailing the coastal waters of north-western Europe in the Middle Ages, different categories of technical information were needed for each port. Sailors had to know the hours of high and low water for each day of the year. Knowledge of the tides, rarely an issue in the Mediterrranean, is vital for sailing around the British coasts; a thirteenth-century fair copy of what must have been a commonplace and much-used document gives the times of flood tide at London Bridge.[13] Sailors also needed soundings – information on the nature of the sea bed as well as depth of water above it. And they needed to know the dates of local, as well as universal, church festivals, for these occasions meant delays in the taking on of food and water supplies.[14]

The age-old practice of passing on way-finding knowledge orally would have posed few problems so long as sailors kept to familiar routes and well-known waters. But by the end of the thirteenth century, changes in ship design had doubled Mediterranean ship capacity and a new type of rudder afforded better manoeuvrability in rough seas. Winter was no longer a bar to trading beyond the Mediterranean and ships could carry larger cargoes.[15] The later Middle Ages accordingly saw the start of a huge increase in foreign shipping in north-western waters. In the fourteenth century, Italian merchants dominated European sea-borne shipment of raw wool and woollen cloth. One route back from England to northern Italy was by way of Bordeaux, thence overland for the short portage to Aigues-Mortes, and from there to a Ligurian port such as Genoa.[16] The alternative was the full sea voyage around the Iberian peninsula. The unfamiliarity of Mediterranean seamen with Atlantic conditions and the problems of safe navigation for rounding Cape Finisterre into the Bay of Biscay and around the Breton peninsula into the English Channel posed a risk not only to their own lives but also all too often to the lives of their Breton rescuers. Traditional seamen's lore thus began to be written down for wider dissemination as route guides or rutters.[17]

The earliest known version of an English sailing direction is a fair copy of one which, if its distant antecedents lay in Brittany, seems from its lack of explicitness to have been made by the compiler for his own use as an *aide-mémoire*. The copy was made in

1486 by William Ebesham for Sir John Paston and concerns the route around Britain and from St. Malo to Gibraltar.[18] Distances are not given but soundings are: 'And ye come out of Spayne . . . till ye come into Sowdyng [the Sorlings that is, the Scilly Isles], And yif ye have a C. fadome depe or els xx/iiij.x. than ye shall go north till the sonde ayen in lxxxij. fadome in feir grey sonde [sand] . . . between Clere and Cillie [Cape Clear and the Scillies]'.[19] These same directions eventually saw print, nearly a century later, as *Richard Proude's New Rutter of the Sea, for the north partes* (1541) and in a typography which seems, with its unpractical unbroken and close-lined text, to have been a deliberate attempt to imitate the layout of Ebesham's manuscript.[20] In 1483–4 the Breton Pierre Garcie produced a book of directions for sailors operating between the Scheldt and the Straits of Gibraltar and around the British Isles.[21] An expanded and liberally illustrated version of Garcie's *Le routier de la mer* was printed in 1520–1.[22] This included 59 woodcut profiles of headlands and prominent coastal landmarks (Fig. 5.2).[23] Garcie also devoted fourteen of 22 chapters to tidal matters.[24] The idea of a pilot guide evidently found approval amongst English seamen. In 1528, Robert Copeland printed *The Rutter of the Sea, with the Havens, Rodes, Soundings, Kennings, Windes, Floodes, and Ebbes dangers and coastes of divers regions. . . .*, in the Preface of which he explained that he himself was no sailor but that he had been given a copy of Garcie's work by a London seaman returning from Bordeaux who asked him to translate it.[25]

Further guides reached England from Brittany in the sixteenth century, notably those of Guillaume Brouscon. Like Garcie, Brouscon appreciated the importance of visual presentation in a guide designed to be easily used by those who could not read or who were in a hurry. Brouscon devised a set of iconic signs for the calendars – a woodman's axe for the month of March, shearing scissors for June, a scythe for July – and tabulated the data for additional clarity. He used circular *cadrans* (dials) for tide tables. And he drew maps of each part of western Europe on which he used compass roses as clocks to show the hour of high tide on the day of a full or new moon for each port (Plate 11).[26] Brouscon's work was immediately adopted in England. A letter from John Marshall to Henry Fitzalan, Earl of Arundel, written between 1544 and 1546, accompanies the British Library's copy of an English version of Brouscon's tide tables and maps.[27] In this, Marshall explains that, as a 'poor traveller in the service of Henry VIII and Edward VI', he spent many years sailing to English possessions overseas and that he had 'collected with care information for seamen not only from around the Kingdom of England but also on the coasts of France, Flanders, Brittany, Wales, Ireland and Spain', which he was now presenting to Arundel.

Despite the obvious interest in England in the work of the Bretons, there seems to have been little attempt to produce further sailing directions or to improve on the imported directions and guides. The situation was changing, though, by the second half of

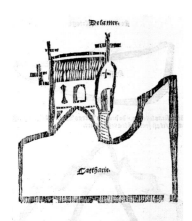

5.2 A woodcut profile from Pierre Garcie's rutter (1483–4), showing the cliff-top church on the [Isle de] Cattharie. Garcie's profiles are characteristically simplified in keeping with their function as readily memorised and easily recognised icons of crucial coastal landmarks. 18.5 × 12.5 cm. British Library C.97.bb.23, Sig. E recto. Reproduced by permission of the British Library Board.

the century, when John Dee is unlikely to have been the only Englishman to produce a seaman's manual. Dee's manuscript 'Queen Elizabeth's Tables Gubernatick' (1576), prepared 'for the British accomplishment of perfect Navigation', is now lost but it is referred to in another manuscript work of his.[28] In 1584, Robert Norman followed his own treatise on the magnetic compass by printing *The Safeguard of Sailors, or Great Rutter*, in major part a translation of a Dutch sea-book printed in 1566.[29] Whether Dee had provided his manual with separate coastal profiles in the manner of Garcie may never be known but the Dutch were certainly illustrating some of their manuals. Already in 1543, Cornelis Anthonisz had included woodcut profiles in his description of navigation in the Baltic, and it was the Dutch who, at a time of commercial expansion, seem to have been the first to produce profiles for the English coasts.[30] The first profiles in an English-language rutter, those in Robert Norman's *Safeguard of Sailors*, were introduced as 'the rising of sundry lands which by no Cart or Platt is expressed or knowne, but resteth only upon the relation of the experimented [experienced] Traveller'.[31] In 1584, the printer Christophe Plantin, then in Leiden, produced what was to become the single most influential sea-book of all, Lucas Jansoon Waghenaer's folio-sized *Spieghel der Zeevaerdt*, translated four years later (the year of the Armada) by Anthony Ashley as *The Mariners Mirrour*.[32]

Waghenaer's book had considerable impact in England, representing simultaneously the rutter tradition and giving a foretaste of the English coastal charts which were soon to appear. Much of the information packed into the 49 chapters of the first part of the *Mirrour* is what one would expect to find in a rutter, except that the reference tables and diagrams are accompanied by exceptionally simple and clear explanations for their use. Then follow two sets of coloured charts. The first group (for the western navigation) provides a 'description and portrait of the greatest part of the sea coasts of Europe', and the second group consists of 'divers perfect plots and sea charts' for the northern and the eastern navigation, that is, the straits between Dover and Calais, the coasts of England, Scotland, Norway, Denmark and the Baltic. Each chart shows a stretch of the coast in plan and details the succession of inlets, headlands, sands and rocks, ports and harbours, marks depth soundings, and indicates the vital wind lines. In addition, there are usually several profiles of that particular coastline as seen from the sea (Plate 12). The charts lack coordinates of latitude and longitude but there were other advantages. A standard scale (about 1:400,000) was used, large enough to allow charts in a bound volume to be of practical use at sea. Estuaries and harbour approaches are depicted at a larger scale. Soundings are given, together with a range of map-marks indicating the position of buoys, beacons, shoals and sands, hidden rocks and anchorages. Waghenaer's copper plates were adapted and copied for the English edition, tables for finding the time of the new moon were adjusted to fit the newly-introduced Gregorian calendar (1582) and some English latitudes were added to the tables. Throughout, the emphasis on clear explanation of the meaning of the map signs and the use of tables confirms that Waghenaer produced his sea atlas not only for seasoned seamen but also for another, relatively new, market – that of the small sea-trader.[33]

Itineraries

Unlike medieval sailing directions, itineraries for land travel have survived in relative abundance and in a variety of formats. Many were simply jotted down wherever space was available at the end of a book, in the margin of a letter, or on the back of a handy document, perhaps as they were dictated. Others are carefully-penned fair copies of a long-vanished original. An itinerary from Boulogne to Orleans was crammed, almost illegibly, into the bottom margin of a letter which arrived in Salisbury in the thirteenth century from an English student in Orleans.[34] Another English itinerary, from somewhere in northern Europe to Florence, was written out on the now-torn fly-leaf of an English translation of the travels of Sir John Mandeville.[35] Similarly informal in context is the note of the six stages of the sea-route between Venice and Jaffa (*Iter a Venetiis ad Joppa*), also with distances, written down on a spare page in another early fifteenth-century volume.[36] In contrast, the group of itineraries written out by an accomplished scribe in about 1400 at the monastery of Titchfield, Southampton, forms part of an impressively bound volume, with oak covers and leather thongs still intact, containing important items such as a list of books in the monastery's library. Twenty-nine routes, between Titchfield and each of the other Premonstratensian houses in England, are also listed and mileages (between each place and the total for the route) are usually given. Further on in the volume, followed by a note of expenses, is an itinerary to Rome, an account of a journey already undertaken but presented here as if to serve future travellers (Fig. 5.3).[37] In fact all the Titchfield itineraries were formally and clearly recorded, as if to be referred to again. In William Worcestre's rather more idiosyncratic account of his travels in southern England, though, the itineraries are secondary to the narrative and to the topographical observations.[38]

Even after printing was introduced, the manuscript itinerary's characteristic succinctness was maintained. The route from Calais through France to Rome, and Naples, occupies less than two pages in a book on the customs of the City of London printed in 1503.[39] By the 1570s, routes were being listed in all major reference works, such as Richard Grafton's *Abridgement of the Chronicles of England* (1571) and Raphael Holinshed's and William Harrison's fuller *Chronicles of England, Scotland and Ireland* (1577) and, more practically, in the humbler almanacs.[40] Handwritten notes also continued to be made on whatever surface happened to be convenient. Those in high places, like William Cecil (Queen Elizabeth's chief minister, later Lord Burghley), added to their own collections of useful routes. Burghley had a secretary make neat

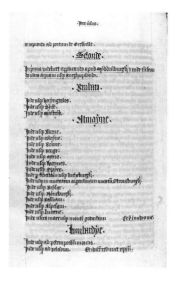

5.3 A page from an itinerary from Orwell, Essex, to Rome (*c*.1400) from Titchfield Abbey, Hampshire, showing the portion between Zealand and Lombardy. The total distance (*summa*, on the next page) has not been supplied. Names of countries are underlined in red in the manuscript, as are all instructions. British Library, Add. MS 70507, fol. 74v. Reproduced by permission of the British Library Board.

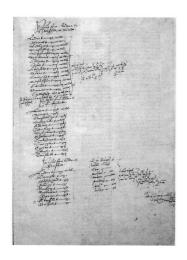

5.4 Manuscript itineraries in Lord Burghley's atlas. Amongst the proof copies of Christopher Saxton's county maps, Burghley and his secretary kept notes relating to postroads and the annual cost of upkeep. Here the notes, concerning the routes from London to Holyhead, Tavistock, and Bristol, and (bottom right) Richmond to Bristol, are in Burghley's hand. Late sixteenth century. British Library, Royal MS 18.D.iii, fol. 4r.
Reproduced by permission of the British Library Board.

lists of the postroads of England, to be kept in the same portfolio as his working copies of Saxton's maps, but he himself made his own notes beside them (Fig. 5.4 and above, p. 66). On another occasion, the only surface handy to Burghley was Lawrence Nowell's map of Britain on the back of which he wrote down the route from Ghent to Bruges.[41] Two hundred years or so later another statesman, Lord Marlborough, would also be listing routes in Flanders, this time for the army.[42] Simplicity and relevance define all these itineraries, from whatever period and context. Elaboration is rare. Physical features are almost never mentioned except for the naming of a mountain pass or the point at which a boat is to be taken. Political frontiers, in contrast, are always indicated.

The road-books which began to appear in the sixteenth century may have started as a printed form of the traditional itinerary but it was not long before their content was expanded to keep up with changes in the market. In essence, a road-book was no more than an itinerary, or group of itineraries, in a leather-bound pocket book, a format which was both portable and durable. From the late sixteenth century onwards, however, the tendency for road-books to become compendia of local historical, anecdotal and social information increased markedly and it is clear that their appeal now lay in a quite different market from that of the traditional well-seasoned traveller. No professional merchant, travelling about England or across western Europe for his pay, would have had much time for seeking out the fine views and points of architectural merit, still less the owners of the great houses which could be seen from the road, as described in the new publications. The leisured voyager, in contrast, travelling for pleasure or to expand his mental horizons, must have welcomed the information as a distraction from the tedium of the journey and as an important part of the educative process of travel.[43] But there was also by now a third type of traveller, the new-style entrepreneur and commercial traveller working largely for himself or for newly-created or expanded industrial or commercial concerns, for whom the dates of market days and local fairs detailed in the expanded itineraries would have been useful.[44]

Like the rutter, the fashion for road-books may have started on the continent, probably in France. One of the earliest is Charles Estienne's *La guide des chemins de France* (1552), written for French users of the post-roads of France but also with a description of the route from Dover to Berwick.[45] The first to deal wholly with English roads was also printed in France. Jean Bernard's *Guide des chemins d'Angleterre* (1579) appeared in a decade remarkable for the number of identical, or very similar, itineraries that were circulating in England (or which survive in greater numbers from this period than earlier). We find itineraries in print not only in specific road-books like Richard Rowland's *The Post of the World* (1576, the first English-language road-book), but also in general books like Grafton's *Chronicles* and Holinshed's *Chronicles*, as well as amongst the papers of statesmen like William Cecil. We have to

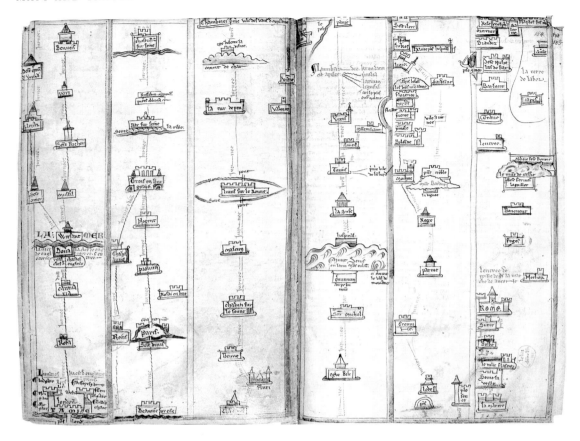

surmise that either a common source had been released in the early 1570s or that the spate of itineraries simply reflects an upsurge of traffic on these mostly long-established and well-known routes.[46] Both trends continued, and by the eighteenth century the amount of information included in road-books meant that the itinerary proper was often lost in the text. Written itineraries have never ceased to be produced. John Ogilby interleaved each set of his strip road maps with the appropriate verbal description in running prose in 1675, and even today most automobile associations produce written itineraries for their members on request. Electronic itineraries for use in a vehicle are now, at the end of the twentieth century, widely available.

Written itineraries had been a commonplace in the thirteenth century, and Matthew Paris' use of one in the construction of his map of Britain has already been noted (Chapter 2, p. 46). A graphic representation of a route, however, was a different matter and maps showing routes were rare even in the late seventeenth century. So the factors which provoked Matthew Paris to portray the route from London to Rome the way he did remain obscure, although it is now generally agreed that the itinerary is unlikely to

5.5 One of the four versions of Matthew Paris's London-Rome itinerary. The route is arranged so as to be read from the bottom left up, starting from London. The French j[o]urnee suggests that each section was supposed to represent a day's travelling, although many intervening stops are missing. 1250s. Each folio 35 × 23 cm. British Library, Cotton MS Nero D.i, fols 183v–184r.

Reproduced by permission of the British Library Board.

have been simply a 'pilgrim map'. Like the other maps destined for his Chronicles, the itinerary was produced in four versions (Fig. 5.5).[47] In each, the itinerary proper ends in a general representation of southern Italy. The Apulian ports of Bari, Brindisi and Otranto, admittedly much used in the thirteenth century by crusaders and pilgrims bound for the Holy Land, are featured, but far more eye-catching, particularly in the most elaborate of all the versions, are the fold-out maps of Rome and Sicily attached to the page at the appropriate points (Fig. 5.6). Paris also shows several alternative routes, for example, between Calais and Paris and over the western Apennines from Fidenza (Borgo Santo Donino) and from Bologna.[48] He included so many off-route places that the

5.6 The final section of Matthew Paris's London-Rome itinerary terminates in a regional map of southern Italy. Separate maps of Sicily (top) and Rome (right) are attached to the page, which measures 35 × 23 cm. 1250s. British Library, Royal MS 14.C.vii, fols 2–5, fol. 4r.
Reproduced by permission of the British Library Board.

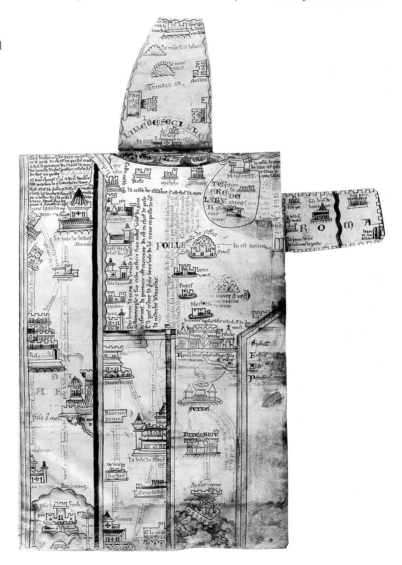

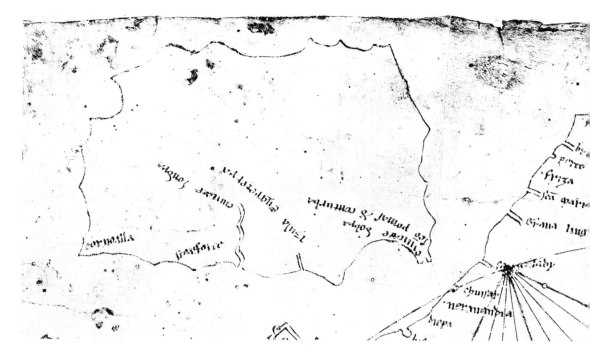

impression is gained that he was not so much concerned with presenting a narrowly-defined, still less a practical, itinerary as with
revealing as much as he knew about the London-Rome-Sicily axis
at a time of considerable diplomatic traffic and trade between
England and the Kingdom of the Two Sicilies.[49] Whatever the
explanation, Matthew Paris's decorative graphic itinerary was
probably unique in its day. It may have remained so, in England at
least, until John Ogilby's *Britannia* of 1675.

Indeed, written itineraries and rutters fulfilled their function so
well that it has to be asked how maps came to be associated at all
with travel. As already stated, the key lies in the traveller's status
and the context of his or her journey, as well as in the distinction
between route-planning and way-finding. A map showing a wide
area of land or sea contains sufficient information for the traveller
to create a route of his choice. The itinerary, in contrast, is not
merely linear but essentially prescriptive. For a medieval merchant
the sequence indicated may have been only advisory but for an
early modern postal official, as for the Duke of Marlborough's
army, it represented a command to travel the recommended route.
At the same time, a recognised itinerary implied a degree of security. Once independent travellers started to pursue their own goals,
setting out for all parts of their homeland and places abroad, more
explicit travel aids were needed. The new maps contained instructions as to their use and gave information that seasoned pilots and
regular travellers carried in their heads. The growth of tourism
from early modern times and the increasingly global scale of British

5.7 Britain on the late-
thirteenth century Pisan chart.
An increase in Mediterranean
trade in western and northern
waters drew Mediterranean
sailors into the Bay of Biscay,
around the Breton peninsula
and into the English Channel.
Better acquaintance with these
coasts led to a marked
improvement in the portrayal of
the coasts of northwestern
Europe on portolan charts;
compare the representation here
with that in Fig. 5.8.
Reproduced from the redrawing
in Yussouf Kamal, *Monumenta
Cartographica Africae et
Aegypti* (Cairo, 1936), vol. IV,
Fascicule 1.) The original map is
in the Bibliothèque Nationale,
Paris (Rés.Ge. B 1118).

Reproduced by permission of the
British Library Board.

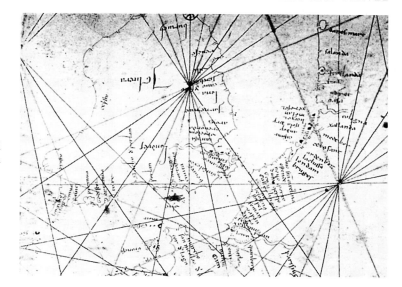

5.8   Britain on the Vescontes's chart of 1325 – to be compared with Fig. 5.7. Reproduced from the redrawing in Yussouf Kamal, *Monumenta Cartographica Africae et Aegypti* (Cairo, 1936), vol. IV, fascicule 1.) The original is in the British Library, Add. MS 27376*, fol. 181r.
Reproduced by permission of the British Library Board.

political and commercial interests from the late sixteenth century onwards can be held responsible for the proliferation of types of road maps on the one hand and for English coastal chart production on the other.

CHARTING COASTAL WATERS: BEGINNINGS

Prior to the sixteenth century, English sailors used information passed on orally. A few may have set eyes on a portolan chart like the one used by the maker of the Aslake *mappamundi*, but such charts came from the Mediterranean, where they supplied 'an amazingly accurate' representation of the coastline and a wealth of place-names, and would have been dismissed by English sailors as a curiosity.[50] Through its network of wind, or rhumb, lines, the portolan chart was a means by which the seaman could travel at will (given favourable conditions) within the entire Mediterranean basin and not just from place to place along a particular coast. Like itineraries, sailing directions date back to early classical times, but portolan charts do not seem to be older than the thirteenth century.[51] Very soon, though, they were recording the northward advance of Italian knowledge of the coasts of Europe, as Mediterranean traders began to ply regularly along the Atlantic seaboard of Europe and into the Channel and North Sea (Figs. 5.7 and 5.8).[52]

No attempt seems to have been made in England to produce charts for navigating British coastal waters prior to the early sixteenth century.[53] Even when the first manuscript charts began to appear, in the second decade, none contained the sort of information on which sailors' lives depended, like the depth soundings given in rutters but which appear on extant English charts only after 1560. The idea of using charts instead of, or as well as, the new pilot books, must have been encouraged by a number of factors. The enhancement of Breton rutters with tidal maps and coastal profiles coincided with an increasing general awareness of distant parts of the world beyond the northern Atlantic Ocean. A rise in the number of maps being made of unfamiliar shores and distant seas would also have led to a better appreciation of the need for charts of home waters, where the narrow seas were becoming busier than ever. The definition of the 'King's Chambers' along the English coasts – that part of the sea enclosed between prominent headlands – may not have appeared on printed maps until after the end of the sixteenth century but sheer congestion of shipping in the estuaries and around the ports in the first decades of the century made the need for guides for foreign ships entering or leaving English ports a matter of urgency.[54] The immediate effect was the creation in 1514 of the Corporation of Trinity House for training pilots and for the improvement of inshore navigation techniques. One of the earliest English coastal charts, showing the southern shore of the Thames estuary from Faversham to Margate dates probably from that same year.[55]

Charts used on board ship rarely survive. As Henry VIII gained enthusiasm for his navy and for the use of maps in strategic and defensive planning, however, large numbers of field surveys were drawn which, from their use as state papers rather than navigational aids, rarely left the office and are thus extant. The treacherous instability of off-shore and estuarine sand and mud banks, like those endangering shipping in the Thames, Humber, Severn and Dee estuaries, and in the southern North Sea near the straits of Dover, meant that repeated surveying was necessary. Fear of invasion from France in the 1530s and 1540s and again at the end of the century, focussed attention on the Channel coast and resulted in a constant updating and extension of coastal defences. Particularly vital harbours and havens acquired a whole series of maps, especially where siltation was a problem, as in the case of Dover, where one of the first extant maps is dated 1531–2. Richard Cavendish mapped the equally vulnerable Thames and Medway estuaries (1544) and Richard Lee the Harwich Road and Orwell channel (c.1543). In about 1539 the coast from Land's End to Exmouth was mapped.[56] True maritime charts (those created specifically for general navigation and characterised above all by depth soundings) were made for the most part only in the 1570s and 1580s, long after the Frenchman Jean Rotz had presented Henry VIII with a magnificent collection of charts covering the world.[57] A map of about 1560 describes 'the river of the Humber & of the sea & seacoast from Hull to Skarburgh'. With its soundings and directions for entering and leaving port or for riding at anchorage in different winds, the Humber map was clearly intended for use as a navigational chart, amongst other things.[58] Another map, dating from c.1590 (and probably by Richard Poulter), shows the mouth of the Tyne with soundings.[59]

5.9   The map of the North Sea in Robert Hitchcock's polemical *A Politique Platt for the Honour of the Prince* (London, 1580) supports his argument that English shipbuilders needed larger fishing boats to be able to compete with the Flemmish 'busses' which were allegedly taking an unfair proportion of the catch. East is at the top. 39 × 48 cm. British Library, C.27.f.3, following sig.E 4. Reproduced by permission of the British Library Board.

While only one map from the two decades of the 1560s and 1570s is known to survive, there are at least eight extant English charts with soundings from the 1580s alone. These were made by engineer-surveyors like Robert Norman (Thames estuary, 1580), Richard Popinjay (Portsmouth harbour, 1584), and Richard Poulter (Humber, 1584; San Sebastian, Spain, 1585; Tyne *c.*1590).[60] William Borough, sometime Master of Trinity House and chief pilot for the Muscovy Company, and one of the most prolific early English chartmakers, was responsible for a wide range of charts. Some were based on voyages of exploration undertaken early in his career, like his chart of the North Atlantic (1576), issued to Martin Frobisher in connection with his search for the north-west passage. Others resulted from his voyages on behalf of the Muscovy Company. Some were of local coastal waters, some, like Borough's plot of the mouth of the River Thames, made in July 1588, were no doubt executed in haste as the Spanish fleet was moving up the Channel.[61] Borough's chart of the passage from England to the Gulf of Finland (c.1579) carries a list of distances from the Thames to Moscow, underscoring yet again the essence of the itinerary which lies at the core of all travel.[62]

As English continental trade reached unprecedented heights in the sixteenth century, so did the exploitation of the resources of the seas around Britain. North Sea fishing had always been as vital to the English economy as to the Dutch and from time to time competition turned into a veritable fishing war. In the 1570s, a decline in the English herring catch was blamed on over-fishing by the Dutch, prompting Robert Hitchcock's lively map of 1580 which highlighted the excessive capacity of the Dutch 'busses' in comparison with English boats (Fig. 5.9).[63] Waghenaer's *Spieghel* (1584) and Anthony Ashley's translation, *The Mariners Mirrour* (1588), also touched on Anglo-Dutch rivalry. For a growing number of interested and educated people, it would be an important work of reference provided they understood how to read the charts and decode the information these contained. Indeed, as more and more maps of all sorts were reaching eager but unfamiliar hands, treatises were printed to instruct readers in map use. Chart use featured especially in William Bourne's *A Regiment for the Sea* (1576) and in Thomas *Blundeville's Exersises . . . [for] all young Gentlemen . . . desirous to have knowledge as well in Cosmographie, Astronomie, and Geographie, as also in the Art of Navigation* (London, 1594).[64] Later came Samuel Sturmy's *The Mariners Magazine* (1669). Yet despite the popularity of the English edition of Waghenaer, Dutch printers continued to assert their cartographical authority and to maintain their hold on the European chart market not only in the late sixteenth century but also for much of the seventeenth.

Seventeenth-century English chart-making went little further than copying manuscript maps of overseas waters for the Muscovy, Hudson's Bay, and East Indies trading companies. Just as county map-making remained in the hands of private map-makers until the creation of the Ordnance Survey, so chart-making remained a largely informal activity prior to the establishment of the Hydrographic Office of the Admiralty in 1795. The parallel runs deeper, for just as the Society of Arts eventually spurred the topographical re-surveying of England in the mid-eighteenth century by rewarding cartographical quality, so the Royal Society (founded 1660) played a similar role in promoting hydrographical map-making through its encouragement of the improvement of surveying instruments, particularly those for use at sea, and through the awarding of premiums.[65] Many privately-produced English charts from the seventeenth and early eighteenth centuries were of high quality. However, Joseph Moxon's 'sea platts' of 1657 were scarcely adequate, as were some in Greenville Collins's *Great Britains Coasting Pilot* (1693), the first systematic survey of the coasts of Britain, made for the Masters of Trinity House. John Seller may have been neither a surveyor nor a navigator, only an instrument-maker and a writer on navigation, but he was the first

5.10   Chart of the 'Channel of Bristol from Silly to St. Davids-Head in Wales', from the first edition of John Seller's *The English Pilot* (1671). The three traditional elements of a sea-book are here represented on one page: coastal profiles (in the deliberately simplified style of the earlier rutters), descriptive text, and chart. Whole sheet 43 × 55 cm; as shown 40 × 27.8 cm. British Library, Maps C.22.d.1 (30). Reproduced by permission of the British Library Board.

A Chart of the Chanell of Bristoll from Silly to St Davids head, in Wales and stretching over to the River of Waterford in Ireland, discovering all the Roads Havens Harbors Depeths and Soundings, upon the said Coasts. newly Corrected and Published by John Seller.

And are to be Sold at his Shop at the Signe of the Marriner's Compass at the Hermitage Stayres in Wapping.

A Scale of English Leagues 20 in one degree

The Land West from *Padstow* sheweth thus when you Sail along by it, three Leagues from you.

Thus sheweth the Land between the Lands End of England and the Island Bresham; and it is about five Leagues Long.

Thus the Land East from *Padstow* sheweth when it is three Leagues East and by South from you.

Thus the Land of *Padstow* sheweth when you Sail along by it three or four Leagues off.

Thus the *English* Shore sheweth from *Axbridge* to *Ilfordcombe*, when you Sail alongst by it.

Axbridge.   Ilfordcombe.

Thus *Wales* sheweth it self from *Wermes-Head* to *Cardiff*.

The Naves;   Cardiff.   Caldy.

To Sail from Londey to Bristol.

He that will sail from *Londey* to *Bristol*, must run alongst the English Coast untill he come within the Point of the *Nass*, for to avoid the *Nass* Sand. And then through between the *Holms*, leaving the *Steep-Holme* on Starboard and *Flat-Holm* on Larbord side, men may also with Ships of small draughts sail about to the Southwards of *Steep-Holm*, but it is there so Shoally that there remaineth at Low-water no more than two fathom water. Under *Steep-Hol..* men may anchor where they will in four or five fathom. He that cometh out from the

Nass, and is bound to *Bristol*, must stand over to *Steep-Holm*, and run alongst to the Northwards of it. It lyeth from the *Nass* E. S. E. about seven Leagues. A little to the Westwards of *Milford Haven* lyeth two little Islands, the Southermost is the smallest, called *Stockholm*, and the Northermost *Scaline*; about two Leagues N. N. W. and N. W. by N. From thence lyeth the Island *Ranasey*; and betwixt these Islands the Land hath a great Bay, called the *Broad-Bay*. *Ranasey* lyeth at the North-Point, and *Scaline* at the South-Point of the Bay; therein alongst by the Shore is good riding for North North East, East and South East Winds, in seven, eight or nine fathom water.

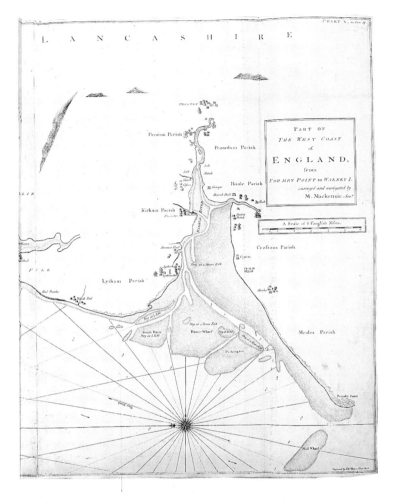

5.11   Part of the Lancashire coast near Preston from Murdoch Mackenzie senior's chart of the west coast of Great Britain (1776). MacKenzie's chart was later adopted as an official Admiralty chart. Whole chart 70 × 115 cm; as shown 31 × 23 cm. British Library, Maps 187.J.1, vol. 1, Chart 10. Reproduced by permission of the British Library Board.

English printer to challenge successfully the Dutch monopoly of sea-chart production by adapting their charts for the English market. It has been said that Seller 'scarcely merits the status accorded him in maritime history', for he merely initiated the London chart trade, leaving it to others to make it successful.[66] However, the charts in his larger folio atlas, *The English Pilot* (1671) were important right through into the nineteeenth century (Fig. 5.10).

About the turn of the sixteenth into the seventeenth century, the Drapers' Company (the so-called Thames school) emerged in London.[67] While Drapers' apprentices probably accounted for most of the charts produced in England in the seventeenth century, their charts were virtually all of oceanic or overseas waters.[68] Accordingly, attention swung back, in the second half of the century, to the harbours and coastal waters of the British Isles, especially those not dealt with in Tudor times. These were now mapped by engineers who, as under Henry VIII, combined marine with military surveying and who, once again, were often not native English. Bernard de Gomme, a Dutchman born in Lille, France, produced charts of Liverpool (1644) and Portsmouth (1665). Christian Lilly, a naturalised Englishman, who came originally from Hanover, was appointed chief engineer with responsibility for Plymouth, and produced a number of harbour maps in south-west England between 1714 and 1720 (for example, Plymouth, Dartmouth, Falmouth, Portland, and the Scilly Isles). Charles Lempriere, a Channel Islander who worked from 1716 onwards at Portsmouth, produced charts which were still being used, with revision, up to 1751. John Desmaretz produced charts of southern harbours from the 1720s through to the 1750s.[69] In parallel with topographical mapping, the eighteenth century brought a demand for more systematic surveys, mostly, though, of the relatively uncharted waters of northern Britain and the Scottish and Welsh coasts.[70] The best of these privately-produced charts were considered sufficiently good to be used by the Hydrographic Office before being replaced by new surveys (Fig. 5.11).

LAND TRAVEL: MAPS FOR WAY-PLANNING

While the shift from written direction to chart took place in England during the second half of the sixteenth century, especially after the publication of Waghenaer's *Spieghel*, a century was to pass before a parallel shift added sheet maps to the land traveller's normal array of aids. While early printed topographical maps contained the same information as an itinerary, a map showing the network of a particular class of road was printed only in 1675, a map of distance lines (ruled lines between places, each line with a figure for the distance, like those of the medieval Gough map) only in 1677, and a map of routes only in 1679. Even so, written itineraries continued to be used, both informally by individuals making their own notes of a route and in printed road-books. From the eighteenth century onwards, sheet maps for a general travelling public began to outnumber the traditional written itineraries, as they overwhelmingly do today, although the ingenuity of map-makers and printers had been stretched since the early seventeenth century to find ways of reducing the mental effort needed in route-planning and map-using and to provide easy-to-use travel aids.

Route planning from topographical maps

We shall probably never know what Henry VIII expected to see on the map which Sebastian Cabot brought from Gascony; presumably features useful in planning political and strategic moves regarding former English territories in France.[71] Given the political circumstances of the English possessions in south-west France, one would expect a topographical map showing towns and villages (perhaps indicating by the nature of the place-sign those which

Essex.

	London	Wanstead	Barking	Rumforde	Hauering	Vnger	Hatfeild Brod	Thaxsted	Dunmowe	Epping streete	Brayntree	Cogshall	Halfted	Maldon	Ingerstone	Burntwood	Billerikay	Raylie	Chelmesforde	Manetre	Harwich	Waltham Ab	Boxsted	Horndon	E. Tilbery	Step-Bumsted
Colchester.	40	34	33	30	28	25	23	19	18	28	13	6	9	10	21	26	22	18	17	6	13	33	4	25	28	16
Step.Bumsted.	35	30	32	28	26	20	15	7	11	22	10	12	9	18	22	25	24	24	17	22	28	25	5	27	30	
E. Tilbery.	18	15	13	11	12	15	20	25	21	17	21	23	25	17	9	8	7	9	14	32	38	19	30	4		
Hornedon.	18	14	12	10	10	12	17	23	18	15	18	20	22	15	7	6	5	18	10	30	36	17	7			
Boxsted.	43	37	37	32	31	27	25	19	19	30	14	8	9	13	24	28	24	21	19	6	13	34				
Waltham Ab.	12	8	10	9	7	8	10	18	16	5	20	27	26	25	15	12	16	22	17	39	45					
Harwich.	55	47	45	42	40	37	36	33	30	21	40	25	19	40	33	37	32	28	28	7						
Manetre.	45	40	40	36	36	32	30	26	25	34	19	13	15	16	28	32	27	23	32							
Chelmesforde.	24	18	18	14	12	9	10	13	8	13	8	12	13	8	5	9	6	8								
Raylie.	26	21	21	17	16	15	18	20	17	19	15	15	17	7	8	12	7									
Billerikay.	19	14	14	10	10	10	14	19	14	13	14	17	18	11	3	5										
Burntwood.	15	9	8	5	5	7	13	19	15	9	17	20	22	16	5											
Ingerstone.	19	14	14	9	9	8	10	16	12	10	13	16	17	11												
Maldon.	30	25	25	20	17	17	16	14	20	9	7	10														
Halfted.	34	29	29	25	28	20	17	10	10	22	5	4														
Cogshall.	35	29	29	25	24	20	17	14	13	23	6															
Brayntree.	28	23	24	20	18	13	9	8	7	17																
Eppingstreet.	13	8	10	8	6	4	8	15	12																	
Dunmowe.	26	20	21	17	16	9	4	5																		
Thaxsted,	32	24	25	22	19	14	8																			
Hatfeild Brod.	20	16	18	14	11	6																				
Vnger.	16	10	12	8	6																					
Hauering.	10	6	6	2																						
Rumforde.	10	5	4																							
Barking.	7	3																								
Wanstee.	6																									

The vse of this Table.

THe Townes or places betweene which you desire to know, the distance you may finde in the names of the Townes in the vpper part and in the side, and bring them in a square as the lines will guide you: and in the square you shall finde the figures which declare the distance of the miles.

And if you finde any place in the side which will not extend to make a square with that aboue, then seeking that aboue which will not extend to make a square, and see that in the vpper, and in the side, and it will showe you the distances. It is familiar and easie.

Beare with defectes, the vse is necessarie.

Inuented by IOHN NORDEN.

5.12 The distance tables for Essex from John Norden's *England. An Intended Guyde* (1625). Instructions for their use are given in the bottom right corner. 10 × 10 cm. British Library, C.77.d.16.
Reproduced by permission of the British Library Board.

were walled or defended), significant political boundaries, and major physical features such as rivers (with permanent bridges), mountains (with passes), hill ranges or escarpments (with gaps), and forests. We would not expect it to show roads, for the army, like all travellers at that time, was expected to make its way from one place to the next by whichever of the local tracks the army's scouts advised should be used.[72] In appearance, Cabot's map may have borne some resemblance to the one produced by Jacques Signot about 1495 in connection with Charles VIII's invasion of Italy.[73] There had been regional topographical maps in the Middle Ages which could have served a similar function: Matthew Paris's separate map of Palestine, for example, has been described as 'a genuine travellers' map.[74] Certainly some of the earliest printed topographical maps were clearly intended to be used in connection with travel. On the continent, Erhard Etzlaub's *Romweg* map of *c.*1500 (with George Erlinger's largely imitative map of the German lands of 1515) and Lazarus Secretarius's map of Hungary (1528) show us how the new genre was to be used in this way.

Etzlaub's objective was to resolve the problem of finding a projection which would permit a flat map from which both distance and direction could be accurately measured and computed. The

5.13 In 1635, Matthew Simmons published his edition of Norden's guide, *Direction for English Travaillers*, in which he replaced Norden's instructions to the user by a tiny map showing each of the towns listed. 3.5 × 3.5 cm. British Library, 291.a.46.
Reproduced by permission of the British Library Board.

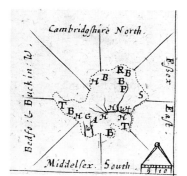

outcome was a map usable by 'Everyman, the traveller accustomed neither to maps nor to travel'.[75] The map shows no roads, only a number of routes connecting northern Europe with Rome, but instructions on the map told the intending traveller precisely what could be learnt from it and how it was to be used.[76] Etzlaub's map was small (28.5 × 40.5 cm) and could have been taken on the journey itself (indeed, the rarity of a map produced in great numbers for the Holy Year of 1500 supports the hypothesis). Not so with Lazarus Secretarius's four-sheet map of Hungary, where each sheet measures approximately 28 × 39 cm.[77] Lazarus's map lacks both roads and routes. No distances are given, although there is a linear scale. Yet Lazarus himself called the map a 'chorography and *itinerary* of the whole of Hungary' (our emphasis), clearly with route-planning possibilities in mind.[78]

The first English printed topographical maps (Saxton's maps of English counties) also show neither routes nor roads, only bridges. However, from Etzlaub and Lazarus we learn how the maps were to be used in the creation of itineraries. We may even imagine somebody like Lord Burghley bending over one of Saxton's county maps, dividers in hand, working out the length, section by section, of one of the postroads for which he had been estimating costs. Burghley was an experienced map-user. Others, planning their first major journey, would have found the exercise beyond their capabilities and would have preferred those of Norden's maps on which a reasonable number of main roads were indicated (as on his maps of Middlesex, Hertfordshire, Surrey, Sussex and Hampshire). After Norden, though, neither William Smith nor John Speed included roads, and new map-users had to learn how to make up their routes from road-less or virtually road-less topographical maps. They also had to learn how to calculate distances and work out directions for themselves, an uninviting task when many had no or little education in arithmetic.[79] Some help came in 1625 with the publication of John Norden's *England. An Intended Guyde*, a pocket-sized ready-reckoner with pages of triangular matrices, one for each English and Welsh county, containing ready-computed distances between any pair of 25 towns in the county (and occasionally some in the next) (Fig. 5.12). Norden's tables were reissued in reduced format in 1635, now with thumb-nail size county maps showing the location of the listed towns, by Matthew Simmons as the *Direction for English Travaillers* (Fig. 5.13). Further editions followed up to the end of the century. For all the apparent popularity of these tables, though, the user still had the task of working out for himself a list of all the places, in correct travel sequence, between point of departure and destination, and the direction to take between each. If these preparatory tasks were to be eliminated, a quite different cartographical aid would be needed. In time, this too was provided, in the form of a map displaying simultaneously every place in question, in its correct geographical position, and the distance between any pair of places.

Distance line maps

A distance line map came on to the English market in 1677, when John Adams printed a twelve-sheet map of England and Wales. Like many topographical maps before the eighteenth century, Adams's map shows no roads at all, only a remarkably large number of places (780 towns and villages). However, a network of ruled lines from one place to another, represents the distance between them, and a number by the line gives the mileage. As Adams announced in the title of his map of England, the distance between any pair of places shown on the map could be conveniently ascertained without scale or compass, simply by identifying the relevant line and reading the figure written against it.[80] On the whole, the resulting computation would be sufficiently close to the real distance to be travelled for shorter distances, although the discrepancies are 'sometimes alarming' on longer routes.[81] However, Adams's distance lines give little hint as to what the traveller might expect on the ground, or even, indeed, whether the implied road was of any use to him at all, for it was by no means always clear whether Adams's informants were giving distances for roads suitable for waggons and coaches or for the more direct cross-country tracks which could be taken by those on foot, on horse-back or with packhorse.[82] Under pressure from friends he had added rivers to his map for the second printing, but – as he himself anticipated – these did not always correct erroneous impressions. For, not only did the straight lines sometimes connect places which had no direct roads between them, but they could also imply bridges where there was none. In the Fenland, for example, the lines on map suggests a number of bridges across the Nene below Wisbech (Fig. 5.14) where there was none.

So, where did Adams get the idea of using distance lines? Had he himself seen or heard about any of the sixteenth-century continental maps with such lines? Could he have seen or heard of, before 1677, the medieval map (wherever it was in his day) that Richard Gough was to acquire in 1774, or any other maps on which distance lines appear?[83] Or is it just that the concept of giving the map-reader a ready-computed distance in this way is self-evident to anyone who was prompted to give thought to the matter?

Unlike the enigmatic Gough map, a fair bit is known about the immediate origins of Adams's map, from the Preface to the *Index Villaris*, a separate listing of the coordinates of some 12,000 places in England and Wales (not all shown on the map). Adams was a Shropshire lawyer working in London. A friend of his had recently established a fishery at Aberdovey, on the west coast of Wales, and wanted to estimate the potential market for his fish within a hundred-mile radius of Aberdovey. Adams sketched out a map on which every market town within the required radius was marked, noting the distances. Back in London, he extended the exercise into a map of the whole country. Once printed and assembled, the new

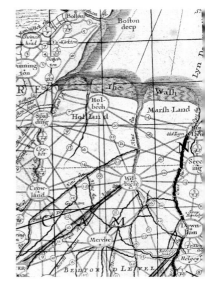

5.14   Detail from John Adams's 12-sheet map of 'notable distances for the use of travellers' to emphasise the role of the lines to indicate distances rather than roads on the ground. In the Fenland, six lines cross a bridgeless section of the River Nene, between Wisbech and the Wash. British Library, Maps 24.e.22, fol.7r.
Reproduced by permission of the British Library Board.

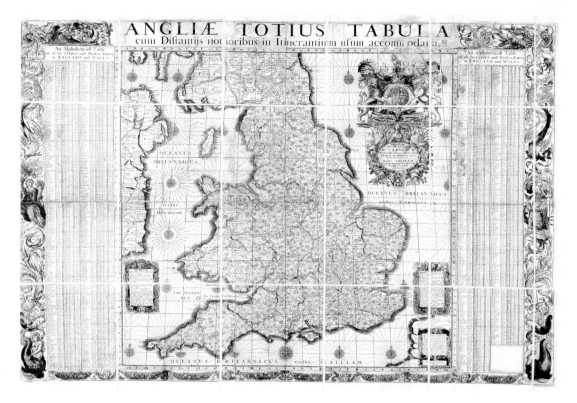

ANGLIÆ TOTIUS TABULA
cum Diftantijs notioribus in Itinerantium ufum accommodata.

5.15   The two-sheet version (1679) of John Adams's distance line map, *Anglia Totius Tabula cum Distantiis Notioribus In Itinerantium Usum Accommodata*, first published in 12 sheets in 1677. 69.5 × 98.5 cm. British Library, Maps 24.e.22.
Reproduced by permission of the British Library Board.

version measured a 'full six foot square' (nearly 2 m square). It was highly successful. Several times reprinted, it was also reproduced as a smaller, more manageable, two-sheet map (Fig. 5.15). In fact, so successful was Adams's map that at least eighteen imitative or derivative maps had been printed by 1750.[84] We today may have difficulty in grasping the practicality of a map which implies a direct road link where there is none, and which ignores major detours and gradients which would affect the actual distance to be trodden or ridden over, and the time needed to complete the journey, arguably highly relevant considerations in a mountainous country like Wales and with a perishable commodity like fish. But regular pedestrian, horseback and waggon travel involves a quite different value of time and notion of punctuality from our own.[85]

Special maps: postroads and turnpikes

Not everybody who purchased printed maps on the Adams model would have necessarily been thinking in terms of their own travel. Some purchasers may have been content simply to display the maps in their homes, to impress as much as for reference. Richard Gough reported how he had seen a copy, 'stretched on rollers at Ashridge

[in] 1767', and Adams wrote to Sir Daniel Flemming that the smaller version could be had for half-a-crown for the map in loose sheets and 'pasted upon Cloath with Rowler and Ledge and the Counties divided with Colours' for five shillings.[86] But others, especially those who provided roadside assistance for travellers (innkeepers, for instance) or who were responsible for regular trade services (such as the organisers of carters and carriers) would have found a map like Adams's essential for general reference. In yet another specialist sphere of activity, maps showing the entire network of the routes used by accredited postal carriers would likewise have been a useful adjunct.

News and knowledge are the essential underpinnings of government at any level. There have also always been differences in the speed of communication, between government news and private news, for instance, or in normal and in exceptional circumstances, such as war. In the Middle Ages, institutions usually provided their own postal systems.[87] In France, the existence of a public postroad system can be documented back at least to a royal decree in 1464. In England, despite evidence for the use of postal relays in the 1480s, the designation of certain routes for accredited carriers of mail and messages, on which post-horses and guides were available, came only in the second or third decade of the sixteenth century. The first national postmaster (*Magister nunciorum, cursorum, sive postarum*), Sir Brian Tuke, was appointed in 1533 to ensure adequate provision of services on routes in unusually high demand for one (official) reason or another.[88] One problem Tuke faced was not only increasing trade and traffic but also the rapidly-accelerating deterioration of the physical state of roads and bridges, partly as a consequence of the dissolution of the monasteries (many of which until then had provided a seemingly inexhaustible source of revenue for the upkeep of roads and bridges), partly through an increase in road traffic, and also from the 1530s the worsening of the weather after the onset of the Little Ice Age.[89] Another of Tuke's problems was the need for a proper road classification and organisation of road maintenance. Legislation in 1555 placed the onus of road maintenance on parishes, giving them a statutory obligation to carry out six days of work on the parish roads annually – but no guidelines as to what constituted acceptable standards. By about 1600, there was a system of nearly ninety 'standing' post stages on four arterial roads (London to Penryn, London to St David's, London to Holyhead, and London to Berwick) with others on branches (for example, to Carlisle, to Ludlow and to Barnstaple).[90] In 1635 Thomas Witherington's sweeping reforms of a motley collection of existing postal services created a self-financing public service.[91] Posts were to travel night and day on designated roads and from designated towns, with several regular services weekly. The key articulations of the seventeenth-century network refined that of the previous century, and comprised eight main routes from London: to Scotland (Great North Road), to Ireland (one by Bristol, the other by Holyhead), to the Welsh Marches (by Shrewsbury), to

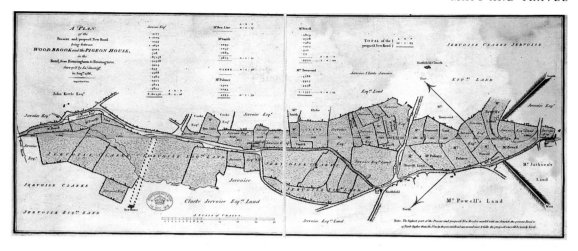

5.16 *A Plan of the Present &*
propos'd New Road lying
between WOODBROOK and
the PIGEON HOUSE, in the
road from Birmingham to
Bromsgrove. Survey'd by Jas.
Sherriff, in Augst. 1786. The
line of the new turnpike road is
printed in yellow. 20 × 47 cm.
British Library, Maps 1210 (8).
Reproduced by permission of the
British Library Board.

Plymouth, to Dover, and to the North Sea ports of Harwich and
Yarmouth. Areas not reached by the main postal system were to be
served by local entrepreneurs.

The conveyance of exceptionally heavy goods, increasing with
the development of large-scale industry in many parts of the country and the ever-expanding pull of the metropolis, made impossible
demands on English roads. Of all attempts to persuade users to
repair the roads they damaged, the most successful in the longer
term was the Act of 1663 which established a system of levying
tolls from road users for the maintenance of individual sections of
the road.[92] The first of these turnpike roads was, appropriately
enough, a stretch of the Great North Road between Wadesmill and
Buntingford, one of the most frequented routes in England but
also, skirting here the Fenland, one of the most low-lying.[93] The
concept of putting barriers across the road to oblige all travellers to
pay gained acceptability only slowly, and half the eighteenth century was to pass before the turnpike road developed into an
essential element in the nation's industrial infrastructure, forming
the system eventually represented by John Cary on his *New Map of
the British Isles* in 1815. Most of the maps of individual turnpike
roads were not made, however, for road users but were produced
as part of the submission to Parliament for permission to set up a
turnpike trust (Fig. 5.16). Their publication, as reduced copies, in
newspapers and periodicals like *The Gentleman's Magazine* or *The
Universal Magazine*, may incidentally have helped raise money for
the engineering work, but they were included primarily to promote
the magazine.[94] Thus, large-scale maps of turnpike roads did not
usually directly concern the eighteenth-century traveller, any more
than the general public today has much use for the large-scale maps
produced for, or used by, highways authorities. Once open,
however, the network of improved roads would have been of great
general interest, and small-scale maps of the system as a whole
provided the best advertisement.[95]

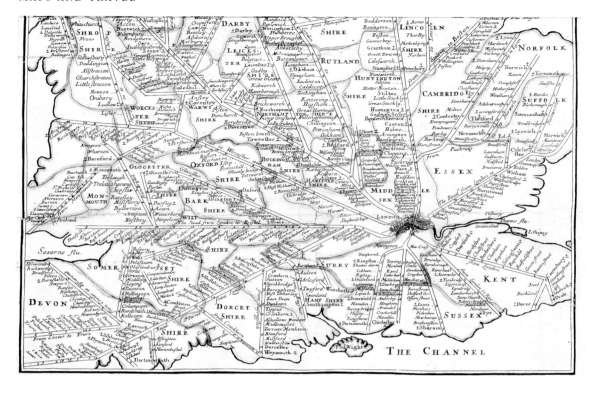

In 1668, three sheet maps showing the postroad network were printed. One was Richard Carr's *A Description of al[l] the postroads in England*, based on drafts made a few years earlier by a senior postal official, James Hicks.[96] In addition to the postroads (marked by double pecked lines, with the mileage noted beside), Carr included rivers, county boundaries and names, towns (marked by a pictorial sign and the name), and many names of coastal headlands. Wind roses and ships adorn the seas and a text at the bottom of the map, in four languages (Latin, French, English and Dutch), notes that the post travels at seven miles per hour in summer and five in winter, and lists the contents of the map. Although each category of information is clear enough in itself, the overall visual effect is confusing and the impact of a road system is lost.

Carr's map reminds us of the fundamental distinction between a topographical map on which roads are represented together with other features, and a special or thematic map on which only routes or roads are shown. Notwithstanding its title, Carr's map belongs to the topographical group, as do the two other 'postroad maps' printed in the same year, one engraved by Wenceslaus Hollar in 1667 but sold only a year later, its title emphasising its novelty, the addition of roads to a general map: *A New Mapp of the Kingdome of England And Principalitie of Wales ... Contayning ... the Highways and principall Roads*'.[97] The third map was Robert Walton's highly decorative *New Map Of England and Wales In*

5.17 The schematic rendering of routes on John Seller's *A New Map of the Roads of England* (*c.*1690) gives the map the appearance of a network of itineraries. Whole map 53 × 42 cm; as shown 27 × 41.5 cm. British Library, *Maps 1205 (2).
Reproduced by permission of the British Library Board.

5.18 George Willdey's map (1713) of postroads and the main cross-roads was based on information from John Ogilby's survey of some 40 years earlier. The lack of roads in the Fenland on this map is striking in comparison with Fig. 5.15. 60 × 60 cm. British Library, Maps K.Top. 5.84.
Reproduced by permission of the British Library Board.

Which the Roads or highways are playnly laid forth.[98] In contrast, John Ogilby's own single-sheet map of 1675, *A New Map of the Kingdom of England & Dominion of Wales. Whereon are Projected all ye Principal Roads Actually Measured & Delineated*, bound into the front of *Britannia*, summarises many more roads than those in the strip-maps which fill the rest of the volume, and shows a network of roads listed on the map as Direct Independents, Direct Dependants, Cross Independents, or Accidentals.[99] No other topographical features are marked.

Different from the maps which show an integrated network of roads, either alone (like Ogilby's) or against a topographical background (like Carr's), are maps of a particular system of routes relating to a specific starting place (usually London) or to a particular operator (such as a certain stage coach company). Such maps are characteristically diagrammatic in style, thematic in content, and structured according to topological relationships, anticipating Henry Beck's famous map of the London Underground system (see Fig. 1.2 on p. 3).[100] These are severely functional maps, designed for efficiency as a guide to, for example, letting passengers know when to get off or how many intervening stops there are before his or her destination. Two examples from the seventeeenth century are William Berry's *The Grand Roads of England Shewing all the Towns you pass thorough* [*sic*] of 1679 and John Seller's *A New Map of the Roads of England Shewing the Reputed distance from one town to another* (*c.*1690) (Fig. 5.17), both printed separately.[101] Two other examples, by Thomas Cross (1682) and by Herman Moll (1695), were printed as book illustrations.[102] As a model of clarity, George Willdey's sheet map of *Roads of England* (1713) is outstanding (Fig. 5.18).[103] Not all route systems maps focussed on London. A hitherto little-known group of maps was centred on Bristol. In 1739, C. Douglas printed a topological *Plan Shewing the Direct Roads* from that city to destinations all around the country; another of his maps, dated 1759, may also have come from a Bristol road-book or, perhaps more likely, a coach timetable.[104] Other towns and cities may well have generated similar maps which remain unrecognised in county archives or private collections.

None of these maps told the traveller anything more about a specific route than he could have discovered from a written itinerary. What maps like Carr's offer is a topographical context for a comprehensive network of itineraries with links to some places not directly on the postroad network. What the route systems maps like Berry's, Seller's, Ogilby's or Willdey's offer is a visual representation of a multiplicity of itineraries mapped out on a single sheet of paper. In either case, as the story behind the production of Carr's map reveals, the original appeal of all such maps was likely to have been to the professionals in charge of the relevant system, serving them in the discharge of their organisational or administrative duties, or to professional travellers whose business took them specifically along these routes.[105] None of these maps, though, gives any real clues about the ground any user would actually have to ride over. They omitted completely the local roads, green lanes and openfield tracks on which the vast majority of the country's rural inhabitants depended daily, especially for their visits to market. And they certainly did not seek to portray the road itself on paper.

JOHN OGILBY'S STRIP-MAPS

Knowing a route to a destination is not the same as knowing which road to follow on the ground. Just as the motorised modern traveller reads road signs, so medieval and early modern travellers

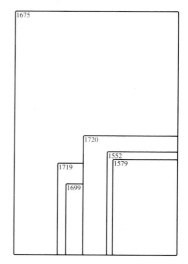

5.19 Comparative sizes of Ogilby's folio *Britannia* (1675), two of the pocket-sized editions of *Britannia* (1719, 1720), and three early road-books (1552, 1579, 1699).

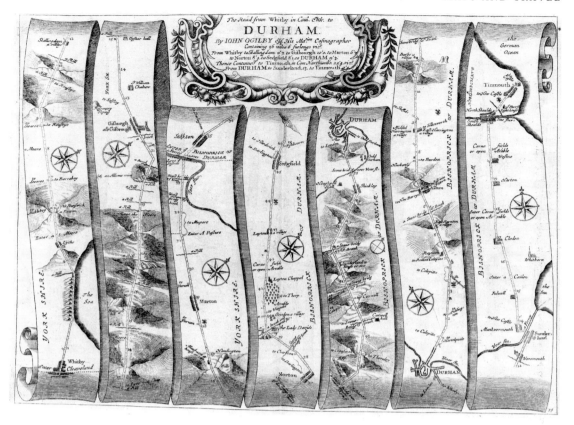

The Road from Whitby in Com. Ebor. to
DURHAM.
By IOHN OGILBY Esqr. His Maties Cosmographer

5.20 A page from John Ogilby's *Britannia* (1675): part of the post-road from Whitby to Tynemouth. While Ogilby made it clear that travellers would cross the river Wear at Wearmouth by ferry, he failed to indicate how they were to cross the much wider river Tyne between South and North Shields. 34 × 46 cm. British Library, Maps C.6.d.8, pp. 196–7.
Reproduced by permission of the British Library Board.

relied on the primeval modes of way-finding, those of looking for landmarks and asking for directions.[106] A wealthy traveller avoided the problem entirely by hiring a local guide or carriage or by employing specialists to organise his journey for him.[107]

At face value, John Ogilby's strip-maps of the postroads of England and Wales might seem to be the answer to the seventeenth-century casual traveller's problem of being sure which road to take and what lay ahead. But did Ogilby intend his maps to be of practical use in this way, to be taken on the road and followed, on horseback or in a carriage? Furthermore, who would need such a map? As single maps, Ogilby's strip-maps could certainly have been folded and carried about, as he portrayed on the title page, and they may have been sold as single maps. But *Britannia* as a whole, a volume of 300 pages weighing nearly 7 kg, can by no stretch of the imagination be called a portable road-guide (Fig. 5.19).[108] So what did Ogilby intend his maps to be used for? Who were the users of a large, heavy folio collection of maps of the postroads of England and Wales; were they those on the road or those in a library?[109]

John Ogilby is usually credited with 'inventing' the strip-road map with the publication of his *Britannia* in 1675, but his claim is

169

worth challenging. The debt he himself acknowledged was to Imperial Rome, and to the written Antonine itinerary and the Peutinger map. Although Matthew Paris's name is nowhere mentioned, however, it is very likely that Ogilby drew his inspiration from Paris, for there is little doubt that Ogilby would have known about Paris's itinerary to Rome even if he had not seen it for himself. Ogilby's maps certainly bear no resemblance to the Peutinger map. The Roman map copied by Conrad Peutinger after it was bequeathed to him in 1508 by Conrad Celtes is not a strip-map in any sense except in its physical dimensions (a dramatic 34 centimetres in height and 6.75 *metres* in length), but a network map, showing a large number of routes throughout the Empire.[110] Unlike the Peutinger map, but exactly like Matthew Paris, Ogilby structured each route in the form of a narrow strip arranged in columns across the page so that, starting in the bottom left-hand corner, the reader can follow the road on paper as if physically riding along it (Fig. 5.20).

Printed editions of Paris's chronicles were readily available throughout the seventeenth century in an edition by William Wats of Matthew Parker's first printing of 1571.[111] Although unillustrated, all of Wats's editions include Parker's preface and his fulsome praise of Paris's 'excellent way of painting and describing the events in their own shape, and with correct proportion of all the limbs and parts' in the map of Britain.[112] Moreover, both Parker and Wats drew attention to the original manuscripts, Parker by pointing out that the 'marvellous proof' of Paris's art was to be found in 'the first copy of his history, written by his own hand', and Wats by advertising on the title page of his edition the manuscripts at Corpus Christi College, Cambridge (to whom Parker had left his own copy), and in the Royal and Cottonian libraries, London. Ogilby could have had access to any of these libraries. He had lived in Cambridge, and in London the Cotton library in particular was well frequented by those whose paths crossed Ogilby's. Moreover, not only were many users of the Cotton library also Fellows of the Royal Society – and Ogilby counted a number of Fellows amongst his intimates – but at least two Fellows (Samuel Pepys and William Petty) had received copies of the subject catalogue of Cottonian manuscripts which had been completed and widely circulated in 1674, the year before *Britannia* was printed.[113]

It could be, of course, that Ogilby's invention of the strip format for a road map was entirely original. However, the fact remains that the existence of Paris's manuscripts, and *ipso facto* the maps which feature so prominently in them, was widely known in early modern times. Plagiarism was common practice, and Ogilby's 'shameless' (to us) pirating of authors has been remarked upon.[114] He never hesitated to consult his scientific friends and acquaintances, especially over the drafting of the maps, and indeed, it may have been his friend Robert Hooke, a Fellow of the Royal Society, who thought up the innovative solution to the problem of distinguishing downhill and uphill slopes on the strip-maps.[115] Like his contemporaries, Ogilby would have been keenly concerned with

5.21 Detail from Fig. 5.20, showing Ogilby's use of pecked and solid lines to differentiate between open and enclosed roads, and his inversion of hill signs to indicate direction of slope. Section of road shown approximately 13 km (8 miles). British Library, Maps C.6.d.8, pp. 196–7.

Reproduced by permission of the British Library Board.

practicality, and pragmatic presentation would have appealed to him at a time when usefulness was 'an Instrument whereby Mankind may obtain Dominion over Things'. Like Hooke, Ogilby would have been quick to appreciate the potential of a format which could be bent to advance his own purposes.[116] The question then arises, if Ogilby had been alerted to the nature of Paris's graphic itinerary, why did he omit to mention the fact? Unlike Imperial Rome, the Middle Ages were held in low esteem in the seventeenth century. A dedication (in this case to Charles II) of a map derived from something as crude as a medieval prototype rather than one dating back to the greatness that was Rome could have been taken as a personal slight.[117]

Although John Ogilby had not in his youth been a professional map-maker, he was by no means unfamiliar with map-making by the time he came to be involved with road surveying. He had already put forward a project for the remapping of the whole country as part of a grandiose scheme for a five-volume atlas which foundered for lack of financial sponsorship and official support. Eventually, he was left with much narrower terms of reference, namely to depict 'the Post Roads for conveying Letters to and from London'. The resulting strip maps are, as J. B. Harley remarked, brilliantly simple in design while containing a wealth of detail relevant to road travel.[118] Ogilby's choice of scale, one inch to one (statute) mile, proved so suitable for general use that it came to be adopted by later county map-makers, including, in due course, the Ordnance Survey. Approximately 7,500 miles (12,000 km) of road, surveyed consistently at 1,760 yards per [statute] mile, and 73 mail roads in England and Wales, are represented on 100 maps and described in 200 pages of written text. On the maps, refinements like the way the inversion of a hill sign warns of the direction of slope according to direction of travel, indications of the nature of bridges (wooden, brick or stone), the presence of gates across roads, the indication (by solid or broken lines) whether a road was enclosed by walls or hedges or open to the surrounding countryside, and land use are all carefully indicated on the maps, together with landmarks like windmills, gallows, water mills, major country houses, and church steeples and towers, and the destinations of crossroads, all of which were relevant to the road-user (Fig. 5.21).

The maps are not without errors, nor are all postroads included. Even so, Ogilby's *Britannia* presented for the first time, in remarkable detail, a cartographical portrayal of the roads themselves. In the final analysis, the great tome is a collection of graphic and written itineraries, destined not for use on horseback, or even in a carriage, and far too costly to risk maltreatment. It sold well, though, and as a delightful library book one has to assume it was read and pored over in the drawing rooms as well as the libraries of those country houses Ogilby had taken pains to include on the maps.[119]

Ogilby died a few months after the completion of *Britannia*. The fact that it was over forty years before the idea of a portable edition took effect emphasises the conceptual gap between whatever Ogilby was trying to achieve and the more practical, or purely commercial, demands that were the concern of later map and road-book printers. Two pocket-sized editions of the maps from Ogilby's *Britannia* were printed in 1719, one by Thomas Gardner and the other by John Senex, to be followed a year later by what proved to be the most successful pocket edition of all, Emanuel Bowen's *Britannia Depicta or Ogilby Improved* (Fig. 5.22). By now, at least, there was a buoyant market for such publications and this remained the case for the rest of the eighteenth century, when an updating of Ogilby's data was long overdue. Two heavily revised editions dominated the market in the last three decades of the century, printed by Daniel Patterson (1771) and by John Cary (1798), respectively.[120] All these small-format editions were designed to be as cheap as possible and to have the widest possible appeal. As the page size was reduced, much of the clarity of Ogilby's folio maps was in danger of being lost and as text and images (mostly the arms of the local gentry) were crowded around the maps, so Ogilby's atlas of a specific road system was well on the way to becoming a general travel guide or companion. These pocket editions were certainly popular, to judge from the number of reprints, reissues and editions put out over the eighteenth century and into the nineteenth. While they would have appealed above all to the 'middling sorts' of people, those who could never have afforded anything like the original folio volume, they would surely have been scorned by the hard-bitten coachmen, postal riders, mail-coach drivers and waggoners who prided themselves, as did their medieval and early modern predecessors, on their road-knowledge.

ROAD MAPS

In early centuries, it was by no means always clear on the ground precisely which was 'the road'. Before enclosure and the 'privatisation' of land, the laws of the ancient common-field farming system could accord travellers way-leave over manorial land, and the tracks which eventually became our 'roads' were in essentials only the most commonly trodden strip of ground. Travellers might deviate from the track, especially where it became impassable in bad weather, or elect to pick their way over the fields – along the headlands and between the furlongs of cultivated openfield – a practice permissible provided no damage was done to the land or to crops. It was only pedestrians and single or small bands of horsemen for whom the option of taking a relatively direct cross-country track instead of the usually more circuitous and over-used main roads was realistic.

The boundaries of unpaved highways were often the subject of litigation. Geoffrey of Monmouth's observation, about 1200, that the roads of ancient England 'were a bone of contention, for no one knew just where their boundaries should be' was still applicable to roads in the remoter districts of England in the later

5.22 The postroad from Oxford to Coventry: page from Emanuel Bowen and John Owen's reduced edition of Ogilby's strip maps, *Britannia Depicta or Ogilby Improv'd* (1720). The information is identical to that on Ogilby's map, from which it was copied (except for the omission or amalgamation of some land-use notes), but it is less easily read. 27 × 25 cm. British Library, C.27.a.13, pp. 218–219. Reproduced by permission of the British Library Board.

to Chipping Norton · Tackley Church

the Hill Hunwell · Goudry Broughton

Comon / enter at Lane / Common / Fields

Atherston

Noted formerly for a religious House of Augustin Fryars, is now a Town indifferently large & well built, having a Chappel of ease. a Free School a Mt on Tues. & Fairs. Mar. 27th July & Sept. 8th & Dec. 4th [at 63]

CommonFields 26

Ladbrook 36

to Woodstock 11

Broughton a Lane

Warwick · to Davenry

Mancester [at 62]
Is supposed to be the Maudnessedum of the Ancients.

a Lane & 25

a Lane 35 & Hedges

10

a Lane & a Brook Stone

Small Bridge

a Farme 34

Pasture

Nuneaton [at 59]
Is said to be so called from a Nunnery foun-ded in 1160 by Amica & Wife of Robt Bosu E. of Leicester. The Town is large & well built having a good Free School. a Mt on Sat. & a Fair on May 3d.

Common Fields to Stratford 24

Watergall house

the London Road

Riding Ground all the Way between Killington and Dedington

to Shipston 23

Water Gall Lane 33

a Farme

a Hedge

Banbury

Wormleighton 32

Pasture Ground

Southam [at 37]
Is a well accommodated Town, having a consi-derable Mt on Mond. & a Fair on St. Peter's-d.

to Woodstock 10

to Shipston · to Blechindon

Comon Fields to Bodicot

Bodicot

to London Boy 21

Common Fields 31

Arrable

a Lane 30

a Lane & Hedges

to Woodstock

to Woodstock · to Ship

to Blechindon

to Shipston

Adderbury · to London

to Twiford

to Somerton

Claydon

enter Warwick Shire

Killington Church

Stone Bridge & Brook

to Ayno

Farmborow 29

to Woodstock · to Ship

Killington Green

to Blackston 19

Common Fields

Killington Green

Arrable a Rill 18

to Wickam 20

Warwick 28

Wormleighton [at 22]
Was formerly honour'd in giving title & Baron to St. Robt Spenser, created by K. James 1st 1mo R. His Grandson Henry Ld. Spencer &c. was 1st Earl of Sunderland by K. Charles 1st lost his life in that Princes cause in ye first Battle of Newbury. The present Ld. Wormleigh-ton is ye noble & Rt. Hon. Cha. Spencer E. of Sunderland, of ye same Noble Family, who has a Seat here & another at Althorp in Northampton Sh.

Stone Bridge & Brook

Cotes[well]

Southam [at 37]

Pasture

to Water Eaton to Cambridge

Arrable to Alse 17

Common Fields 27

to Woolvercote

Woolvercote

to Hemton

Deddington

a Rill

The Manor of Hogs-Norton als. Norton Juxta Tivecross [at 66] Was given by K. Eldred ye Saxon to Elseth his Servt, by Deed dated 951 which we shall take ye opportunity of inserting here, as a Specimen of ye ancient Forms of such kind of Grants, for ye Rea-ders satisfaction. The Te-nor of it is as followeth. being thus signed. —

Wormleighton [at 22]

Deddington [at 7]
Is a Town of great An-tiquity, govern'd by a Bay-liffe, has a Mt on Sat. & Fairs on St. Lawrence & St. Martins day.

Arrable on both Sides

to Woodstock

Stone Bridge & Archers Brook

15

to Dunstew · to North Aston

Dunstew 14

to Bicester

Steple Aston 13

Rousham Church

W. Domino Dominoru dominante in seculo

OXFORD · **OXFORD** · **WARWICK** · **OXFORD**

"seculorum Ego Eldred Rex, Anglorum Gubernator et Rector cuidem "mihi fideli∬imo Ministro, Æl∫eth uni Ca∬atos perpetualiter concedo in il- "lo loco, ubi jamdudum Soticolæ illius Regionis nomen indiderunt, Et Northtu- "ne, ut habeat et po∬ideant quamdiu vivat, Et po∫tquam univer∫itati via "adierit, cuicunque voluerit Hæredi derelinquat in æternam hæreditatem, Si "autem prædictum Rus liberum ab omni mundiali tam in magnis quam "in modis, rebus, Campis Pascuis, pratis, Silvis, sine expositione, et pontis "Arci∫ve instructione. si qui denique mihi non obstante hanc re liberta- "tis Charitam livore depre∬i violare satagerint, agminibus atra caliginis lapsi Vocem au- "diant. Discedite a me maledicti. si non ante Mortem Emendaverint Penitentia. Istis "terminis ambitur prædicta Tellus. This lant Uthalangemara to Northtune Hæc Charta Carax- "ata est. A.D. 951. Indict.9 Ego Eldred Rex. Ego Eadgiva Regi Mater. Ego Odo Dorovernensis Episcopus &c.

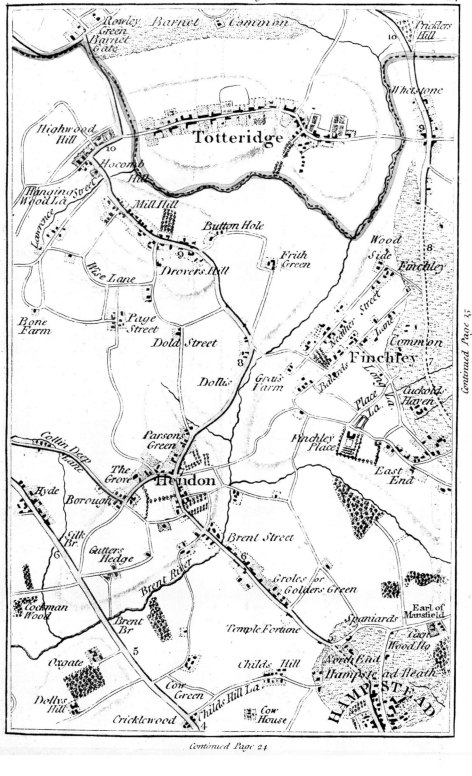

Rowley Green

Barnet Common

Barnet Gate

Pricklers Hill

10

Whetstone

Highwood Hill

10

9

Hocomb Hill

Totteridge

Hanging Street Wood La

Mill Hill

Button Hole

Wood Side

Finchley

8

Lawrence

Wise Lane

Drovers Hill

9

Frith Green

Mother Street

Lane

Bone Farm

Page Street

Dold Street

Common

7

Finchley

Balords

Dollis

Grass Farm

8

Cuckold Haven

Long La

Place La.

Parsons Green

Finchley Place

Collin Deep Lane

The Grove

East End

Hyde

Borough

Hendon

Silk Br

Gutters Hedge

Brent Street

6

6

Groles or Golders Green

Brent River

Earl of Mansfield

Cockman Wood

Brent Br

Spaniards

Cage Wood Ho

Temple Fortune

5

5

North End

Oxgate

Childs Hill

Hampstead Heath

Dollys Hill

Cow Green

Childs Hill La

Cow House

HAMPSTEAD

Cricklewood

4

Continued Page 45

seventeenth century.[121] Not only was each track liable to become braided as passers-by tried to avoid potholes and deep ruts, but there were often alternatives. Writing from Cumbria in November 1676, Adams complained: 'Many of ye wayes are so ruff as a man can scare meet with two persons who agree exactly in their distances; besides there being severall wayes from many places, one often fancy[ing] one way [but] another person another way to be nearer.'[122] The confusion of 'severall wayes' was not confined to distant regions. The Great North Road (Watling Street) offered travellers three quite different alignments for the twelve mile (20 km) stretch between Alconbury, Huntingdonshire, and Wansford, Northamptonshire.[123] Which 'road' was followed depended on the status of the traveller, the size or weight of load or baggage, and the nature of the waggon or conveyance. Inevitably, the best defined, and best maintained, highways were those permanently reserved for those the business of the realm, the king's or government's business, messengers, accredited carrier services, and such-like.[124] Independent travellers using them had to take their mounts from regulated post-inns.

From the mid eighteenth century onwards, though, a number of factors promoted the definition of the physical road with a precision rarely seen since the Romans constructed and administered their paved highways. Enclosure affected roads as well as fields, leading to stipulated road widths and the containment of many lowland and also upland roads between field walls, fences, ditches or hedges.[125] Turnpiking led to a further narrowing of the trodden surface, encouraging travellers to use only the regularly maintained portion of the highway. By the end of the eighteenth, and in the first decades of the nineteenth century, John MacAdam's recipe for metalling roads was taken up as he visited one part of the country after another to advise turnpike trusts on road construction.

As the distinction was created between good roads and bad roads, so it was worth indicating this on maps. Individual map printers like Robert Morden and Philip Lea had been adding selected roads to county maps in the second half of the seventeenth century. In the mid eighteenth century, winners of the Society of Arts awards for large-scale county maps were indicating cross roads as well as main roads. By the end of the eighteenth century, the increasing scale of road improvement encouraged publishers to advertise the fact that their new topographical maps included the road network, and their maps now allowed travellers to select a physically more comfortable route, a point warmly approved by reviewers. In April 1786, Cary's map of Middlesex was praised as 'particularly convenient for occasional consultation' – both as a county map and as a road directory'.[126] Cumbersome titles designed to ensure that the prospective purchaser appreciated the wealth of information contained in each map, almost always mentioned the road first. In 1768, Cary burdened his small octavo booklet containing, in 28 separate plates, a one inch to one mile map of Middlesex with just such a comprehensive description:

5.23 Roads on topographical maps: one of the 50 sections into which John Cary's map of the country 15 miles around London was divided for binding into an octavo road-book. 1786. 15 × 9 cm. British Library, Maps C.24.a.13 (14).
Reproduced by permission of the British Library Board.

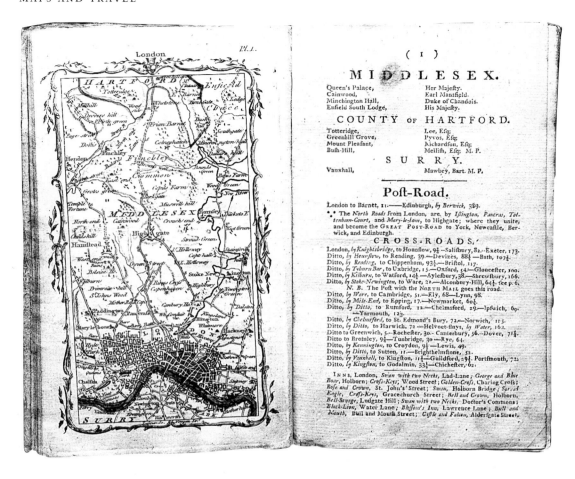

(1)

MIDDLESEX.

Queen's Palace,   Her Majefty.
Cainwood,   Earl Mansfield.
Minchington Hall,   Duke of Chandois.
Enfield South Lodge,   His Majefty.

COUNTY OF HARTFORD.

Totteridge,   Lee, Efq;
Greenhill Grove,   Pyvos, Efq;
Mount Pleafant,   Richardfon, Efq;
Bufh-Hill,   Mellifh, Efq; M. P.

SURRY.

Vauxhall,   Mawbey, Bart. M. P.

Cary's Actual Survey of Middlesex ... wherein The Roads, Rivers, Woods & Commons; as well as Every Market Town, Village Etc. are distinguished; & Every Seat shewn with the name of the Possessor, Preceded by a General Map of the County, Divided into its Hundreds. To which is added, An Index of all the Names contained in the Plates.[127]

The topographical map of Kent printed in 1769 by John Andrews, Andrew Dury and William Herbert is no less detailed:

... in which are Expressed all the Roads, Lanes, Churches, Towns, Villages, Noblemen and Gentlemens Seats, Roman Roads, Hills, Rivers, Woods, Cottages & everything Remarkable in the County, Together with the Division of the Lathes & and their Subdivision into Hundreds.[128]

Cary's titles were often identical from map to map. The lengthy title of his *Actual Survey of the Country Fifteen Miles Around London* (1786), printed in 50 sections, each 15 × 9 cm (Fig. 5.23) is identical to that of his Middlesex map, printed eighteen years previously. Roads were also marked on atlas maps. On the county

5.24 Two pages from Mostyn Armstrong's uncoloured London-Edinburgh road-book, showing part of the Middlesex section. The Great North Road itself is scarcely discernible amongst all the other topographical information and it is difficult to match the cross-roads listed on the opposite page with the map. 16.5 × 9 cm. 1776.

Reproduced by permission from a private collection.

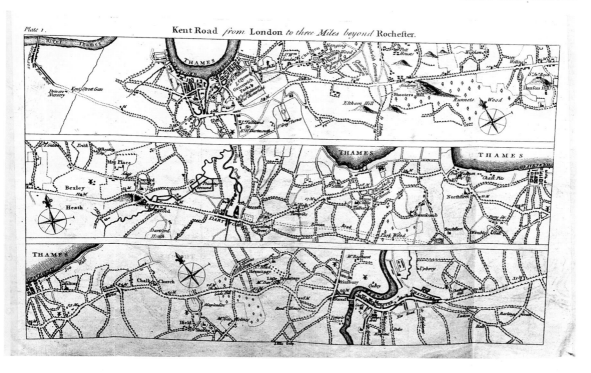

Plate 1.   Kent Road *from* London *to three Miles beyond* Rochester.

5.25   Mostyn Armstrong's map of the Kent road. The road network within the area shown on each strip is depicted. 15.5 × 28 cm. 1760.
Reproduced by permission from a private collection.

maps in Cary's *New and Correct English Atlas* (1787–9), turnpikes are identified, their starting and finishing points named, and some distances are given together with – as advertised on the titlepage – 'Directions for the junction of the Roads from one County to Another'.

Increasingly, maps in road-books were no longer maps of the itinerary but effectively topographical maps, like those in Mostyn John Armstrong's *An Actual Survey of the Great Post-Roads between London and Edinburgh with the Country of Three Miles on each side* (1776) (Fig. 5.24) and in his description of the London-Dover road, printed the following year. In 1818, Cary put out a reduction of his six-sheet map of the British Isles on which were shown 'the whole of the Turnpike Roads'.[129] Two years earlier, Aaron Arrowsmith portrayed what appears to have been a meticulously correct network of roads (in three classes: post roads, main and minor roads) on his great $16\frac{1}{2}$-sheet map of England and Wales.[130] At the same time, graphic itineraries continued to be produced. Cary's first publication in the road-book genre, his *Actual Survey, of the Great Post Roads between London & Falmouth, including a branch to Weymouth, as well as those from Salisbury to Axminister, . . .* (1784), included a summary map of the whole route as well as 50 topographical maps. Later on, Cary found yet another way of enhancing his itinerary maps, namely by adding sight lines, as on his *Survey of the High Roads from London . . .*

(1790) where lines ruled from the road showed 'the points of sight from where the Houses are seen' (Plate 13). As Fordham noted, some of these houses were visible from three or even four points.[131] Maps in general travel books, like *The Kentish Traveller's Companion* of 1776, attributed to Mostyn Thomas Armstrong, were essentially topographical maps, however cleverly arranged on the page (Fig. 5.25).

Later came specialist maps, such as those for cyclists (from the 1880s), and motorists (from the first years of the twentieth century). The topographical map, however, has remained the paramount way-finding aid for a large majority of independent travellers.[132] Even professional and business organisations, whose scale of operations fits the local or regional area which can be shown on a conveniently-sized assemblage of sheets, continue to find the topographical map essential for reference. Only within major urban areas is the street map more useful. For general use, though, where Christopher Saxton laid the foundations for county surveys, and John Ogilby showed the suitability of a scale of one inch to one mile for topographical maps, enterprising map printers and publishers have continued to exploit, with great success, the market for the small-scale travel aid needed for long-distance, high-speed modern travel, namely the thematic road map and road atlas.

1. Plan of the tabernacle in Solomon's temple, Jerusalem, in the Codex Amiatinus. Double folio. Late seventh century, Jarrow. Biblioteca Laurentiana, Florence, Codex Amiatinus I.c.II, fol. 2v–3r.

Reproduced by permission of the Biblioteca Laurentiana.

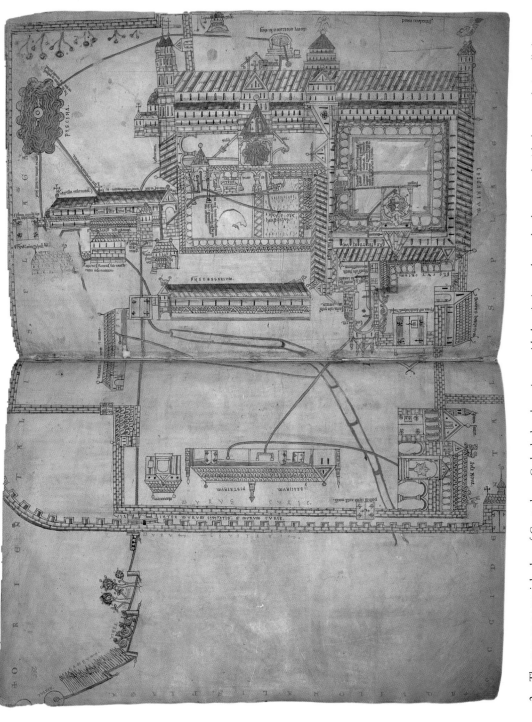

2. The map as memorial: plan of Canterbury Cathedral and precincts, with descriptive and anecdotal texts and with the recently-installed waterpipes and drains highlighted in red, produced in honour of Prior Wibert, the instigator of these and other improvements. Originally approximately 56 × 34 cm. East is at the top. *c.*1153–61. Trinity College, Cambridge, MS R.17.1, fols 284v–285r.

Reproduced by permission of the Master and Fellows of Trinity College, Cambridge.

3. Plan in a processional reminding participating clergy where they were to stand for, in this case, the Easter Vigil litany. Deacons, sub-deacons and priests are indicated by double circles representing tonsured heads in plan, others by the vessels they held. 21.5 × 13 cm. Late fourteenth or early fifteenth century, Norwich. British Library, Add. MS 57,534, fol. 62v.

Reproduced by permission of the British Library Board.

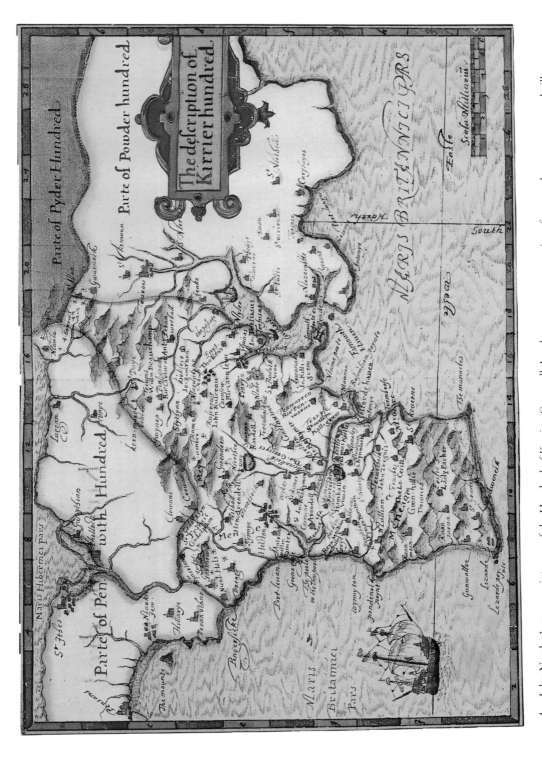

4. John Norden's manuscript map of the Hundred of Kerrier, Cornwall. Landowners names are given for great houses, towns and villages are distinguished, and lead mines marked (cross with four dots). 23.5 × 34.5 cm. c.1597. Trinity College, Cambridge, 0.4.19.

Reproduced by permission of the Master and Fellows of Trinity College, Cambridge.

5. Manuscript draft of William Smith's map of Hertfordshire, showing corrections which were adopted for the printed version (see Fig. 3.19). The original date of 1601 has been changed to 1602 and a Latin key has been added, for instance. The white circles marking the location of each place are holes punched through the paper in the process of transferring the map onto the copper plate. 41 × 49 cm. British Library, Maps C.2.cc.2 (13).

Reproduced by permission of the British Library Board.

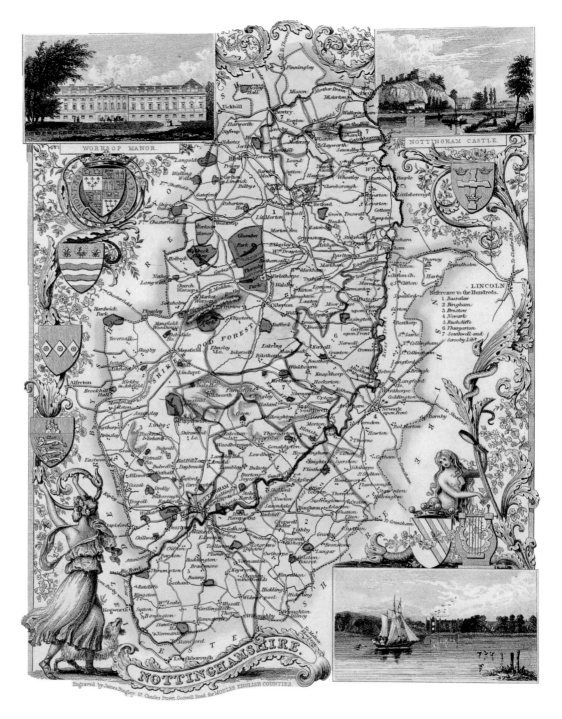

6.   The map of Nottinghamshire in Thomas Moule's *English Counties* (1837). At a time of industrialisation and intensive urban growth, vignettes of country houses and other decorative features are used to evoke a nostalgic rural image. 25 × 19 cm. British Library, C.29.b.2 (xvi).

Reproduced by permission of the British Library Board.

7. Lanhydrock, Cornwall, estate atlas, 1693–9. The Robartes family owned properties widely dispersed throughout the county of Cornwall. Individual maps of the dispersed parts were bound together to form an estate atlas with this elaborate title page.

Reproduced by permission of Cornwall Record Office.

8. Detail from the title cartouche on Mark Pierce's manuscript map of Laxton, Nottinghamshire. Significantly, the globe presents the Atlantic, plied by the ships which brought William Courten, the new owner of the Laxton estate, his wealth. 1635. Bodleian Library, MS C 17:48 (9a).

Reproduced by permission of the Bodleian Library.

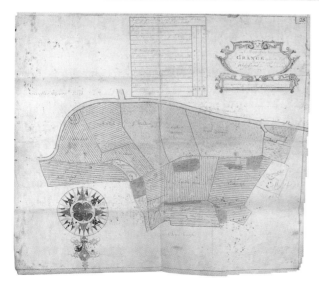

9. Estate map of Bideford East the Water, Devon, by Joel Gascoyne, c.1699. Some highly decorated estate maps, such as this one, can be counted as minor works of art; the map boasts an elaborate cartouche, compass rose, and vivid colouring. 32 × 46 cm. 1 inch to 10 perches (1:1980). Devon Record Office, NDRO 2379A/Z38/12.

Reproduced by permission of Devon Record Office.

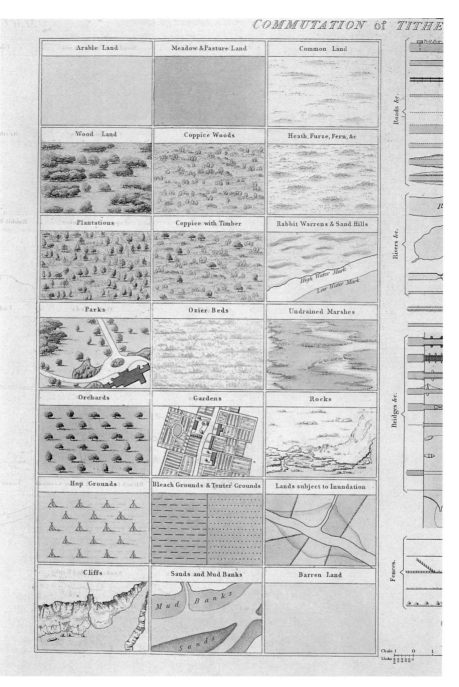

10.   Some of Robert Kearsley Dawson's recommended tithe map symbols, 1836. Dawson hoped mapping for tithe commutation would be extended to cover the whole country so that England would be provided with a comprehensive mapped cadastral survey on the continental model, but the Treasury refused additional funding and only a minority of tithe maps were produced using the full range of symbols. 'Copy of Papers Respecting the Proposed Survey of Lands under the Tithe Act', *British Parliamentary Papers (House of Commons)*, 1837, XLI.

Reproduced from a copy in a private collection.

11. Two of the charts in Guillaume Brouscon's manual for pilots sailing around the coasts of southern Britain and Ireland. The lines indicate the hours of high tide at each port. Each map 12 × 8 cm. 1540–50. Manuscript. British Library, Add. MS 22,721, fols 10v–11r.

Reproduced by permission of the British Library Board.

12. Lucas Jansoon Waghenaer's chart of the English coast between Dover and Orfordness. Sandbanks are clearly outlined, the nature of the coast is indicated both on the map and in the profiles along the top, and soundings are given. Each chart is presented from the seaman's perspective; here, west is at the top. 32.5 × 50 cm. British Library, Maps C.8.b.4. (23).

Reproduced by permission of the British Library Board.

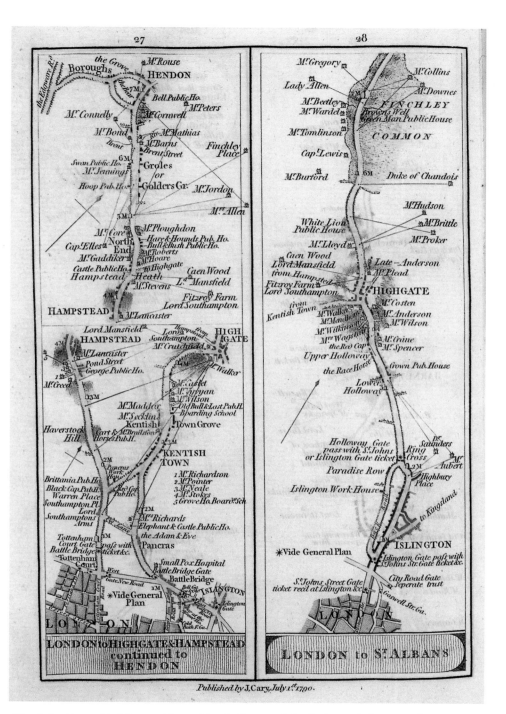

13. John Cary added lines to his road-book, *Survey of the High Roads from London . . .*, to 'shew the points of sight from where the Houses are seen'. Dimensions of the page reproduced here, showing the Hampstead to Hendon road, north of London, 13 × 19 cm. 1790. British Library, Maps C.24.c.4.

Reproduced by permission of the British Library Board.

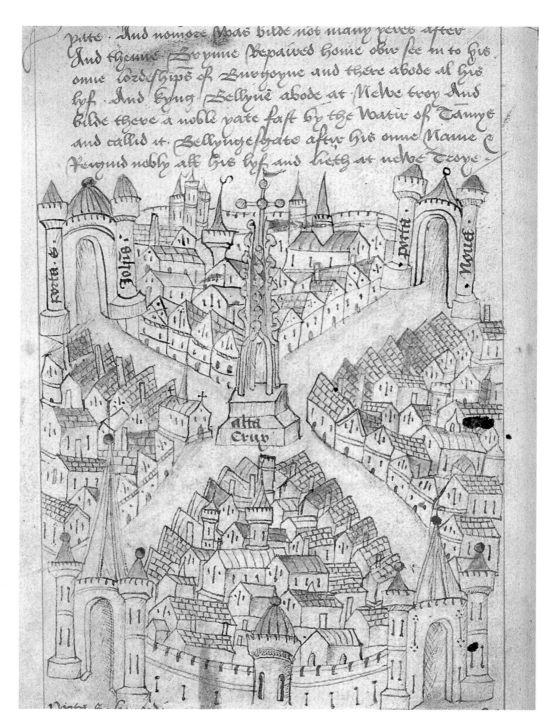

14.   Robert Ricart's plan of Bristol, *c.*1480. This bird's-eye view of the central part of the city is from his unpublished history of Bristol. Like Chaundler's map of Wells (see Fig. 6.3), Ricart's Bristol is also highly selective in the features portrayed. 15 × 12 cm. Bristol Record Office 04720, fol. 5v.

Reproduced by permission of the Bristol Record Office.

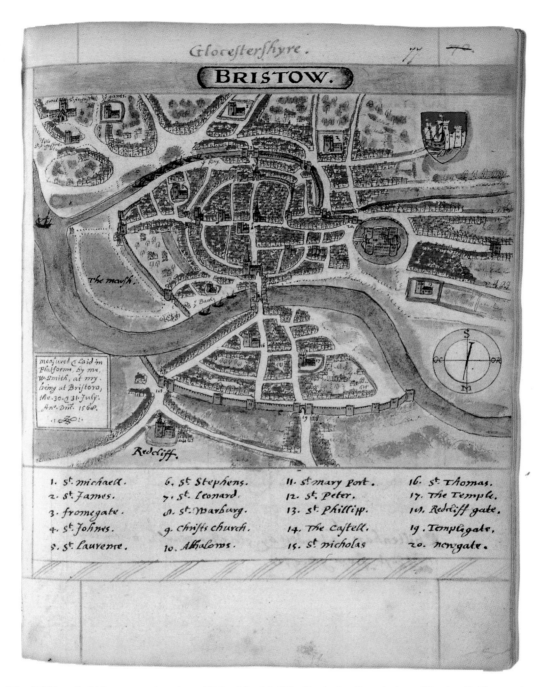

15.   William Smith's manuscript plan of Bristol for his 'The Particular Description of England'. The text in the cartouche states that the town was 'measured and laid in a Platform by me W. Smith, at my being at Bristoro the 30 & 31 July Ano Dm 1568', although the title page is dated 1588. 16.5 × 15 cm. British Library, Sloane MS 2596. fol. 77r.

Reproduced by permission of the British Library Board.

16. John Hooker's plan of Exeter, c.1587. Hooker was City Chamberlain from 1555 and his map depicts something of the economic life of the city (note the mills, shipping and fishing on the River Exe) but is also interpreted by historians as a celebration of the wealth and power of this place in Tudor England. 37 × 54.5 cm. British Library, Maps C.5.a.3.

Reproduced by permission of the British Library Board.

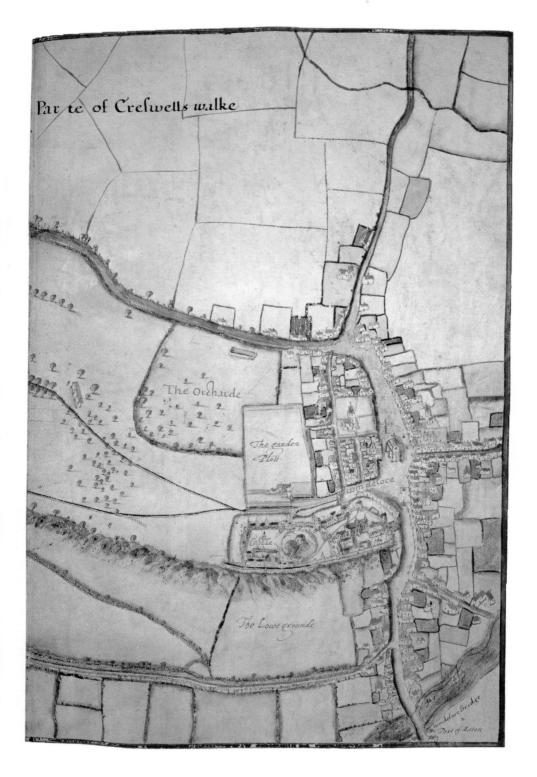

Parte of Creswells walke

The Orcharde

The garden Plott

wmdelove

CastLe

The Lowe grounde

windelove Bridge
Part of Eton

17.   The town of Windsor appears on one of the eighteen maps in John Norden's 'Description of the Honor of Windsor' (1607), an apparently unsolicited survey of Crown property offered to King James I. 37 × 48 cm. British Library, Harley MS 3749. fol. 5r.

Reproduced by permission of the British Library Board.

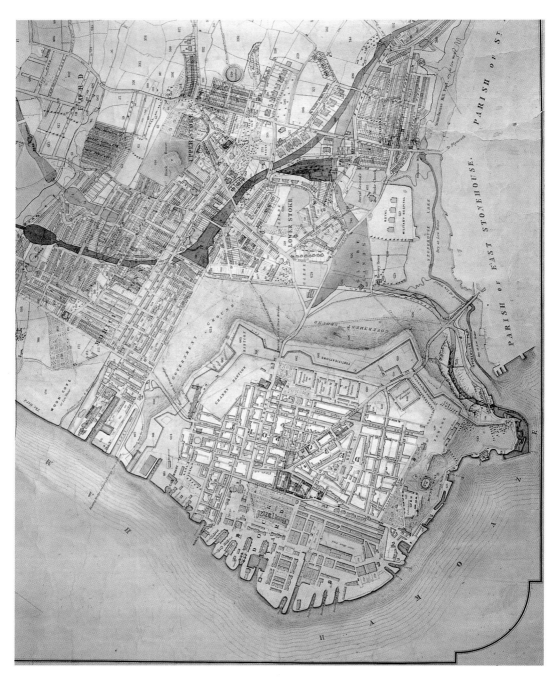

18.   Tithe map of the parish of Stoke Damerel, Devon, 1842, by J. H. Ruttger, lithographed by Standidge. The built-up area of Devonport is not mapped in great detail, but closer attention is paid to both the dockyard and to suburban houses and their gardens (on which tithe was payable). 76 × 105 cm; 1 inch to 6 chains (1:4752). The area illustrated is about 2.3 × 2.1 km (1.4 × 1.3 miles). Public Record Office, IR 30/9/383.

Reproduced by permission of the Keeper of Public Records.

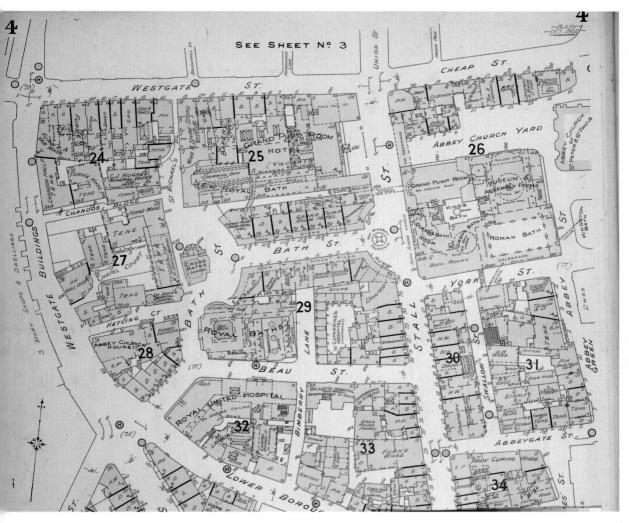

9. Charles Goad, fire insurance map of Bath, 1902. The so-called 'Goad maps', of which this is an extract from one sheet, were produced to satisfy the demands of insurance underwriters who needed large-scale town plans in order to identify particular buildings and to assess insurance risks. The maps are highly detailed portrayals of the physical fabric of towns. This particular extract covers an area of about 300 × 250 m (328 × 273 yards). British Library, Maps 145.b.9.(1).

Reproduced by permission of the British Library Board.

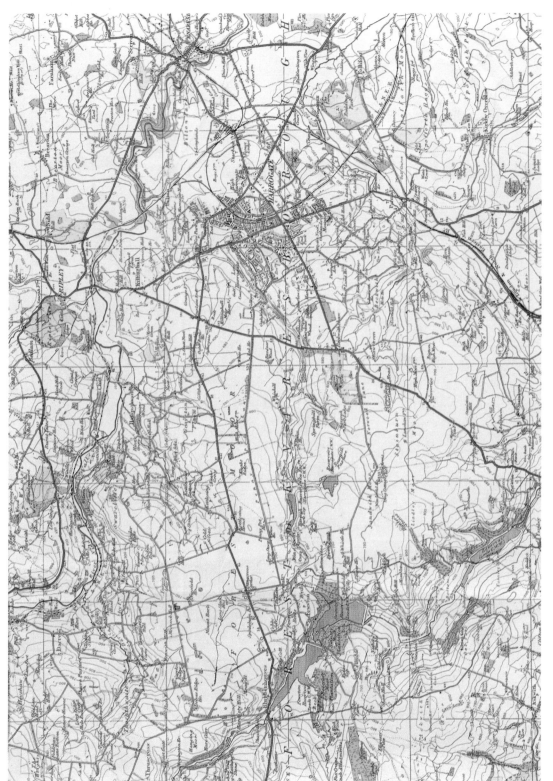

20. Ordnance Survey one-inch Popular Edition (1:63,360) sheet 26 (Yorkshire), 1925. The Popular Edition was the first of several generations of modern-day Ordnance Survey small-scale maps, with relief shown by contours and roads prominently coloured to serve wayfinding needs. Each sheet 82 × 67 cm. The area covered by this extract is about 13 × 18 km (8 × 11 miles).

Reproduced by permission of Department of Geography, University of Exeter.

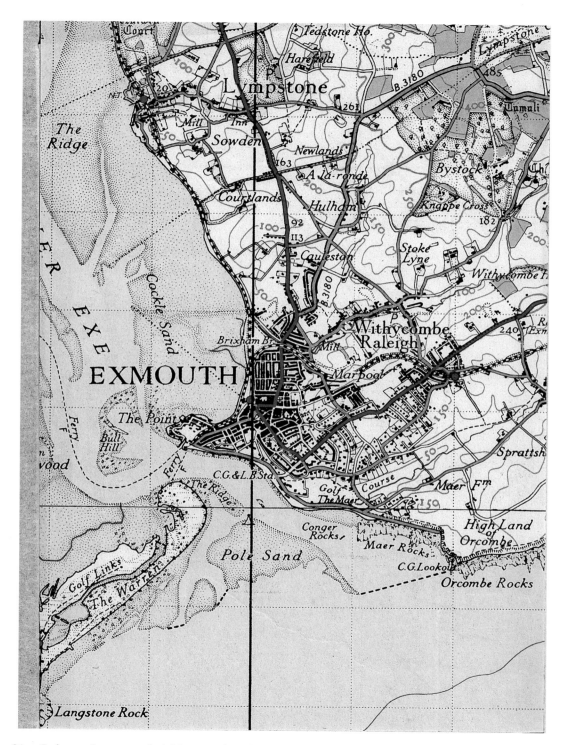

21. Ordnance Survey one-inch New Popular edition (1:63,360) sheet 146 (Exeter), 1946. This edition of the one-inch map was the first to carry the metric National Grid for locating places by co-ordinate references. Each sheet 82 × 67 cm. The area of Exmouth covered by this extract is about 7 × 5 km (4.3 × 3 miles).

Reproduced by permission of Department of Geography, University of Exeter.

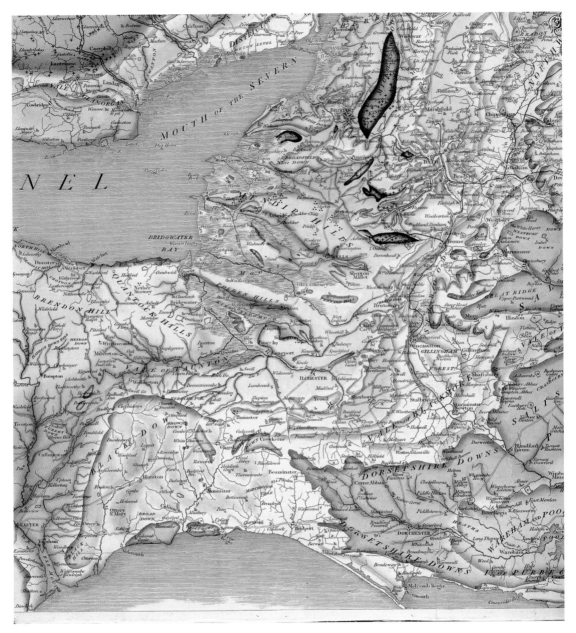

22.   William Smith's *Delineation of the Strata of England and Wales*, 1815. Smith's map was printed in 15 sheets at a scale of 5 miles to 1 inch (1:316,800) by John Cary. A sophisticated system of hand-colouring outcrops is used, for example, different intensities of tone of the same colour are employed to distinguish scarp and dip slopes. Detail showing southern England from the Severn estuary to the Dorset coast. British Library, Maps K.Top. 5.76.6.Tab.

Reproduced by permission of the British Library Board.

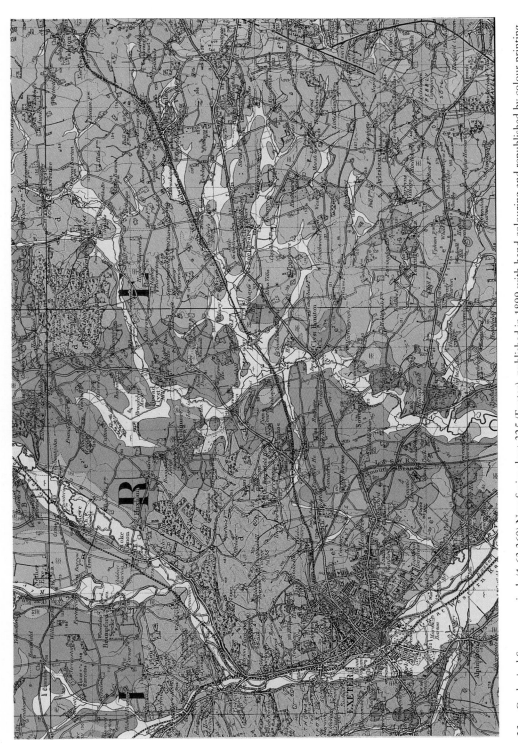

23. Geological Survey one-inch (1:63,360) New Series sheet 325 (Exeter), published in 1899 with hand-colouring and republished by colour printing in 1912. The area covered by this extract is about 13 × 18 km (8 × 11 miles).

Reproduced by permission of Department of Geography, University of Exeter.

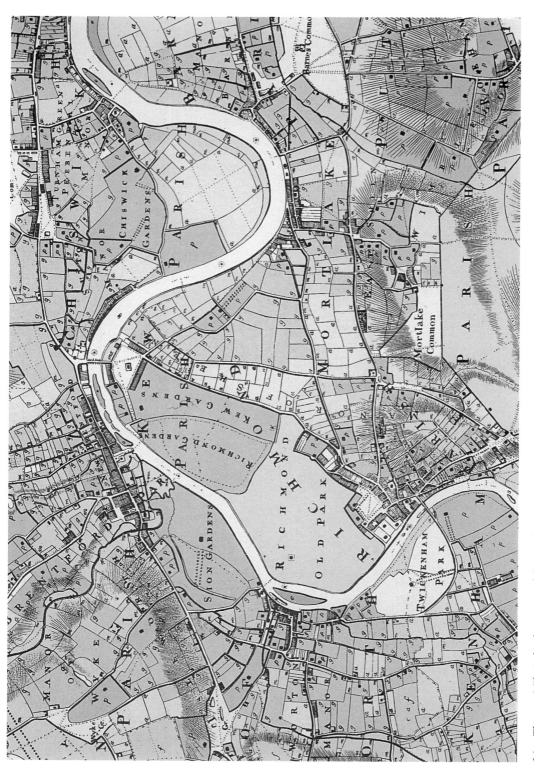

24. Thomas Milne's land use map of the London area, 1800. This is the first English example of a special purpose land use map of a large area of country (some 673 sq. km (260 sq. miles) around London). Not only is the area mapped large, but the land use classification is much more elaborate than anything previously attempted. Milne's map is published on an Ordnance Survey Old Series one-inch base reduced to two inches to one mile (1:31,815). The extract covers an area of about 6 × 10 km (3.7 × 6.2 miles).

Reproduced from a copy in a private collection.

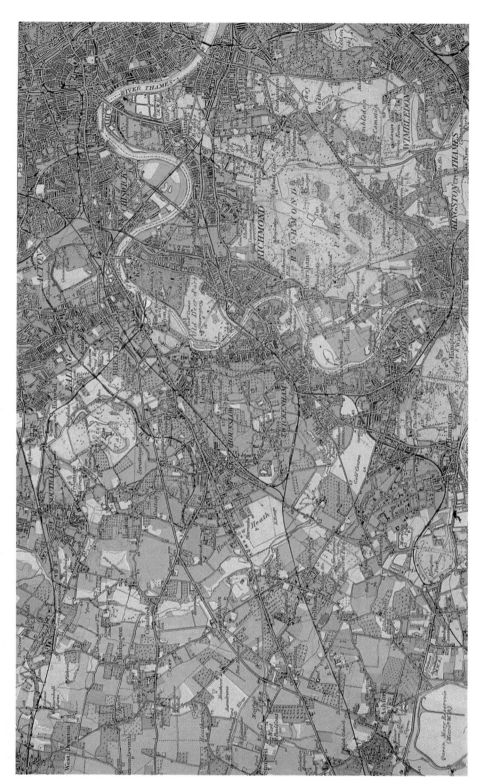

25. Land Utilisation Survey of Great Britain, sheet 114, parts of Middlesex and Surrey, published 1931–2. Land use was plotted by school pupils and teachers in the field on 6-inch maps and then reduced for publication on 170 1-inch maps. The area covered by this extract is about 6 × 10 km (3.7 × 6.2 miles).

Reproduced by permission of Department of Geography, University of Exeter.

26. Charles Booth's *Descriptive Map of London Poverty*, 1889. Several versions of the map were printed between 1889 and 1903 and published as separate sheets by Edward Stanford and also inserted cut up in an appendix to Booth's *Life and Labour of the People of London* (first published by Williams and Norgate, 1889–1891). Later versions, some extended, were published through to 1903. Originally on four sheets, six inches to one mile (1:10,560); the area shown is about 2.5 × 4.0 km (1.5 × 2.4 miles).

Reproduced by permission of the British Library Board.

Mapping Towns

Orthodoxy has it that there are no medieval maps of towns, and that the first extant cartographical representation of an English urban area is probably the bird's-eye view of part of the city of Bristol (*c*.1480). Were this to be the case, it would be yet another aspect of a Romano-medieval hiatus, for there were urban maps, as well as town views, in Roman Italy and the Empire.[1] At the same time, it may be that we are simply failing to acknowledge what has survived.

REPRESENTING TOWNS

There can be no serious doubt that Roman surveyors worked in England, although whether they were exclusively from Italy and Gaul or whether there were also trained Britons is another question. The English landscape is rich in examples of Roman grid-pattern street layouts, and nowhere more impressively than at Bath, Chester, Chichester, Colchester, Exeter and York.[2] Less clear is the extent to which the process of surveying was accompanied by mapping, although there are plan representations of a small number of Italian cities and their associated centuriated lands (Tarracina, Hispellum and Minturnae) in the didactic manuscripts of the *Corpus agrimensorum*.[3] That there is no English example in the *Corpus agrimensorum*, and that there is no surviving English map evidence, does not necessarily mean that maps were never made of the new towns of Roman Britain. Marcus Vitruvius Pollio's treatise *De architectura* (first century BC) not only describes the practical tasks which have to be undertaken by Roman architects and surveyors but also refers to *ichnographia* or the drawing of a ground plan (presumably on papyrus, parchment, or wax tablets) by rule and compasses.[4] It is also known that maps were used in the administration of Roman cities, and fragments of such maps have survived, ranging from the so-called official plan of Rome, the *Forma Urbis Romae*, down to a small piece portraying a few properties in Isola Sacra, near Ostia.[5] If only a few fragments of a major stone-cut map of the Imperial capital city survive, the chances of our knowing about less durable records of all the urban and rural surveying in Roman Britain are slim indeed. Exactly what happened to towns in England after the withdrawal of the Roman army in 410 AD is unclear. There was certainly destruction and decay but archaeological investigations suggest that this was much less widespread and of more unequal intensity or duration than

6.1 Symbolical plan of York in bird's-eye perspective. The historical narrative is set within a walled city, through which flows a river, crossed by a stone bridge. Early fifteenth century. 22 × 16.5 cm. British Library, Harley MS 1808. fol. 45v. Reproduced by permission of the British Library Board.

was formerly supposed. The fact remains, however, that while it is difficult to point to graphic documentation from an urban context much before the twelfth century, it is even less easy to explain why this should be so.

As in other categories of medieval mapping, surviving representations relating to towns are striking for their variety. This applies to subject matter as well as to style. There are written descriptions of the urban landscape, like William Fitz Stephen's account of London from sometime between 1173 and 1175, and the monk Lucian's description of Chester of about 1195.[6] There are town signs on topographical maps, like Matthew Paris's maps of Britain and Palestine, on Paris's graphic itinerary, and on the Gough map of England. Such signs tend to be conventionalised icons, but occasionally the portrayals seem to include characteristics or features that the map-maker happened to know about. Thus Paris's depiction of the profile of London, the island site of Lyon, and the pine trees of Pontremoli on his itinerary, and the portrayal of the newly-built walls at Coventry on the Gough map, involve a degree of realism.[7] Medieval manuscript illumination in general contains an abundance of pictorial representations of towns. The New Jerusalem was a standard illustration in commentaries on the Apocalypse, from Beatus of Liébana onwards.[8] A late thirteenth-century copy of Geoffrey of Monmouth's History of the kings of Britain (Historia regum Britanniae) has had added, early in the fourteenth century, a dozen marginal drawings of various English towns illustrating the text at appropriate places.[9] Most are in profile (side-view) with gable-ends giving an effect of perspective but one of the more elaborate shows a meticulously-drawn London (Trinovantum) as a circular walled city, flanked by a crenellated tower, in low oblique. Within the walls is a large church surrounded by houses. In the foreground is the water gate mentioned in the text.[10] An early fifteenth-century copy of the same work also contains a depiction of a walled city from a high-oblique angle, probably York. One of three full-page, boldly-coloured illustrations found in the manuscript (another is the map of Britain also described in Chapter 2), the function of the York plan was plainly symbolical rather than topographical (Fig. 6.1).[11] Also notional rather than realistic is the illustration of Constantinople in the Luttrell Psalter (1345), thought to have been modelled on the city of Lincoln.[12] Without doubt, many other examples could be added to a corpus of medieval urban representation. For example, depictions of Rome and of York are amongst a number of drawings of walled cities illustrating the chronicles of the Percy family (c.1485) and of Peter of Icham (late fifteenth century).[13]

Towns were also represented in plan. In many continental copies of Beatus's Commentary on the Apocalypse, the New Jerusalem is represented in this way, albeit in a high degree of abstraction. Copies of the work made in England in the thirteenth century also portrayed the visionary city this way (see Fig. 2.3, p. 11).[14] Matthew Paris drew Rome in plan for his itinerary (Fig. 6.2), showing an unnamed river running across the city (con-

tained within crenellated walls) and portraying pictorially the two
key gates (north to Lombardy and south to Apulia) and several
named churches (including St John Lateran and St Peter's). On the
map of the Levant which follows the itinerary in the St Albans
chronicle, Paris likewise depicted two walled cities in plan – Acre
(together with its separately walled suburb) and Jerusalem.[15] We
also find plans of individual urban features in other documents.
These tend to show either a single house-plot, small groups of
plots, or a structure within the walled area, like the cloth-stretching
tenter-racks of Exeter, *c.*1420 (see Fig. 2.19, p. 26).[16] Such
non-book plans appear to date only from the fifteenth century, a
tardiness which could be significant. Alternatively, the lateness of
such plans could be little more than an accident of survival.
Amongst these urban plots is the mid fifteenth-century plan of the
London Charterhouse, which lay immediately outside the city walls
across the parish boundary between Clerkenwell and Islington; the
four plans from the muniments of the wardens of London Bridge,
representing tenements in Deptford and Lambeth, from 1470–8,
the plan from Durham city, showing sixteeen burgage tenements
along the street now known as Old Elvet but then part of the 'new
borough' of Elvet, from 1493; and a plan showing tenements hard
up against the city wall at Exeter (the latter portrayed in side view)
which was attached to a notarial document, concerning a property
dispute in the parish of St Mary Arches dated 1499.[17] There may
also have been printed town maps in the last decades of the fifteenth century. We hear of a map of London, engraved on copper,
which was supposed to date from 1497 and to have been bought
by Christopher Columbus's son Ferdinand.[18]

6.2 Plan of Rome as a fold-out
on Matthew Paris's London-
Rome itinerary. Virtually
identical to that in the copy of
the itinerary in Corpus Christi
College, Cambridge, MS 25
except that the latter identifies
the Tiber as La riviere. 1250s.
5.5 × 9 cm. British Library,
Royal MS 14.C.vii, fols 2–5,
fol. 4r.
Reproduced by permission of the
British Library Board.

TYPES OF URBAN MAPS

If nothing else, and irrespective of the perspective from which they
were rendered, all these surviving representations of towns should
alert us to the possibility that towns were indeed depicted, in one
way or another and for one reason or another, in the English
Middle Ages. We should also be prepared to accept that medieval
urban representations fulfilled different, and an undoubtedly more
limited range of functions from their successors. Finally, we should
accept that way-finding was not one of the functions of a medieval
town map. A detailed plan on which individual streets can be identified is unlikely ever to have been needed in medieval, or even early
modern times, especially in Britain where (London excepted) most
towns were scarcely larger than many southern European villages.
Town plans for way-finding came into use on the continent
towards the end of the sixteenth century as north European tourists
began to visit the celebrated urban centres of Germany, France
and, above all, Italy, seeking the great buildings of classical times,
and not infrequently engaging in some discreet spying for their government in the course of their sight-seeing.[19] Having said this, there
are medieval city plans from the continent which do show the dis-

position of major architectural features (although not streets) in sufficiently correct spatial relationship to each other to have been of aid to a foreigner seeking one of them. Just three of a number of cases in point are: a plan of Milan in a chronicle written by Galvano Fiamma (c.1330); the Limburg brothers's celebrated version of a plan of Rome, taken from a Sienese mural for the 'Très Riches Heures du Duc de Berry' (c.1442); and the various bird's-eye views of Milan, Venice, Florence and Rome and other cities by Pietro del Massaio in the Venice manuscript of Ptolemy's Geography (1472).[20]

On reflection, we realise that towns present the map-maker with the most complex landscapes of all. Buildings within a town tend to be of different ages, styles (vernacular or designed) and functions. They are arranged on their plots along a street in different ways, with a large, small, or non-existent space surrounding them. The streets may be broad and ruler-straight or narrow and tortuous, or they may be formally-created terraces, crescents, circles, or squares. There may be great squares, gardens and parks. Towns also contain a range of land uses – residential, commercial, industrial, administrative, ecclesiastical, agricultural – which sometimes contribute to pronounced spatial differentiation. They may be sited on hilly or on relatively level terrain. And, underlying all, invisible in the landscape but a key to the urban texture, is the property ownership cadastre.[21] Concomitant with the heterogeneity of the townscape is a variety of cartographical genres subsumed by the general designation 'urban map'. A guide is clearly needed amongst all these different maps from different times and different contexts. Our classification attempts to convey something of the range of purposes of town mapping from the Middle Ages onwards. As in all typologies, few categories are mutually exclusive, for most maps served more than a single purpose, and most maps could be described under more than one heading. So only broad categories are identified below, presented, as far as possible, chronologically. Full justice to the wealth of town maps in England still awaits specialist treatment.[22]

Urban topography in chronicles and books

By far the majority of the earliest known English representations of towns are illustrations in medieval manuscripts, a role continued in the printed book long after the sixteenth century. In fact, for medieval and early modern times, this is the most important single category of urban maps. The map of the town and precinct of Wells (Fig. 6.3), for instance, was one of four commemorative illustrations in a book compiled by Thomas Chaundler in praise of William of Wykeham's founding of New College, Oxford, and his other good works. Drawn between 1463 and 1466, it presents in a clear manner from a low oblique viewpoint the two parts of the town, the 'beautiful and ornate [bishop's] palace' and the walled urban area.[23] More widely known is Robert Ricart's map of Bristol, drawn twenty years later (c.1480) (Plate 14). Ricart was

6.3   Thomas Chaundler's bird's-eye view of Wells. The circular walled area contains the bishop's palace. In the foreground is the town within its rectangular walls, gateways, market place and extra-mural housing, which two figures are admiring. 28 × 17.5 cm. Early 1460s. New College, Oxford MS C.288: on deposit in the Bodleian Library.
Reproduced by permission of the Warden and Fellows of New College, Oxford.

the author of an unpublished history of Bristol which he illustrated with a map of the central part of the town, also in an oblique perspective, as if from a vantage point on a nearby (but imaginary) hillside. Just as Chaundler's map of Wells was highly selective as regards which elements of the townscape were portrayed, so Ricart too omitted several major features known to have existed at the time, notably the castle and the Bristol Bridge at the end of the High Street. Such selectivity has led some modern commentators to suggests that 'Ricart must have been thinking more in terms of a picture than a map', as if to deny the essential cartographical nature of the representation, forgetting that all maps have to omit something of the landscape and that the issue is only a question of degree.[24] That said, it is surprising that features as important as the castle and bridge were omitted.

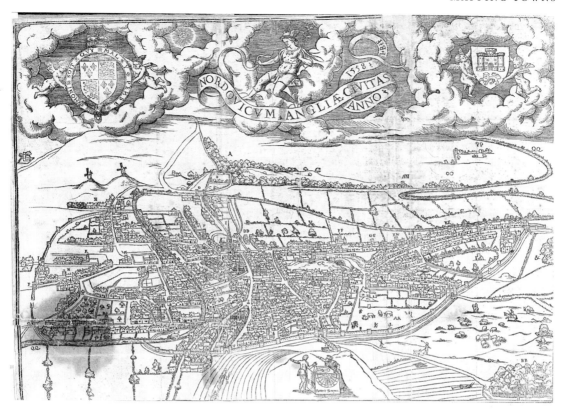

6.4 William Cuningham's plan of Norwich, 1558, the earliest surviving printed map of an English town, illustrates his surveying and map-making treatise *The Cosmographical Glasse*. Cuningham is portrayed at work with his surveying equipment in the foreground. 31 × 41.5 cm. 1559. British Library, 59.i.28, folded between fols 8 and 9.
Reproduced by permission of the British Library Board.

Early English printed books contain a number of maps of towns. A recent survey of sixteenth-century books shows town maps to be particularly characteristic of bibles, news-sheets, pamphlets, and books on travel and exploration.[25] Nearly all are of places abroad. We find maps in English books, or in English-language books printed on the continent for the English market, of towns such as Athens, La Rochelle, Bordeaux, Venice, Geneva and Jerusalem and even of places in the New World such as the native settlements captured by Francis Drake in his raid on the Spanish Main and the West Indies, 1585–6. The earliest surviving printed map of an English town, that of Norwich (1558), is in William Cuningham's surveying and map-making treatise *The Cosmographical Glasse* (1589) (Fig. 6.4), on which Cuningham himself is portrayed together with some of his surveying equipment, probably the 'Geographicall plaine sphere' described in his text.[26] The ancient university cities of Oxford and Cambridge were also mapped in the sixteenth century as they vied with each other for pre-eminent status. Only one of these university town maps was a book illustration, the copperplate map of Cambridge by Richard Lyne (1574) which accompanied Dr John Caius's *De Antiquitate Cantabrigiensis Academiae* (Ralph Agas's map of Oxford was printed separately, see below).[27] In the same decade, William

Harrison was formulating his project of including a set of maps of English and Welsh cathedral cities in Raphael Holinshed's *Chronicles*. In the event, these maps were not available in a sufficiently complete form by the time of publication and the *Chronicles* appeared in 1577 with only a map of the siege of Edinburgh castle in 1573.[28]

Both William Smith and John Norden produced a number of maps for their respective chorographies (written descriptions combining topography, history and geography). In their work we note an increased emphasis on the ichnographical, or plan, perspective. From 1568 onwards – the date of the plan of Bristol which he himself plotted (as he tells us) – Smith prepared a number of town plans and views for his *The Particuler Description of England with the Portratures of Certaine of the Cheiffest Citties & Townes*.[29] The book contains (amongst a good deal of heraldic matter) eight profile views of towns, and seven town plans (London, Bath, Bristol (Plate 15), Cambridge, Canterbury, Norwich, and Rochester) rendered from an oblique angle of some degree. The angle adopted by Smith varies from plan to plan. On that of London, as a result of the foreshortening inevitable from a low-angle view, the City itself is shown by densely-packed houses and churches, rather than in plan, and only in Southwark in the foreground and in Westminster and around the Tower of London to west and east respectively is there a clear sense of urban layout. In contrast, Canterbury, like the others in the plan group, is presented from a much higher viewpoint, allowing one to see comfortably into the intramural area where all the salient elements of the town's structure are clearly revealed in their relative locations even though all upstanding features (walls, fences and buildings small and large) are shown pictorially, as in all oblique views. In 1594, long after Smith had returned from his four-year spell in Nuremberg, he prepared (or completed) a historical account of that city, illustrated by three maps, one a plan and one a view of the German city, the third a topographical map of its surrounding territory (Fig. 6.5). In what is thought to be the original manuscript, the one today in Nuremberg, all three maps are on separate pages. For the copy now in Lambeth Palace Library, however, Smith squeezed the two urban maps together to present them on a single page, slightly cropping the foreground of the view to make the fit. He also omitted to label as many of the churches in the view as he had done on the original. In other respects, the two versions of all three maps are similar in their essentials.[30]

Besides admirably demonstrating even more effectively than do the illustrations in the *Particuler Description* the respective advantages and disadvantages of the two ways of representing towns, Smith's Nuremberg drawings (1594) offer a rare insight into contemporary attitudes towards the representation of towns. His depiction of Nuremberg (Fig. 6.6 top) is similar to his view of the City of London, inasmuch as it is the skyline which predominates, much as it would be seen from a hill away from, and high above,

6.5   Map of the 'territory of Norinberg': detail from one of the two known copies of William Smith's manuscript 'A Breef Description of the Famous And Beautifull Cittie of Norenberg'. The three dedicatory epistles in Lambeth Palace MS 508 (thought to be a copy of the original now in Nuremberg) imply that there was also a third presentation copy. 1594. Whole map 20 × 29.5 cm.
Reproduced by permission of Lambeth Palace Library.

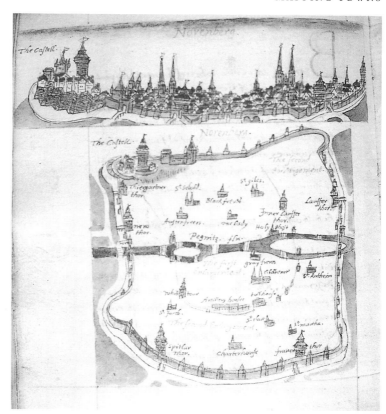

6.6 William Smith's view and plan of the city of Nuremberg. In Smith's first version, these are on separate pages and all churches and buildings on the skyline are named in the view. In all other salient respects, however, both view and plan are identical in the copies he made for presentation. View 4 × 14 cm, plan 11 × 10.5 cm. Lambeth Palace MS 508.
Reproduced by permission of Lambeth Palace.

the city. Although all major buildings are clearly delineated and identified (in the original version each upstanding spire and tower is named), no glimpse of the ground between any of them and the close-packed houses surrounding them is to be had. Only their relative disposition helps correlate the features of the manuscript view with those on the plan and, no doubt, with the real buildings on the ground. But while the view of Nuremberg is of similar nature to that of London, Smith's plan of Nuremberg (Fig. 6.6 bottom) is quite different from any of his English town plans, even though some of these were created after Smith had left Nuremberg. Smith, it has to be inferred, was consciously tailoring his cartographical format according to the context or purpose of his work. If the Nuremberg plan looks 'academic', it is because – as part of his historical account of the town – it was so intended. Apart from the castle on its rock in one corner of the walled area, and other features along the outer wall, only twenty intra-mural structures are shown and named. Smith's focus was clearly on showing the stages of urban development, indicated by an inner line separating 'The first Enlargement' from 'The second Enlargement'. From Smith's town plans we learn that the choice between view and plan, and between the various ways of presenting the latter (from bird's-eye view to ichnographical), may have been far more frequently a deliberate and rationalised decision in medieval and early modern times than is usually recognised.

John Norden was a friend, or at least a close acquaintance, of William Smith and it has never been made perfectly clear to what extent the cartographical features introduced by Norden (key, road signs, town plan insets on his county maps, a reference grid) came to England from Nuremberg by way of Smith. Norden, author of a celebrated early seventeenth-century text on land surveying, had had in mind an ambitious project, the publication of a set of county topographical descriptions to cover the entire country under the title *Speculum Britanniae*. Each county volume was to include maps of towns as well as of the county. In the preamble to the (unpublished) manuscript of his pilot description of Northamptonshire, Norden addressed Lord Burghley '. . . may it please your honour to consider whether it might be expedient, that the most principall townes, Cyties and castles within every Shire, should be briefly and expertly plotted out . . . as at this day they were'.[31] After references to the plans of Peterborough (now lost) and Higham Ferrers (now in the British Library), Norden concluded with a clear indication that the latter was not his own creation but his redrafting of a pre-existing map.[32] Norden's *Northamptonshire* was an experiment with which he was not entirely satisfied, and he began the *Speculum* proper with the county of Middlesex, which includes two maps of the cities of London and Westminster (1592–93), followed by the volumes for Surrey, Essex and Hampshire.[33] For the latter, general county maps were produced but no town plans. For the Sussex volume, however, Norden inserted a bird's-eye view plan of Chichester as an inset on the general county map, usefully filling up space otherwise occupied just by sea (Fig. 6.7). He went on to complete only two further county volumes, Hertfordshire and Cornwall (the latter left unpublished), but again these lacked town maps and, once the manuscript for Cornwall was completed in 1604, Norden set the *Speculum Britanniae* aside as a failed project.[34] In Norden's town plans, the angle of view is always oblique, with the result that, although the buildings are represented pictorially, they do not wholly obscure the street pattern.[35]

Despite Norden's abandonment of his own chorography in the early seventeenth century, the genre itself did not disappear, continuing through to the nineteenth century. From 1669 until his death in 1676, John Ogilby was promoting one of his characteristically monumental (and over-ambitious) publication projects. Starting out in 1669 with the idea of a five-volume 'English Atlas' covering Africa, America, Asia, and Europe as well as the British Isles, three years later the hard reality of lack of financial support obliged Ogilby to offer a revised project, a six-volume 'Britannia', the final volume of which was to be 'A New and Accurate description of the famous City of London, with the perfect Ichnography thereof, according to its six and Twenty Wards, in a fair Volume illustrated with the Scenography of all Eminent Buildings and Places belonging thereunto' and which would include a plan of the City of London and a plan of the City of Westminister. In 1675, a further revision of the project foresaw a modest three-volume 'Britannia', the first book of which was to comprise 'A Description of the 25 cities with particular charts of each of them, but more particularly those of London and Westminster', the second 'A Topographical Description of the Whole Kingdom', and the third a book of the roads of England and Wales.[36] In the event, only three large-scale town maps and the road book were published. The maps were of Westminster (surveyed by Robert Felgate and Gregory King and published in 1674), Ipswich (completed in 1674 but engraved only in 1698) and London (completed 1676 and put on the market early in 1677 by William Morgan, Ogilby's step-nephew, to whom the business passed when Ogilby died in September 1676). The twenty-

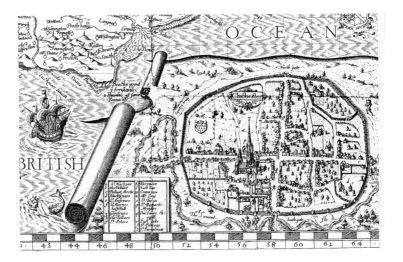

6.7 John Norden's plan of Chichester, 1595. Norden was the first English map-maker to inset a map of an important town within the borders of a county map. On his map of Sussex, the plan usefully fills space that would otherwise be occupied by sea. Whole map 29 × 52; inset 10.5 × 18 cm. Reproduced, by permission of the Estate of Harry Margary, from a facsimile printed by Harry Margary, Lympne, Kent, 1970.

sheet London map could not have made a greater contrast with earlier (and many later) town plans similarly originating as part of a written topography. Surveyed by William Leybourne, this remarkable achievement was on a scale of 100 feet to one inch (1:1200). It shows every plot in the city, to a substantial degree highly accurately.[37]

In contrast, the seven town plans included in Daniel and Samuel Lysons's ambitious, much later, but similarly unfinished county topographies (*Magna Britannia*, 1806–1822), are modest, small-scale affairs, derived from existing work and produced to a uniform format to fit the broad folio pages in six of the nine county volumes. The seven town plans in the *Magna Britannia* are of Cambridge, Chester, Reading, Carlisle, Whitehaven, Derby, and Exeter. Each volume has a county map. There is also an 'Ancient Plan of the City of Carlisle' taken from a manuscript plan and two extracts from the *c*.1539 manuscript map of the southern coasts of Devon and Cornwall (see below), showing the Havens of Exmouth and Plymouth. Similarly uniform in style are the 21 town plans engraved between 1804 and 1810 by J. Roper, from G. Cole's drawings, for *The British Atlas* (1810).[38] This last, which also contains 53 county maps and a general map of navigable rivers and canals, was issued by Vernor, Hood and Sharpe as an accompaniment to the 18-volume *Beauties of England and Wales*.[39] In addition to the towns already depicted for such works, or as insets on county maps, there were by this time plans of the newly-industrialising towns of Manchester and Liverpool as well as places like Bedford, Durham, Winchester and Worcester, as in Cole's atlas.

Printed town maps on separate sheets or in atlases

However well integrated into the argument, maps in books are normally adjuncts to a text. Maps printed on separate sheets – for use as separate maps, for display or for binding into an atlas – reflect a different order of priority. In Italy, for example, Jacopo de' Barbari's great woodcut of Venice (1500) was printed on six sheets to be assembled and displayed, like Francesco Rosselli's large-sized views of Pisa (woodcut) and Florence, Rome and Constantinople (copperplate).[40] We know of nothing similar in England until the 1550s and the 'lost copperplate' map of London.

The 'lost copperplate' map has been dated to between 1553 and 1559, with an inclination towards the latter.[41] The author of this remarkable survey of what was by then a large, densely-packed and busy city of some 100,000 inhabitants has so far eluded all research, as has almost everything else about its production. Its scale works out at about 34 inches to one mile (1:1863), and its assembled size must have been about 112 × 226 cm.[42] The three known surviving plates (out of a possible twelve or fifteen) are worn, implying considerable use, although whether printing took place in London or abroad is another unresolved question.[43] The plates show how the map depicted the city's streets to scale, portraying buildings in elevation and with, it would seem, attention to architectural detail. Idiosyncrasies in street-name spelling may reflect a foreign engraver's unfamiliarity with English or, it has recently been suggested, vernacular pronunciation of local names in various parts of the city at the time of the survey.[44] The 'lost copperplate' map is also known from a rather rough woodcut copy of the original, for long erroneously attributed to Ralph Agas. This version is dated to between 1561 (when the spire of the old St Paul's fell down) and 1570 (when the new Royal Exchange was completed) (Fig. 6.8).[45] Smaller than the original (about 28 inches to a mile and measuring 71 × 183 cm), and extending much further north of the city to show the hills of Highgate and Hampstead, the woodcut map was printed on 15 sheets.[46] The map of London in Georg Braun and Frans Hogenberg's *Civitates orbis terrarum*, a six-volume collection of city views and plans printed in Cologne between 1571 and 1617, appears to have been based not on the woodcut version but on the original copperplate map. In 1562/3, the printer Jean Godet registered a 'Carde of London' with the

6.8 The 'lost copperplate' map of London, c.1558, showing the area of Three Crane's Wharf from one of the three plates (out of a possible 12 or 15) known to have survived. Map-maker unknown. The map depicts the city's streets to scale and portrays buildings in elevation and with attention to architectural detail. Assembled size of the whole would have been about 112 × 226 cm; 34 inches to 1 mile (1:1863). Reproduced from *The A to Z of Elizabethan London* (London Topographical Society, Publication 122, 1979) by permission of the Society.

Stationers Company: was this the copperplate map, or something quite different, yet another map of London?[47]

There were sheet maps of other towns in sixteenth-century England. In 1578, four years after Cambridge, Oxford got its map. Ralph Agas took his view from the north of the city explaining, in doggerel verse, his choice: 'For there the buildings make the bravest show, And from those walks the scholars best it know'.[48] Agas was an experienced estate surveyor, later author of a surveying treatise, *A Preparative to Platting of Lands* (1596), and well-accustomed to large-scale plan drawing, which could explain why he chose to draw his map of Oxford at the very large scale of 50 inches to one mile (1:1267). The map was engraved some ten years later by Augustine Ryther, who had worked for Saxton and who, in 1592, also engraved a second map of Cambridge, one by John Hammond.[49] Both Agas's and Hammond's maps are characterised by a true-to-scale ground plan while showing the elevations of the buildings in considerable detail.[50] Nothing much is known about Hammond. However, a John Hammond was a student at Clare Hall, Cambridge, at exactly the same time as Edmund Rudd. Rudd died while still at college, in the spring of 1576, leaving a drawing table and an unfinished map, the latter valued for probate at the astonishingly high price of nine shillings.[51] Speculative as it may be, it is tempting to conclude, in default of evidence about any other John Hammond or about Hammond's precise relationship to the printed map, that the Hammond who was at Clare Hall until 1579 was the one with whom the map of Cambridge – engraved only just over a dozen years later – is usually associated; and that the map had been mostly the work of Edmund, son of John Rudd the map-maker.

John Hooker's original map of Exeter (Plate 16) has remained in manuscript, although, like so many of these early town plans, various copies were made and engraved for printing.[52] It is not surprising that Exeter finds itself in the company of London, Cambridge, Oxford and Norwich as one of the earliest English towns to be surveyed; these were amongst the wealthiest towns in the country. Exeter was a mercantile centre for the wool trade of south-west England and acknowledged to be among the half dozen towns at the top of the late sixteenth century English urban hierarchy. Nearly all these early town maps were published at a reduced scale by Braun and Hogenberg.[53] In his preface to Volume II (1575), Braun had invited people to send him 'portraits' of English towns, to be engraved by Hogenberg and published in the succeeding volumes of the *Civitates*. By the time the last volume had been completed, the *Civitates* included nine plans of English towns (Cambridge, Bristol, Chester, Norwich, Exeter, York, Lancaster, Shrewsbury and Canterbury as well as the one based on the 'lost copperplate' map of London), one of a Scottish town (Edinburgh), two of Irish towns (Dublin and Cork), and a number of views.

So far, the surveying and mapping of English towns was proceeding in an *ad hoc* manner. It was John Speed who produced the first comprehensive set of town plans. His *Theatre of the Empire*

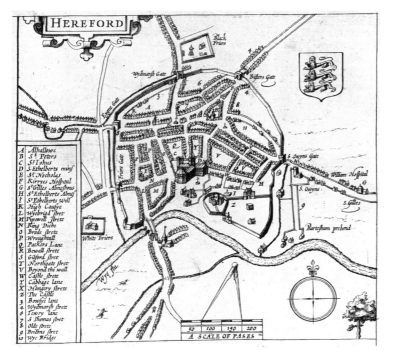

The legend on the map reads:

A | Alhallowe
B | S.t Peters
C | S.t Iohns
D | S. Ethelberts minf
E | S.t Nicholas
F | Kirryes Hofpital
G | S.t Gilles Almeshous
H | S.t Ethelberts Almg
I | S.t Ethelberts Well
K | High Caufye
L | Wychriad ftret
M | Pipowell ftrete
N | King Ditche
O | Brode ftrete
P | Wroughtall
Q | Packers Lane
R | Bewall ftrete
S | Gilford ftret
T | Northgate ftret
V | Beyond the wall
W | Caftle ftrete
X | Cabbage lane
Y | Monyery ftrete
Z | The Caftle
3 | Bowfye lane
4 | Wydmarfh ftrett
6 | Iewry lane
7 | S Thomas ftret
8 | Olde ftret
9 | Brifons ftret
10 | Wye Bridge

6.9 John Speed's plan of Hereford, 1610. Speed produced the first comprehensive set of English town plans. His *Theatre of the Empire of Great Britaine* (1611) contains town maps as insets on his county maps after the manner of John Norden (see Fig. 6.7). 15 × 16 cm. British Library, Maps C.7.c.5. Reproduced by permission of the British Library Board.

of Great Britaine (1611) contains town maps as insets on his county maps after the manner of John Norden's map of Sussex. The usually rather self-deprecatory Speed was justly proud of his town maps, pointing out that, while 'some have bene performed by others, without Scale annexed, the reste by mine own travels, and unto them for distinction, the Scale of Paces . . . five foote to a pace I have set . . .'.[54] The implication is that Speed was in Ireland surveying Dublin and obtaining a map of Cork, for these town maps were amongst those used by Braun and Hogenberg (together with copies of the maps of York, Lancaster and Shrewsbury). Speed's map of Hereford (1610) typifies the way Speed gave streets their names and showed clearly the bridges, walls and churches of each town (Fig. 6.9). A variety of features important to the economic and social life of early seventeenth-century burgesses are also shown with some realism on Speed's town maps: gates, crosses, conduits, drying frames for fishing nets, tenter-frames for textiles, windmills and water mills, lime kilns, gallows, stocks, cockpits, and maypoles. Houses, though, are highly stylised. Confusingly for the unwary modern historian, Speed did not hesitate to indulge in antiquarian extravagance at the expense of townscape veracity. His map of Gloucester, for example, portrays the town walls in a pre-seventeenth-century state.[55] Notwithstanding such shortcomings, Speed's county maps contain a valuable collection of English and Welsh town maps, most of them the earliest-known depictions of particular towns. His use of a measured plan as a base for plotting topographical information set a pattern for others to follow. In other respects, such as in his hierarchical classification of places, reserving pictorial signs for market towns and cathedral cities 'to provide for his customers a visual aid to the urban scene', he was observing contemporary cartographical practice.[56]

Town plans continued to be added to county maps in the eighteenth and nineteenth centuries. The townscapes recorded on maps by people such as Agas, Cuningham, Hooker, Lyne, Smith, Norden and, above all, Speed were readily plagiarised and many of the later maps 'continued to portray these archaic images ... certainly until the 1770s, possibly later, and in a few cases much beyond'.[57] New town plans were being made for county maps, however, even before the 1770s. A few of the county maps in Emanuel Bowen and Thomas Kitchin's *Large English Atlas* had inset 'ichnographies' of towns: Lewes and Chichester on the Sussex map (*c.*1749), Bath on the Somerset map (*c.*1750), and Plymouth on the Devon map (*c.*1754) besides a plan of Kenilworth Castle (Warwickshire, *c.*1751) and a number of views or prospects.[58] Benjamin Donn's award-winning *Map of the County of Devon with the City and County of Exeter...* (1765) had plans of Plymouth and Exeter (Fig. 6.10). Peter Burdett added a plan of Derby to his award-winning map of Derbyshire (1767), and one of Chester to his map of Cheshire (1777). However, John Chapman and Peter André's *trompe l'oeil* plans of Colchester and Harwich Harbour (*County of Essex*, 1777) may well belong to the plagiarised category. From the mid nineteenth century, we can cite John Rapkin's map of Liverpool which was printed in about 1845 by John Tallis as part of the latter's collection of British town plans. This map is set across the top of a print of a view of the busy harbour, around which vignettes of major buildings 'drawn and engraved by H. Winkles' complement the deliberately pictorial attractiveness of the sheet as a whole.[59]

Irrespective of their pedigree, by piggy-backing on county maps town plans were being assured of a wide currency. England has never been a land of cities in the same way as other western European countries but by the mid seventeenth century English county towns, spas, and river- and sea-ports, had risen to pre-eminence as the seats and symbols of organised social life as well as the focal point of trade and industry to a degree unknown since Roman times.[60] Not only townspeople, still a distinct minority of the total population, but many others in a county would have turned with interest to the plan of 'their' town displayed on the map of the county. Just as architectural insets graced Cole and Roper's town plans for the *Beauties of England and Wales*, so town plans would have enhanced the commercial appeal of county maps, whether these were small, like Speed's or large, like Donn's and Burdett's wall maps.

6.10 *The Town and Citadel of Plymouth* by Benjamin Donn, 1765. The map shows the street layout and extent of the built-up area with admirable clarity, but reveals almost nothing of the pattern of building. Inset on Donn's map of the county of Devon. 29 × 38 cm; 1 inch to 6 chains (1:4752).
Reproduced by permission of the Department of Geography, University of Exeter.

Maps for planning fortifications and urban defence

While there may not have been much practical need for detailed or scaled maps of English towns in early Tudor times, Henry VIII quickly appreciated their value in planning national defence. The technicalities of warfare had changed considerably from the relatively static and defensive strategies of the Middle Ages, when sieges were usual and the field of action compact. In the Renaissance, however, the arrival of firearms shifted the balance of power to the attacker, so that if the walls of a town or a castle were to withstand the impact of cannon, a rebuilding programme was needed to ensure the adequate protection of each town.[61] The Italian system of fortification outworks (later taken up and developed by French engineers, notably Marshal Vauban in the seventeenth century), was based on carefully calculated lines of fire, designed to minimise the damage caused by incoming cannon fire and to maximise the possible angles of outgoing lines of fire (Fig. 6.11). Maps were the ideal medium on which to protract these lines and to record, inform and plan a nation's defence and in the 1530s at the latest, maps started to play an important role in the military affairs of England. Unlike continental Europe, where the fortified town was one of the outstanding charateristics of the late medieval and early modern periods, island England enjoyed a more settled political situation and increasingly diverged from the continental pattern of fortification. In keeping with continental practice, the towns of Roman Britain had been walled and their walls maintained and, where necessary, rebuilt or extended throughout the Middle Ages.[62] By the sixteenth century, however,

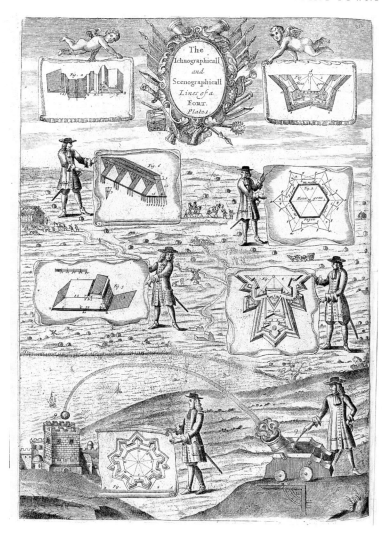

6.11 Renaissance fortification: Richard Blome's exposition of the 'Iconographicall and Scenographicall Lines of a Fort' from his *The Gentlemans Recreation* (1686). 32 × 22 cm. British Library, G 7428.
Reproduced by permission of the British Library Board.

most towns in the interior of England and away from the Scottish and Welsh borders were not thought to be at risk from external attack and attention was concentrated on coastal towns. By the second half of the seventeenth century, after the Civil War, town walls were largely obsolete for defence, although many were maintained in good repair throughout the early modern period as a way of controlling access and facilitating the collection of taxes and excise duties.[63]

Thus it was after 1538 that Henry VIII, fearful of a French invasion, embarked on a massive fortification building and mapping programme along the vulnerable coasts of Britain, as far north as Berwick in the east, along the south coast to Land's End and north to Carlisle, and also around Calais, England's last possession in

6.12   Gian Tommaso Scala's manuscript plan of proposed fortifications at Tynemouth, Northumberland. The early medieval monastery of Jarrow is shown in the foreground. 66 × 64 cm. 1545. British Library, Cotton MS Augustus I.ii.7.
Reproduced by permission of the British Library Board.

France. Italian personnel were active in England, working along-side English military engineers in surveying sites and advising on the ways these should be defended.[64] In such a context, scale is a key factor, and the henceforth regular practice of drawing maps to scale in England has been linked directly to this Tudor phase of mapping for fortification and defence.[65] Amongst the many extant examples of such threat-of-war mapping we can point to the maps for Hull and for Scarborough, both drawn in 1539–40, and Gian Tommaso Scala's map of Tynemouth (1545) (Fig. 6.12). The choice of examples is considerable, for few if any of the large numbers of the surviving English military maps from this period saw active service, and those we can turn to in the map archives are in effect well-preserved state papers – maps or copies of maps made

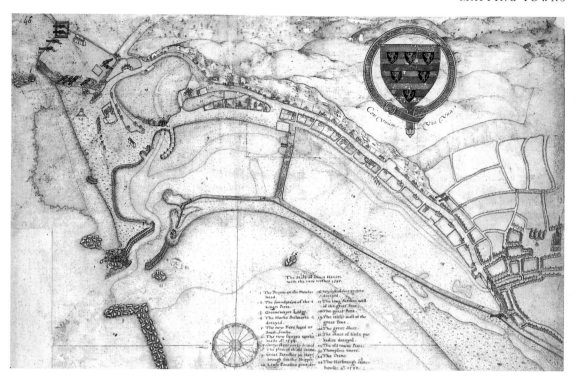

6.13 Map of Dover harbour, showing 'the state of Dover haven with the new workes 1595'. Much of the construction indicated had already been effected, replacing 'decaied' groynes and piers. The projects shown here laid the foundation of modern Dover, most strikingly with the 'long earthen wall of the Great Pent' breakwater (no. 17 on the plan) and urban development along the strand. 36 × 46 cm. British Library, Cotton MS Augustus I.i.46.
Reproduced by permission of the British Library Board.

for the king and his ministers.[66] Particularly important or vulnerable places were mapped time and time again. One of the earliest of Henry's military maps is the first extant plan of Dover (1532). Drawn by the Italian artist Vicenzo Volpe, it shows a scheme for the improvement of the harbour, the idea of which was to extend this inland along the narrow valley which offers the only break in the high chalk cliffs which famously characterise this stretch of coast.[67] A few years later, in September 1538 (immediately after England's break with Rome and perhaps in fear of imminent repercussions), another view of Dover Harbour was presented to the king. Probably drawn by an Englishman, either John Thompson or Richard Lee, it shows that the earlier ambitious scheme was never wholly effected. Maps of Dover continued to be made throughout the century, as danger threatened or as siltation necessitated (Fig. 6.13). Another English military map-maker of this period was the surveyor-spy John Rogers, who was connected with preparations for the negotiations over the Anglo-French frontier around Calais and who was responsible for a number of maps, including at least three of the fortications at Boulogne (1545–6).[68]

Back home, in England, local concerns centred on the effectiveness of measures taken to protect coastal towns in the event of a raid. There were few illusions about what could happen. A now-lost map evidently portrayed an attack by the French on the town of Brighton in 1514, for a copy has survived made in 1545 for a

report on the state of the town's defences when a new attack was threatened in 1539–40 (Fig. 6.14).[69] The copy shows the raiders' big ships at anchor out to sea, while galleys bring enemy soldiers to the shore. Local reinforcements are seen marching down the lanes from neighbouring villages, called in by the burning beacons outside the town. Houses in the town are on fire.[70] Another map shows Portsmouth (1545). Drawn to a consistent horizontal scale of one inch to 100 feet (1:1200), this is the earliest known map of any British town drawn wholly in plan, on a uniform scale, and entirely without pictorial elements (Fig. 6.15).[71] It also shows 'by far the earliest scheme for the defence of an English town by means of a fully-flanked bastioned system in the Italian style'.[72] Streets and buildings are depicted but urban layout was secondary to the main purpose of the map, which was to illustrate the intended improvements to the defences of this strategically important port and naval city. The continuing importance of Portsmouth is reflected in the fact that nearly all known subsequent maps of Portsmouth, down to 1800, concern the development of its defences.[73] The long, exposed, coasts of Devon and Cornwall also needed protection. One remarkable map is a 3.5 metre long panorama, dating from between 1538 and 1540, which shows the coast from Land's End to the River Exe and which contains notes about the distance from the shore to nearby towns and the capacity of beaches as enemy landing places (Fig. 6.16).[74] Forty years later,

6.14   The French attack on Brighton, Sussex, shown here, took place in 1514 but the map is probably a copy of one of several maps drawn up in 1539, in the context of renewed threats, to underline the vulnerability of the town's defences. The map records the enemy landing from lighters from big ships anchored in the roadstead and reinforcements marching down the country lanes, summoned by the burning beacons. 62.5 × 92 cm. 1545. British Library, Cotton MS Augustus I.i.18.
Reproduced by permission of the British Library Board.

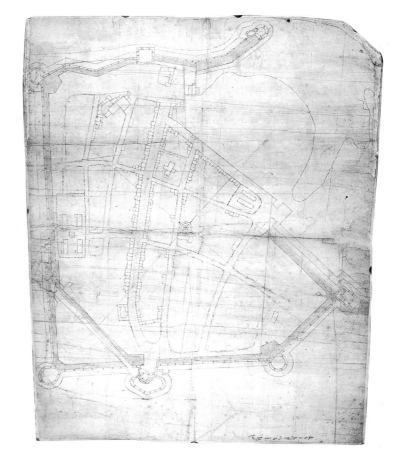

6.15 Plan for the fortification of Portsmouth, 1545. Drawn to a consistent horizontal scale of 1 inch to 100 feet (1:1200), this is the earliest known map of any British town drawn wholly in plan, on a uniform scale, and without pictorial elements. 58 × 78 cm. British Library, Cotton MS Augustus I.i.81. Reproduced by permission of the British Library Board.

6.16 A panoramic map of the coast from Land's End to the east Devon border was prepared in 1539 or 1540 in the course of planning national defence. The detail reproduced here depicts Land's End, Mount's Bay and the Lizard. Some forts are labelled 'not made', like the one between Mousehole and Penzance, while others were only 'half made'. The written notes comment on the suitability of the beaches for enemy landings and the distance to the nearest towns and salient points. Whole map 80 × 360 cm; as shown here 70 × 90 cm. British Library, Cotton MS Augustus I.i, 35,36,38,39. Reproduced by permission of the British Library Board.

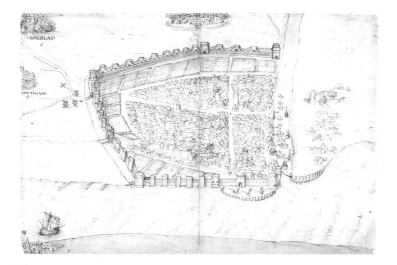

6.17 The vulnerability of Hull-on-Humber, on the east coast of England, is underlined in this plan of 1539. Walls surround the town on three sides, but there was no protection on the east, the side from which an attack might be expected, other than the River Hull, across which a chain guarded the port against intruders and a small battery of four cannon at its mouth. 57×77 cm. British Library, Cotton MS Augustus I.i.83.
Reproduced by permission of the British Library Board.

Lord Burghley annotated a printed map of the same area with information of a similar nature, also remarking on the exceptional length of the beach of Slapton Sands.[75] Four centuries later, actual landings on Slapton Sands in the course of Allied practice for the Normandy landings of World War II, resulted in tragic disaster. On the night of 27–28 April 1944 a serious lapse in communication by Coastal Command allowed a flotilla of German E-boats to penetrate the naval defences and in the subsequent attack almost a thousand American servicemen on board landing craft were killed.

Away to the north-east, where the broad Humber estuary also offered a potential enemy easy access far inland, town defences were also found to be wanting in the 1530s. A plan of 1539 shows Hull with its unprotected river frontage and only a chain across the mouth of the river to deter alien shipping (Fig. 6.17). Henceforth, the Humber estuary and the town of Hull were repeatedly mapped for military purposes. In the 1540s, John Rogers produced a very different-looking map from the first, drawn to scale, probably from his own field sketches, and depicting projected improvements for the town's defences. These included a wall with bastions along the hitherto unprotected eastern side of the river. Like the Humber, mapped again in the 1560s, the Thames estuary had rendered London vulnerable to penetrative sea-borne raids from time immemorial and whole series of maps record the improvements to defences along the lower Thames, the north Kent coast, and the fortifications of the Medway towns of Rochester and Chatham as well as along the Channel.

Maps for urban administration

Most maps can be used in a number of ways – some far removed from the map-maker's original intention – and it becomes difficult, especially in the absence of any comprehensive study of English urban maps, to single out which map in particular might have been used in the course of administration of some aspect of urban life or landscape. There is the occasional hint. In the sixteenth century, Richard Lyne said his map of Cambridge was made 'principally for [the] cause' of showing how certain river diversions would 'be a singular benefite for the healthsomnes both of the Universities and of the Towne . . .'.[76] His map of 1574 carries a text in which the physical location of the town is described with particular reference to the King's Ditch. This ditch, shown on the map, had been originally part of the urban defences but in Lyne's day was 'now found convenient for the cleansing of the dirt from the streets, and for washing filth into the Granta'. It would seem that Lyne's map was made principally to advance the project of diverting a particular brook. A full transcription of the text is given by Willis Clark and Arthur Gray, who also cite a letter written on 2 November of the same year by Andrew Perne, Master of Peterhouse and Vice Chancellor of the University, to Lord Burghley.[77] Perne is concerned about plague in Cambridge. His letter pinpoints the context, if not the express purpose, of Lyne's map, and the extract is worth quoting at length: 'I do send to your honor a brief note of such as have died of the plage in Cambridge hitherto, with a mappe of Cambridge, the which I did first make principally for this cause, to shewe howe the water that cometh from Shelford to Trumpingtonford and from thence nowe doth passe to ye mylles in Cambridge, as appearith by a blewe line drawne in the saide mappe to Trumpingtonford (without any comodities) might be conveighed . . . into the King's Ditch, the which waie as appearith by a red lyne drawne from the said Trumpingtonford to the King's Ditch, for the perpetual scouring of the same, the which would be a singuler benefite for the healthsomnes both of the Universities and of the Towne, besides other commodities that might arise thereby'.

A few years later, Ralph Agas explained how he set out a map of 'a Cittie, Borough, and Towne . . . [with its] streets, waies and allies, as may serve for a just measure for paving thereof, distance between place and place, and such other things of use: the buildings of all sorts in their number, measure, forme & proportion, as each mans interest, claim, and demand may truly appeare . . .', implying that maps could be produced to help with urban management.[78] Such explicitness, however, is all too rare and we have to turn to analogy with better-studied maps from other parts of Europe to see how maps of towns with, sometimes, their immediate environs could serve a wide range of regional functions (for example, military, judicial, economic).[79] Environs maps, as they are known, betray their urban connection in their titles, as for example, George Perry's *Map of the Environs of Liverpool (1769)*. Robert Morden's *Map Containing the Towns, Villages, Gentlemens' Houses, Roads, Rivers & other Remarks for 20 Miles Round London* (1686) was overlain by a rectangular grid, indicating its intended use as a reference map, possibly by shopkeepers whose customers came from the towns, villages and great houses shown on the map.[80] John Cary's *Actual Survey of the Country Fifteen miles round London* (1786) was published both as a single-sheet 'General Map' and as a road-book in which the map had been cut to form 50 page-sized plates (see Fig. 5.23, p. 175).[81]

The use of maps in urban administration in England, as elsewhere, seems to have developed late. As in the case of military maps, it was need which prompted the tool. Peter Barber draws attention to the fact that central government called for town maps in 1540–41 in connection with the abolition of ecclesiastical liberties and immunities.[82] Maps for use within towns came about with the growth of the towns themselves, and with an expanding urban population. It is difficult to separate local way-finding needs from the needs of outsiders and, especially, of those we would today class as tourists. The statistics we have been able to study – for maps of the various parts of London (the City, Westminster, Southwark), 1550 to 1850 – are arresting.[83] The plotted data break naturally into three periods, each of a hundred years. In the first period, from 1550 to 1649, only seven maps of London are known to have been produced (roughly equivalent to one map every fourteen years); in the second, from 1650 to 1749, 67 maps are known (equivalent to approximately one map every fifteen months); and in the third, from 1750 to 1849, 300 maps were produced (an average of three a year). The exponential upswing of this last period, however, contains two surges: the first, as just noted, in the 1750s, and the second about 1800, when the average rate of production of new maps of London rose from two per year to four. Moreover, existing maps were being reissued at an unprecedented rate. Before the 1780s, it was rare that a map of London was reissued more than ten or a dozen times, at most. After 1787, however, certain maps were outstanding for their durability. John Cary's map of 1787 was reissued 21 times between that date and 1825 and his map of 1790 was reissued nineteen times up to 1836; Charles Smith's 1801 map was reissued 28 times before 1843, and G. F. Cruchley's 1826 map went through no fewer than 32 editions and reissues before 1847. Many other maps continued to be reissued on a large number of occasions after 1878. These figures only reinforce the sharpness of contrast between the early and the later years of this third period of London town plans. The factors involved are complex and overlapping, but there can be no denying the astonishing rate of production of maps of London from the last fifteen years of the eighteenth century and throughout the first half of the nineteenth century compared even with the second half of the seventeenth century, let alone anything previously.

Many provincial centres and the smaller market towns were rarely mapped in their own right, especially before the eighteenth century, but they did sometimes appear on large-scale property and cadastral maps as part of a particular landowner's estate. In this context, the urban part of the map would have shared the same fiscal function as the rural part.[84] It was while working as an estate surveyor that Agas included the market town of Toddington on the manuscript map he made for Lord Cheney in 1581, in much the

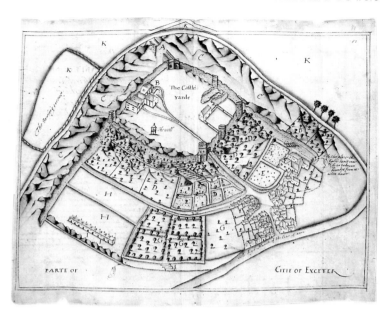

6.18 John Norden's plan of Exeter Castle, 1617. Norden's map of the 'Castel of Exon' is one of a number of maps bound into a book containing details of 60 surveys of Duchy of Cornwall properties which he undertook in summer 1617. Letters on the map relate to an explanatory table: 'B', for example, is 'the house where the assises and sessions are helde'. 30 × 36 cm. British Library, Add. MS 6027, fols 80v–81r. Reproduced by permission of the British Library Board.

same way as Saxton soon after mapped the small towns of Manchester (1596) and Dewsbury (1600).[85] Likewise, the town of Windsor appears on the manuscript map of 'Windsor Castle, Park and Town' which John Norden made in 1607 for James I (Plate 17).[86] Chelmsford was mapped by an estate surveyor, John Walker senior, in 1591 as part of the manor of Bishop's Hall, and Exeter Castle and its precincts were surveyed by John Norden in 1617 as one of the many estate mapping commissions he undertook for the Duchy of Cornwall (Fig. 6.18).[87] Among later examples, we can cite the plan of Worksop, in Nottinghamshire, on an estate map made for the Duke of Norfolk in 1775, and two plans of Newark in the same county included on maps made for the Duke of Newcastle in 1788 and 1790.[88] Small towns often appear on tithe maps because unbuilt land within them was tithable: a particularly striking example is Devonport in the parish of Stoke Damerel (Plate 18). There are also estate and cadastral manuscript maps which show a small part of a town (or, even more frequently, individual plots within a town or city) and which were produced in the course of some aspect of urban administration. In the late sixteenth century, Ralph Treswell was working for St Bartholomew's Hospital, London when he was surveying their property in county towns like St Albans in 1595.[89] Although only a third of the surveys undertaken by Treswell were in, or close to, the City of London, just two commissions account for over 50 separate plans showing single properties or small groups of properties for rent valuation or disputes. The plan of property at Limehouse Dock (1588), for instance, was executed in connection with St Bartholomew's Hospital's argument with the tenant over overdue rent while, in 1612, another of his maps was used to fix tenancies and rents of properties owned by the Clothworkers' Company in Fleet Lane, at the western edge of the City of London (Fig. 6.19).[90] Nearly a century later, William Leybourne was engaged in the service of the City of London.[91] One of his duties was to produce maps of certain of the city's markets, notably those at Leadenhall, Newgate, Woolchurch and Honey Lane which were selling meat, fruit, vegetables and dairy produce.[92] The markets were subject to strict regulation, but the main objective in asking Leybourne for maps was to ascertain the rents from the stalls meticulously marked on his plans.[93]

The mapping of an entire metropolis like London, however, called for a very different level of organisation and finance and was unlikely to have been willingly embarked upon had the need for maps not been relatively assured. Town planning was the prime stimulus for such maps of London in the seventeenth century following the devastation caused by the Great Fire of 1666. By the eighteenth century, a number of increasingly accurate, large-scale topographical surveys lent themselves admirably to use in general administration. John Rocque started his first map of the metropolis in 1739 – using an up-to-date theodolite to take bearings from steeples and to measure angles from street corners and a chain for ground measurement – but interrupted his work to map Bristol (1743), Exeter (1744) and Shrewsbury (1746), and to complete a second map of London (1745). His first London plan was eventually published in 1747 as the 24-sheet *Plan of the Cities of London and Westminister and Borough of Southwark* on a scale of 200 feet to the inch (about 26 inches to one mile or 1:2436).[94] The map, a contemporary catalogue advised, should be assembled, mounted on canvas, and hung from rollers 'from the cornice of the wainscot' in order to 'let it down' for examination at pleasure – or it could be made 'into a beautiful & useful screen'.[95] That the map had more serious uses is inferred from the fact that it was reissued in a reduced, eight-sheet, version in 1755, updated to show 'all New Roads that have been made on account of Westminster Bridge, and the New Buildings and Alterations to the present year . . .'.[96] Like the larger original, the 1755 map carried a lettered grid so that the same street index could be used. Rocque's 1745 map, started in 1741, was *An Exact Survey of the City's of London, Westminister ye Borough of Southwark and the Country near ten miles round . . .*, on 16 sheets and a scale of 1000 feet to the inch ($5^1/_2$ inches to one mile, or 1:12,000) and engraved by Richard Parr.[97] A yet smaller map on a single sheet was produced in 1754 under the title '*A Plan of London on the same scale as that of Paris: In Order to ascertain the Difference of the Extent of these two Rivals*'.[98] A fascinating glimpse of the variety of uses to which town maps could be, and were, put is gained from annotations made later on a copy of the 1762 edition of Rocque's one-sheet map by the Quartermaster General (Major-General George Morrison) to record, in different colours, the location of encampments, the positions of cavalry and infantry, and the routes of the army

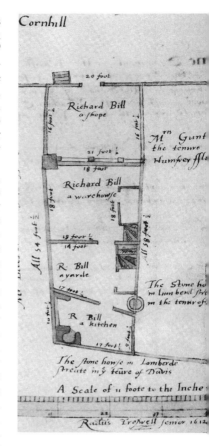

6.19 Ralph Treswell's survey of 16 Cornhill, opposite the Royal Exchange, City of London, 1612. One of a number of highly detailed plans undertaken by Treswell for the Clothworkers' Company. Each part of the building was measured and described. 20 × 9 cm; 1 inch to 11 feet (1:150). Clothworkers' Company, Plan Book, 'Book of Treswell Surveys', no. 21.
Reproduced by permission of the Clothworkers' Company.

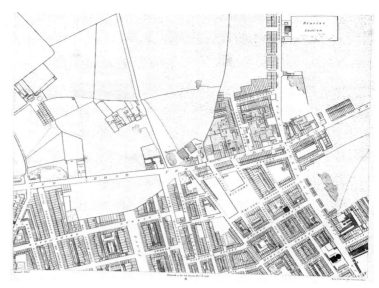

6.20   Richard Horwood's Map of London, 1792–9. Part of one of the 32 sheets which make up his *Plan of the Cities of London and Westminster the Borough of Southwark, and Parts adjoining Shewing every House.* Horwood's map is one of the most richly detailed maps of London ever produced. Each sheet 50 × 55 cm; 26 inches to 1 mile (1:2376). British Library, Maps 148.e.7.
Reproduced by permission of the British Library Board.

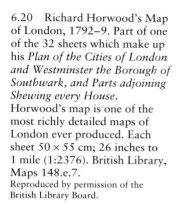

patrols under his command in quelling riots against Catholics in 1780.[99]

While other European capital cities, like Rome and Paris, were still having to make do with bird's eye views, John Rocque had given London a superb ichnographical tool, albeit a not wholly comprehensive one (a number of courts, alleys and yards were omitted). It was left to Richard Horwood to create 'one of the most richly detailed maps of London ever produced'.[100] Horwood's 32-sheet map, *Plan of the Cities of London and Westminster the Borough of Southwark, and Parts adjoining Shewing every House* was published over the years 1792–9 (Fig. 6.20).[101] And indeed, every individual property was shown on Horwood's map, unlike Rocque's. In the prospectus to his work, Horwood describes some of the uses to which his map might be put. For lawyers, he thought it would provide an accurate set of parish boundaries, while for those in commerce it would serve as a directory to locate the premises of suppliers and customers. The map could be used for route-finding and even, because of the accuracy with which it was drawn and its large scale, to measure the length of road journeys in order to check the charges demanded by hackney carriages, as Horwood himself thoughtfully pointed out on the map. The plates for Horwood's map passed eventually to William Faden. Perhaps rather surprisingly, for Faden was a canny businessman and not given to wasting time or money on uncommercial ventures, Faden revised the map before reissuing it on three separate occasions (1807, 1813, 1819).[102] Another pointer to the usefulness of Horwood's map to those responsible for London's fabric and inhabitants is the fact that he had been given money by the Phoenix Insurance Company, for whom he may have formerly worked and which presumably relied on his map in the delicate business of settling fire claims.

205

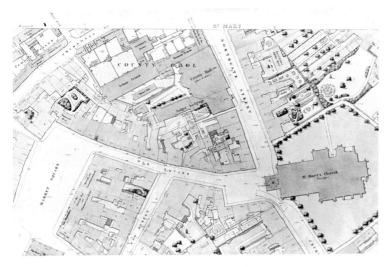

6.21 Plan of Warwick made for the Local Board of Health, 1851, by the Ordnance Survey. Mapping such as this, with close attention paid to building projections, garden layouts and basement levels, represents the apogee of manuscript urban mapping. 68 × 98 cm; 1 inch to 10 feet (1:528). The area covered by the extract is about 200 × 300 m. Warwick Record Office, CR 1618/unnumbered. Reproduced by permission of Warwick Record Office.

The proliferation of special maps relating to London and the other major English cities testifies to the increased variety of uses that maps were put to in the nineteenth and early twentieth centuries. At the same time, general town maps like Horwood's were also pressed into, or adapted for, a continually broadening functional spectrum. In 1851, London's population was still a relatively modest 2.5 million inhabitants. By 1891 it was 4.3 million. Administration of this great mass of people occupying a steadily expanding built-up area was not just a matter of convenience but one of sheer survival. Maps were needed in the struggle to ensure a healthy and sanitary environment and safe water supplies – and not just in London but in cities throughout the country. For instance, the Public Health Act of 1848 enabled Local Boards of Health to produce maps of the municipalities over which they had jurisdiction to help plan proper drainage and sewerage systems.[103] Many of these so-called Board of Health maps were produced privately for a particular Board, but a number of the Boards subcontracted their mapping to the Ordnance Survey who, by the early 1850s, had judged a scale of 1:1056 (60 inches to one mile) inadequate for sanitary planning and were mapping a number of towns at twice that scale (120 inches to one mile or 1:528).[104] Some of the Board of Health maps were published, others remain in manuscript, and a number have been lost (Fig. 6.21).[105] Maps were also used in connection with providing water and sewerage. In 1855, a Metropolitan Board of Works was established for London. A few months later, it was decided 'that a map of the whole area within the jurisdiction of the Board be provided, defining the boundaries of the various vestries and district boards'.[106] A number of proposals for this map were put to the Board both by commercial map-makers and by the Ordnance Survey, which had already made (1849–50) a skeleton survey of London for sanitary purposes. In the event, an existing commercial map, Edward Stanford's 24-sheet *Library Map of London and its Suburbs* (1862), was used. This was on a scale of 6 inches to one mile (1:10,560) and showed a wealth of potentially useful detail, including the boundaries of a dozen administrative bodies which only had to be coloured by purchasers (or by the publishers on demand) to create the required administrative map.[107] Stanford's *Library Map* plates were also used to provide a base map for a variety of other purposes, such as the 'School Board Map of London', and Charles Booth's 'Descriptive Map of London Poverty'.

Yet another aspect of maps being used to meet the ever-increasing demands of urban administration is the way they began to be produced in the eighteenth century to help deal with the increasingly devastating and expensive problem of fire. Insurance companies were prepared to sell fire insurance, but their underwriters demanded accurate large-scale town plans in order to identify the location of a particular building and to assess the risk attached to the related insurance proposal. Charles Goad was by far the major company in England producing such fire insurance maps from the last decades of the nineteenth century (Plate 19).[108] These maps portray the form of towns in incredible detail. Not only are individual buildings depicted in plan and their commercial uses indicated, but also characteristics such as the type of building materials, interior arrangement of staircases, party walls, doors (with their direction of opening) and windows are shown.

Maps for town planning

There has been town planning in England since Roman times. Not the least important phase – as recent archaeological investigation demonstrates – was in the Middle Ages.[109] Like so many contemporary villages, medieval new towns were laid out in an orderly and logical, if not always perfectly geometrical manner for kings, bishops and feudal lords hoping for fiscal gain through the conversion of income from agricultural to urban and commercial uses. No plan of a medieval new town has survived and there are but few documentary references to the work of survey: only 'the dumb witness of the sites in England, Wales and Gascony that have the simple rectilinear grid of streets' stands in testimony to the necessary process.[110] The simplicity of a design based on right angles and the symmetry of a chequer-board layout arguably makes a plan hardly necessary and it has been suggested that the medieval 'rectilinear grid made no more demands on techniques of measurement than the ability to set out a straight line, to divide it into equal proportions, and to set another line at right angles to it'.[111] So, as in Roman England, the landscape evidence of medieval new towns and extensions to smaller burghs fails to inform us whether the layout was planned on paper first or recorded as a map immediately after building. In contrast, Renaissance town planners focussed attention on the drawn plan. The 'ideal cities movement' of fifteenth- and sixteenth-century Italy developed new concepts of

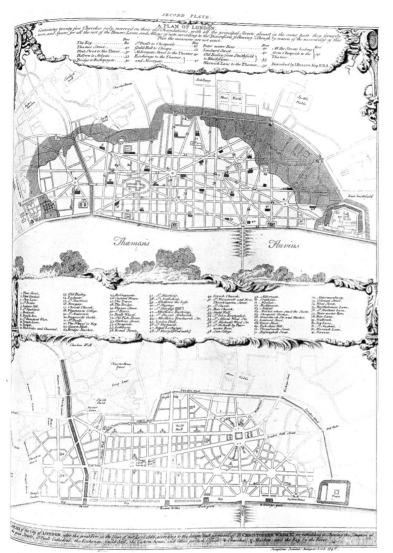

6.22 Rebuilding London: plans by Christopher Wren and John Evelyn, 1666. Both plans show clearly the influence of continental, classically-inspired, baroque town planning with their elements of symmetry, radial and grid pattern streets, circular 'ronds points', and great plazas. 32 × 21 cm. Wren's and Evelyn's plans were published by the Society of Antiquaries of London in 1789 (*Vetusta Monumenta*, Vol. 2).
Reproduced by permission of the Devon and Exeter Institution.

urban space and town design which often remained purely abstract and theoretical, the end product being mostly a paper plan rather than a constructed town. The movement began in Italy with the early fifteenth-century rediscovery of Vitruvius's treatise on architecture and with the reworking of classical planning ideas late in the century by Leon Battista Alberti and Antonio Filarete.[112] Vincenzo Scamozzi, Leonardo da Vinci and others developed the theory and practice further and editions of Vitruvius were published in Italian, French, and German. Inigo Jones was the first to introduce these Renaissance town-planning ideas into England when he laid out the piazza of Covent Garden for the Duke of

Bedford in 1630, but the best opportunity to judge the influence of continental ideas on English practice and to assess the role of maps in the planning process occurred in association with the rebuilding of London after the Great Fire of 1666.

From 2 to 6 September 1666 fire raged through the City of London and consumed some 60 per cent of its fabric, including 13,000 houses, St Paul's Cathedral, the Royal Exchange, the Customs House, the halls of 44 livery companies, and 87 parish churches. Some 177 hectares were laid to waste.[113] It was the most devastating and destructive fire ever witnessed in England. Urban fires were great tragedies, but at the same time they provided unique opportunities for rebuilding towns in new styles. Four days after the fire, while the burnt-out area of London was still smouldering, plans were active for a comprehensive rebuilding of the city. When the city authorities submitted their proposals to King Charles (March 1667), he asked for these to be plotted onto a map.[114] A manuscript skeleton survey of the burnt-out area, in six sheets, was forthcoming within a week from the team of surveyors.[115] A two-sheet reduction of their map, sold by Nathaniel Brook, was engraved by Wenceslaus Hollar and titled *An Exact Surveigh of the Streets Lanes and Churches contained within the Ruines of the City of London First Described in Six Plats by John Leake, John Lennings, William Marr, Will. Leybourn, Thomas Streete & Richard Shortgrave in Dec[ember] 1666*. The map was published in 1667.[116] This showed the area devastated by fire, the lines of the burnt-out streets, and the sites of the ruined medieval churches and other major buildings.

At about the same time, Robert Pricke produced two small maps on a single sheet. The main map, showing the whole of London and Westminster and suburbs was drawn so that it could be 'judged what proportion is burnt and what remains standing'. The second, *An exact Map representing the condition of the late famous and flourishing City of London as it lyeth in ruins . . .*, was flanked by lists of the destroyed churches and livery halls keyed to the map.[117] More practically, maps were produced to show proposals for the rebuilding of the city. Quickest off the mark was Sir Christopher Wren, who presented his *Londinum Redivivum* to Charles II a week after the conflagration.[118] Wren happened to be freshly returned to England from a long visit to Paris, where he had studied that city's classical buildings and formal street designs, and the continental influence of baroque planning dominates his proposals. Both his and John Evelyn's designs invoked symmetry with radial street patterns, circular piazzas and the promise of great vistas along broad avenues (Fig. 6.22).[119] They had been created too hurriedly, however, and without regard for the irregularities of terrain, and neither the King nor Parliament would accept either Wren's or Evelyn's schemes. Other proposals were received from Valentine Knight, Richard Newcourt, and Peter Mills, the City Surveyor.[120]

In the event, none of these radical plans was adopted. The loss of an opportunity to create a new, monumental city is part of the

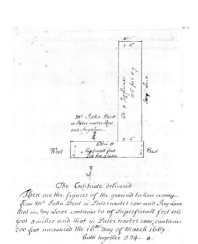

6.23 Rebuilding London: John Oliver's Certificate Survey, 1667. London was not rebuilt after the Great Fire to one of the several grand plans made at the time, but piecemeal and plot by plot. Before building could start on any site, however, claims to ownership had to be confirmed or 'certificated', as shown on this plan of the property of John Bent in Ivy Lane and Paternoster Row. 22 × 16 cm. Guildhall Library, Corporation of London, MS 84, p. 97. Reproduced by permission of Guildhall Library.

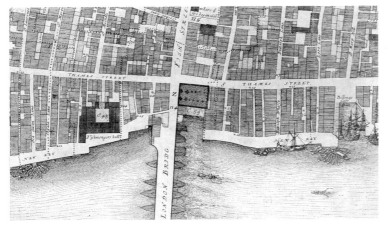

6.24 John Ogilby and William Morgan, *A New and Accurate Map of the City of London*, 1676. The rebuilt City is pictured on Ogilby and Morgan's map; this extract includes Fish Lane and the monument 'where ye fire began'. 152 × 248 cm, or about 52 inches to 1 mile (1:1188). British Library, Maps C.7.b.4. Reproduced by permission of the British Library Board.

old story of ambitious but unrealistic idealism in urban planning.[121] Other maps record what actually happened; pragmatic and piecemeal rebuilding, plot by plot, along the (often widened) old street lines. Claims to ownership had to be certificated before building could start on any site, however, which meant that one of the six commissioners appointed in 1666 (Sir Christopher Wren, Hugh May, Roger Pratt, Robert Hooke, Edward Jarmon and Peter Mills) had first to survey each plot individually and sign the relevant piece of paper (Fig. 6.23).[122] The results of their work are reflected indirectly in the 20-sheet map of the City of London made a decade later by John Ogilby and William Morgan (1676) (Fig. 6.24). Here on Ogilby and Morgan's map, we see, as Ralph Hyde puts it, 'not an ideal City with straight streets and noble

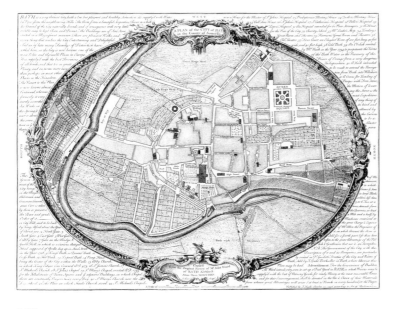

6.25 John Wood's plan of the City of Bath, 1735. A cartographical celebration of the beginnings of the classical transformation of the city. The mapped area is about 1.0 × 1.3 km (0.8 × 1.0 miles). British Library, K.37.14. Reproduced by permission of the British Library Board.

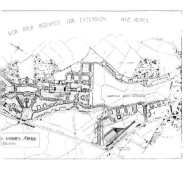

6.26   Plan for Hampstead
Garden Suburb, by Raymond
Unwin and Barry Parker, 1907.
The garden city movement of
the turn of the twentieth century
is the ideological opposite of the
style of town planning
epitomised at Bath and pictured
in Fig. 6.25. Here the map
portrays a vision of low-density
building and informal, irregular
street lines. Mapped area is
about 1.2 by 2.0 km
(0.7 × 1.2 miles).
Reproduced from H. R. Aldridge, *The
Case for Town Planning: A Practical
Manual for the Use of Town
Councillors, Officers, and Others
Engaged in the Preparation of Town
Planning Schemes* (London, National
Housing and Town Planning Council,
1915), p. 407.

vistas as Christopher Wren had initially envisaged, but a fairly
practical, improved City, the best that could be sensibly hoped
for'.[123]

While London lost the opportunity of rebuilding in the grand
manner, the city of Bath and its extension and embellishment in the
following century provides one of the best examples of the English
interpretation of classical urban design.[124] Map representations of
intended schemes played a central role in its planning. John Wood
the Elder, the architect who set the pattern for the transformation
of the city from a small provincial town to the most fashionable of
all English spas over the half century from 1728, had produced at
least one rural estate map of Bramham Park Estate, Yorkshire, at
some time between 1722 and 1727.[125] Printed in 1731 the
Bramham map depicts the geometric tree planting and ruler-
straight avenues of a formal landscape garden and park in the
classically-inspired manner that John Wood was to translate into
an urban context at Bath where, in 1728 he set out Queen's Square
and the surrounding streets in a way which effectively established
the form and direction that was followed by the city's later extensions. In 1730 a plan of a proposed scheme by Wood for a grand
Royal Circus was 'transmitted to a Person at that Time in London',
while other of his plans were published in the three volumes of his
An Essay towards a Description of Bath (1742–3). The Essay
includes Wood's schemes for a Royal Forum but it was his map
of the whole city (1736) which seems to have been produced
specifically to celebrate his achievement of the Queen's
Square development, since this features prominently on the map
(Fig. 6.25).[126]

The garden-city movement of the turn of the twentieth century
is the ideological opposite of the style of town planning epitomised
by eighteenth-century Bath. Perhaps the most significant English
contribution to urban planning theory, the garden city was based
on the idea of combining the best of city life with the best of rural
living in a carefully controlled low-density, 'garden' environment.[127] While the maps showing proposed developments in
London and Bath had been used to emphasise the formality of
straight avenues, *ronds points*, and squares, garden-city planners
used maps to emphasise a quite opposite concept and to underline
their vision of the sort of low-density, informal arrangement of
buildings and streets eventually constructed at Letchworth and
Welwyn, Hertfordshire and at Hampstead Garden Suburb, in
north London (Fig. 6.26).

Maps in guide books, street atlases and town directories

As we saw in Chapter 5, maps were not generally used for finding
the way about England or across western Europe much before the
eighteenth century. Finding the way about a strange city, though,
was a different matter. As the travelling class expanded, printed
maps of towns for the tourist at home or abroad were made available. Maps of this genre display a diversity of form and include

generalised bird's-eye views of towns on the small-format pages of early tourist guide books, large-scale, multi-sheet or bound atlas maps with named streets, and maps which locate individual building addresses in town directories. In the case of tourists, the issue was not so much to locate a house or street as to find the particular feature they were to admire, paint or (mindful of the hidden agenda of many early travellers) spy on for their government back home. As the built-up area of towns expanded and the urban fabric increased in density and complexity, both the amount of spatial information communicated on a map and the range of town guides increased. On the maps, streets are named and these names are often be listed in an accompanying index keyed to the map by reference to a grid.

One of the early forerunners of the modern 'A to Z' type of street atlas is dated 1653. This anonymous *Guide for Cuntreymen in the famous Cittey of London, by the help of which plott they shall be able to know how farr it is to any street* was a re-issue of John Norden's map of London printed over 50 years earlier in the Middlesex volume of the *Speculum Britanniae*, to which an enlarged list of streets had been added.[128] Later in the century, while John Ogilby failed to produce the volume of *Britannia* which was to have included views and descriptions of English cities, both his map of London and his nine-sheet map of Ipswich were accompanied by explanatory booklets listing street and place names shown on the maps. Ogilby's example was followed in the eighteenth century by John Rocque and others.[129]

It is difficult to define guide books unexceptionally, since a wide range of books can serve as travellers' guides. Some of the earliest guide books to include town maps were of early seventeenth-century Italian cities, while from the eighteenth century, oblique views and plans were commonplace in guide books.[130] By 1723, the idea of a foldable 'pocket map' had been established, a format which became the norm for town maps in the nineteenth century.[131] Along the south coast of England, the mild climate of Devon was attracting an important tourist trade from the middle of the eighteenth century onwards. By the first part of the nineteenth century, tourist guide books to individual towns like *The Tourist's Companion being a Guide to Plymouth*, (1823), *The Panorama of Torquay* (1832), and *A Descriptive Sketch of Sidmouth* (c.1836), were including maps as a matter of course (Fig. 6.27).[132] By the mid nineteenth century, the guide-book genre was well established, and publishers began to specialise. John Murray, the first publisher to use the term 'handbook' for a continental guidebook (1836), started his English series with Peter Cunningham's *Handbook of London* in 1839 and went on to offer the tourist a series of handy volumes containing short descriptive inventories of localities and monuments arranged as itineraries and equipped with maps 'to facilitate their use . . .'.[133] In his *New Picture of London and Visitors' Guide to Its Sights* (1847), Edward Mogg also elaborated on the touristic role of maps: 'to avoid the inconvenience of taxing

6.27 Sidmouth, Devon, and the neighbourhood, 1858. This is an example of a map aimed at the tourist market. It was copied closely from the Ordnance Survey one-inch map with some minor additions, notably the 'Paths'. Area shown covers about 6 × 6 km (3.7 × 3.7 miles). Peter Orlando Hutchinson, *A New Guide to Sidmouth and the Neighbourhood . . . illustrated with Map, Views and Diagrams* (Sidmouth, 1858).
Reproduced by permission of the Devon and Exeter Institution.

6.28 Plan of Penzance: inset on the county map in Kelly's *Directory of the County of Cornwall*, 'with maps engraved expressly for the work', 1897. A good example of a functional street map of a type common since the mid nineteenth century: little is shown other than streets, public buildings and field boundaries. These last suggest derivation from an Ordnance Survey source. Scale of original, 6 inches to 1 mile (1:10,560). Area shown covers about 1.7 × 1.6 km (1 × 0.9 miles).
Reproduced by permission of the Devon and Exeter Institution.

his friend to an attendance upon him in his peregrinations, it is indispensable that he provide himself with a good plan . . . like the clue in Ariadne, they will conduct him through the labyrinth, and occasionally consulted, will enable him unattended to thread with ease the mazes . . .'[134]

Away from the increasingly lucrative tourist market, most town and trade directories were also being provided with maps as an aid to identifying a specific street listed in the directory. From its first edition (1790), *The Universal British Directory of Trade and Commerce* included a general plan of London.[135] Many of the larger publishers of directories, such as James Pigot of Manchester, commissioned purpose-produced maps and plans. In the 1820s, Edward Baines's Lancashire directory contained maps of all the principal settlements in the county and from 1843 onwards, a map of London was inserted into Frederic Kelly's directory of the city.[136] By the end of the century, maps were so integral a part of Kelly's directories that every street entry was accompanied by its map reference (Fig. 6.28). Just as the directories themselves had to be frequently re-issued to keep up with changes in the town and new developments on the outskirts, so too the maps were regularly updated to ensure their continued usefulness.

Power in cities was often contested and elaborate maps could be both a symbol of power and the means by which that power was exercised. The cartographical and non-cartographical contents of a map, particularly the inscriptions and dedications which it bore, and the decorations in borders and cartouches were all important manifestations of power: of a monarch, a noble, or a town council. Just occasionally, as we have observed, a pre-modern town map was made with a specific practical function in mind – as in the case of Richard Lyne's map of Cambridge (1574) and as Ralph Agas reminded readers of his surveying treatise (1596) – but such evidence is rare. Aaron Rathborne was surely articulating the common motive when he and Roger Burges were applying for a twenty-year patent for their intended portfolio of town maps in 1617. In their lengthy application no hint is given of the possible utility of the maps.[137] Instead, we are told by Rathborne that he was intending to emulate foreign practice in printing maps 'to the great honor and renowne of those princes in whose dominions [the maps] are'. This, then, is the light in which we should see, for example, John Hooker's map of Exeter (1587, see Plate 16), a map much reproduced, adapted, and copied by contemporaries and near contemporaries, even as decoration on a leather screen.[138]

Hooker's portrayal of the city in which he was Chamberlain from 1555 has been studied from many angles and analysed for what it can tell us about Exeter's architecture and townscape, but important questions remain – why was this map made at all? Why was a considerable sum of money expended on it? What was its intended purpose? While no straightforward answers to such questions are yet forthcoming, we can be certain that Hooker's map is more than just an anodyne topographical description of the walls, streets and buildings of a cathedral city set in a rural hinterland. It tells us something of the economic importance of the River Exe from the details of fishing, milling and shipping shown on the map and we also see how 'people walk the streets, horses draw carts, market stalls are on the cathedral green, cows are being driven across Exe Bridge and even a game of hockey is in progress'.[139] It portrays the city as a late sixteenth-century fact, certainly. But its iconography conveys deeper messages. The arms of the City of Exeter and the Bishop of Exeter acknowledge the patronage of the city by Queen Elizabeth I, and the map is a celebration of the wealth and power of this important late sixteenth-century town under her government. By citing in turn the Roman, Celtic, Saxon, Latin and English names by which the city has been known, the map also parades the lineage and antiquity of the city. Reflecting the political significance of a concentration of inhabitants, their activities and their wealth on which not only the local region but also the nation depended, town maps, perhaps more than any other of the map genres discussed in this book, serve many interests beyond the strictly practical.

The Spirit of Modernity: Maps in Everyday Life

That the map is today a medium of everyday communication is a truism easy to illustrate. It is difficult to find an issue of a major newspaper that does not contain at least a small scale map to pin-point a little-known, and perhaps distant, place where a newsworthy event has occurred or to illustrate the political geography of a region of the world. Larger-scale maps illustrate the reconstruction of disasters – the last moments of a train, a ship or an aircraft, or the last route taken by a victim of violent assault. On 13 September 1997, every British daily newspaper carried a detailed map of the route through central London of the Princess of Wales's funeral cortège. Everyday we see the weather forecast in map form on television.

It is equally hard to imagine an entirely map-less household today as it is to call to mind a map-less mass media. There is but a minority of households that has no need of a road map for planning a car journey, while there can be few people who can live their lives without a street map of their own city or of the town to which they turn for shopping or services. A pocket diary map of Henry Beck's map of the London Underground network (see Fig. 1.2, p. 3) is an essential way-finding aid for all visitors (and many Londoners). Anyone who walks for pleasure in England will know and use those Ordnance Survey large-scale maps which portray footpaths; those who sail in English estuaries and inshore waters should have access to a hydrographical chart, if only for their safety. Children are taught how to make and read maps as part of their primary school education. As adults, we might draw a sketch map to help the first-time visitor find our home.

The unremarkable use of maps in everyday life in England is mirrored also in the world of scientific inquiry. Professional geographers use maps to reveal spatial patterns in their data and to seek spatial correlations between phenomena. Geographical Information Systems (GIS) are founded on electronically stored maps. Useful as both maps and GIS are in geographical and allied research, they also have wide applications outside academia. GIS have been developed as 'package programs' so that individuals and organisations (both public sector and commercial) responsible for administering territory can easily create and superimpose maps of many different variables to assist their decision-making.

In short, the academic, political, economic, social and personal utility of maps is now a demonstrable *sine qua non*. But it was not ever thus. While we have found no reason to talk in this book about map-less periods, except in respect of some specific types of maps at particular times (for example, the lack of estate maps in the Middle Ages), we are comfortable with the assertion that 1,000, 500 or even 200 years ago, society *at large* was not as aware of the utility of maps as it is today. We do not imply a progressive development of maps and map use through time, but, taking a chronological view, we can see how the history of map use in England and Wales is punctuated by apogees, eclipses, nadirs and renaissances. When we come to the last period, that of the nineteenth and twentieth centuries, better documentation makes it easier to analyse the process of diffusion of map awareness and to see more clearly how maps have come to touch almost every social stratum. For example, the institutionalisation of basic topographical map-making was both a function of increasing demand for maps and charts in the nineteenth century and also of the economies of scale generated by centrally-organised surveying, cartography and printing which reduced the unit cost of maps and promoted their wider dissemination and use. Another nineteenth-century development was the increasing use of maps within the scientific community to investigate relationships between phenomena across space, and to help solve social and other problems. Work in the natural sciences and in the emerging social sciences, not least in the establishment of 'modern' geography, demonstrated to a wider public the powerful heuristic and didactic roles of maps that had been appreciated centuries before by scholarly medieval monks.

In this final substantive chapter we explore some facets of this nascent 'spirit of modernity'. We review, first, the emergence of the Ordnance Survey, Great Britain's national mapping agency and the originator today of almost all new British topographical mapping. Second, we turn to the parallel development of the Hydrographic Office of the Admiralty and that part of its work concerned with charting the coastal waters of England and Wales. Finally, we trace the development of thematic maps to illustrate the value of maps for investigating, recording and communicating spatial phenomena in modern scientific inquiries.

Mapping the nation: the Ordnance Survey of England and Wales

The story of the origins and development of the Ordnance Survey as England's national mapping agency is one of the changing role of an institution. From being a handmaiden of the military at the end of the eighteenth century, the Ordnance Survey has acquired a wider role as provider of maps for the public at large.[1] The establishment of the Ordnance Survey as a government-funded, official mapping agency did not herald the demise of privately-produced, commercial map-making in England. As we have seen in Chapter 3, there was initially a productive two-way relationship between the producers of commercial maps of English counties and

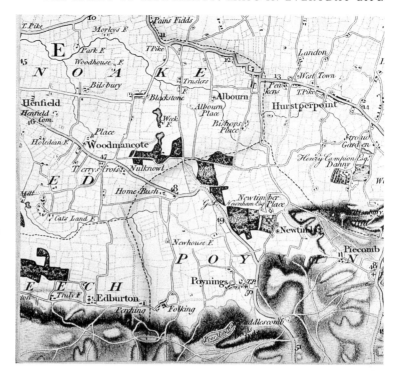

7.1   North-west Kent, published by William Faden in 1801. Once Ordnance Survey geodetic data became available, they were used by many private commercial cartographers, as for example by William Gardner for this 1 inch to 1 mile (1:63,360) map of Sussex. Although published in colour, it is stylistically similar to the Ordnance Survey maps which were to follow in the first years of the nineteenth century. British Library, Maps 5395.9.
Reproduced by permission of the British Library Board.

the Ordnance Survey.[2] Indeed, men such as Christopher and John Greenwood and their contemporaries who were mapping counties up to the end of the 1830s as speculative, money-making, ventures, expected to be able to employ Ordnance Survey geodetic data to ensure accuracy for their maps. The two inches to one mile map of Sussex, left unfinished by Thomas Yeakell and William Gardner in 1783, was completed in 1795 by T. Gream with geodetic data supplied by the Ordnance Survey, and was then printed and published by William Faden in 1795 (Fig. 7.1). Although in colour, this final edition of Yeakell and Gardner's map is stylistically similar to the Ordnance Survey's own maps which were to follow in the first years of the nineteenth century.

From about 1829, private firms like that of the Greenwoods were gradually eclipsed by the Ordnance Survey as the originators of new maps at the one inch to one mile scale, although private firms continued to publish road maps and atlases at smaller scales, and street guides to towns, as they do today. But even they relied on Ordnance Survey data. The last point is important. Right from its beginnings, the Ordnance Survey was a provider not only of maps but also of the geodetic skeleton on which other maps could be based. But while at the start of the nineteenth century, private publishing firms were losing their hold on the printed topographical map, private surveyors continued to be the main source of large-scale manuscript maps of estates and farms until the

Ordnance Survey commenced surveying at the much larger scale of 25 inches to one mile (1:2500) in the 1850s.

The Ordnance Survey of England and Wales had two points of departure, neither of which was in England. The earlier was in Scotland in the 1740s, just at the time that the re-mapping of England by private individuals was about to embark on its most active phase. In Scotland, by contrast, it was the government which initiated a military survey as part of the political aftermath of the Jacobite rebellion of 1745–46.[3] The battle of Culloden, which terminated the Highland uprising, brought the English army rudely face to face with the reality that its task was being made more difficult for the want of reliable maps; and so the survey of Scotland was established. One of the officers employed on the survey, William Roy of the Royal Engineers, tried twice (in 1763 and in 1766) to persuade the government to make an official survey of the whole of Britain which would incorporate the earlier Scottish work with a new survey of England and Wales, for publication at a scale of either one inch or 1.25 inches to one mile (1:63,360 or 1:51,138). The rationale behind Roy's suggestion was to provide the nation with a map which would serve as a tool for the defence of the realm but which had been made, in Roy's words, 'during times of peace and tranquillity'.[4] Periods of peace, however, are not the most propitious times in which to try to persuade governments of the value of funding precautionary mapping and the authorities in London were not moved to finance Roy's proposals. The result was that no equivalent map to that of Scotland was made in England in the eighteenth century. Instead, the topographical mapping of English counties continued in the hands of private, commercial map-makers.

Another factor involved in the origin of the Ordnance Survey was not a mapping project but an international geodetic exercise undertaken to determine the precise relative positions of the two observatories of Paris and Greenwich. In 1784, a base line had been set out on Hounslow Heath for a system of triangulation which was eventually extended to link London and Paris. In England, the project was under the general direction of the civilian Royal Society, but day-to-day work was directed by Roy, by then the country's leading geodesist, assisted by a group of men from the Royal Artillery. By the time of Roy's death, in 1790, the London-Paris triangulation was complete. Roy had planned to extend the triangulation to the whole of Britain, and this process was begun in 1791 when the Trigonometrical Survey of the Board of Ordnance was founded to complete that work.

The 'Old Series' one inch to one mile maps

In 1784, at the same time as the London-Paris triangulation was begun, the Duke of Richmond set a group of surveyors to map the Plymouth naval dockyard and surrounding country at a scale of six inches to one mile (1:10,560) to provide a base map for planning new fortifications. In about 1788, the same group of surveyors

7.2 Ordnance Surveyor's drawing, Fowey Estuary, Cornwall, 1805, by Robert Dawson senior. The drawing shows field boundaries and other minor details not included on the published map illustrated in Fig. 7.3. Scale of original is 3 inches to 1 mile (1:21,120); area covered is about 4.5 × 6.0 km (2.8 × 3.7 miles). British Library, OSD 9 (also published on microfilm by Research Publications, Reading, 1989). Reproduced by permission of the British Library Board.

went on to map other areas of military significance along the south coast, including the Isle of Wight and a substantial part of the county of Kent. The surveyors were still working in Kent when, in 1795, the decision was taken to extend the entire mapping project to the whole country. England was now at war with France and the need for reliable maps was pressing. For the initial manuscript maps, known today as 'Ordnance Surveyors' Drawings', the original scales of six inches to one mile (preferred for areas of military significance, notably along the south coast) and three inches to one mile (1:21,120) were reduced after about 1805 to two inches to one mile (1:31,680).[5] Figure 7.2 is an extract from one of the Drawings of part of the Fowey estuary, Cornwall, produced at the three-inch scale, and showing details which were eventually eliminated when the final map was published at a scale of one inch to one mile (Fig. 7.3). Exactly whose initiative it was to implement Roy's idea to publish the survey at this scale is unclear, but the facts are that in 1801 the survey of Kent was published by the independent map-maker and publisher, William Faden, 'Geographer to the King', and that thenceforth all engraving and printing would be undertaken by the Board of Ordnance itself.[6] The modern name 'Ordnance Survey' first appeared in 1810 but for many years the organisation was better known either as the 'Ordnance Map Office' or the 'Ordnance Trigonometrical Survey'.

By reducing the scale of field survey, the entire process was speeded up so that by the end of the Napoleonic Wars, in 1815, the southern half of England and Wales had been mapped, and one

7.3 Ordnance Survey Old Series sheet 30 (1813). This indicates the amount of generalisation necessary (compare with Fig. 7.2) for the published map at the smaller 1 inch to 1 mile scale (1:63,360); the area covered in the extract is about 5.5 × 7.5 km (3.4 × 4.6 miles).
Reproduced from a copy in a private collection.

7.4 Hachuring on the Ordnance Survey one-inch Old Series map, sheet 75 NW: the Moelwyn mountains, North Wales, 1840. This shows engraved hachuring at its most sophisticated and expressive. Scale of original, 1:63,360; area covered is about 7.5 × 7.5 km (4.6 × 4.6 miles).
Reproduced from a copy in a private collection.

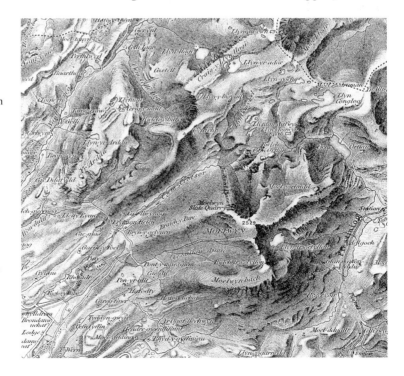

inch to one mile maps were published for about a fifth of England. It took 69 years, from 1805 to 1874, for the 110 sheets that make up the complete coverage of England and Wales at this scale to become available. The immediate advantage of these Ordance Survey one inch to one mile maps (known to us as the 'Old Series') compared with the earlier, privately-produced county maps, lay in the greater amount of topographical detail they offered the user and in the way hachuring was used systematically to show relief, a feature important to the military (Fig. 7.4). More minor roads and minor place-names are included on the Ordnance Survey one-inch maps by comparison with county maps and careful attention was paid to the outlines of buildings.

The 'Battle of the Scales': 1:2500 becomes the standard map scale

A perhaps quirky feature of the early history of the Ordnance Survey is the way that what happened in England was the result of initiatives outside the country, either elsewhere in the kingdom or across the Channel. In the debate over scales, the activities of the Ordnance Survey in Ireland were to have a profound influence on developments in Ordnance Survey mapping in England. The problem was the selection of an appropriate scale for the basic survey of the country. The Ordnance Survey mapping of Ireland had originated from a need to update the basis on which the land tax was collected. This required a survey of the boundaries and the contents of the taxation units of local areas known as townlands.[7] The task fell to the Board of Ordnance.[8] As in England, triangulation of the country was the first step; in Ireland this was completed in 1841. Parallel with the geodetic work, the Ordnance Survey began mapping Ireland at six inches to one mile.

Following the success of the Irish experience, it was argued in London in the 1830s that the scale of six inches to one mile should be introduced into England and Scotland once primary triangulation, suspended in 1823 as a result of the survey of Ireland, was resumed. Accordingly, in October 1840, the Treasury authorised

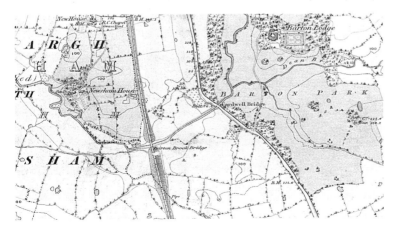

7.5 Ordnance Survey six-inch map of Lancashire, sheet 53, 1848. This is a good example of the style of the early Ordnance Survey six-inch mapping of Britain. Each sheet 75 × 100 cm. 6 inches to 1 mile (1:10,560); area shown in the extract is about 2.5 × 1.5 km (1.5 × 0.9 miles).
Reproduced from a copy in a private collection.

the mapping of northern England and Scotland on the six-inch scale. Work began in the following year in Lancashire, with many of those same surveyors who had worked in Ireland being employed (Fig. 7.5). In 1843, the six-inch survey began in Scotland. Progress in both England and Scotland was slow, delayed in part by the amount of intricate urban surveying (much greater than in Ireland) that was needed. By 1851, only Lancashire and Wigtonshire had been completed. Frustration voiced in Scotland led to the appointment in 1851 of a House of Commons Select Committee. This recommended the abandonment of the six-inch scale in favour of the former policy of field survey at two inches to one mile with publication at one inch to one mile. This recommendation effectively launched what became known as the 'Battle of the Scales', in which technical considerations of mapping were taken over into political debate. The suggestion that the six-inch scale was to be abandoned drew protests from the counties of northern England and Scotland, in which surveying at this scale was about to start, and it became apparent that the whole issue needed thorough review. A consultation exercise in 1853–54 suggested that a scale of about 1:2500, or 25 inches to the mile, would be best for surveying and that this might provide the standard scale from which smaller scale maps would be derived and published. An official investigation in 1850 had already indicated that maps on the six-inch scale would be inadequate for land registration purposes. It had also decided that the recently completed tithe surveys were too inaccurate to be an acceptable substitute. Moreover, tithe surveys did not extend into Scotland, the part of Britain most vociferous in its demand for an official, large-scale survey.

In the event, the 1:2500 scale (25.54 inches to one mile) was adopted for surveying and for published maps not because of positive enthusiasm for it – expert opinion was about evenly split between the merits of 24 inches to one mile (1:2640) and the traditional scale of one inch to three chains (1:2376) employed for many private estate maps, enclosure maps and tithe surveys – but because the scale of 1:2500 had been recommended as an international standard by an International Statistical Conference held in Brussels in 1853. Furthermore, the 1:2500 scale was already being used for the maps of several European cadastral surveys, most notably the Napoleonic *cadastre* of France.

There may have been no government demand for large-scale maps for cadastral purposes in England and Wales, but there was certainly strong demand for large-scale maps from commercial map-users in England. The middle years of the nineteenth century were times of unprecedented social and economic change; towns were growing fast and industrial activity was burgeoning, especially on the coalfields of midland and northern England. Railway companies required maps for planning new ventures, and the fast-developing railway network was itself transforming the pattern of economic activity and the way land was being used. Regional specialisation in agriculture was intensifying. Urban-

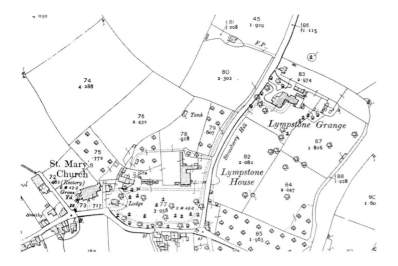

7.6   Ordnance Survey 1:2500 of Devon, sheet 93.9, 1905. This exemplifies the clean-cut appearance of the mature 1:2500 (about 25 inches to 1 mile) style. Different symbols indicate land cover by orchards and scattered trees. Each sheet 75 × 100 cm; the area shown in the extract is about 450 × 250 metres (490 × 273 yards). Reproduced by permission of Department of Geography, University of Exeter.

based industry was developing in conjunction with the railways as the latter provided a means for the assembly of raw materials and the distribution of manufactured products. All these processes brought a demand for printed maps on the largest possible scale, administrative tools which would be relatively cheap and available in sufficient numbers to serve this growing and very varied market demand for mapping.

Surveying for 1:2500 mapping began in County Durham and in the Scottish counties of Ayr and Dumfries in the spring of 1853. The decision to map at this scale was opposed by various sectional interests and the Battle of the Scales was not finally resolved until 1858, by which time a pattern of mapping had been set which was to make late nineteenth-century Britain the best-mapped country in the world. All cultivated and settled areas were first surveyed and then mapped at 1:2500 (Fig. 7.6), larger urban areas (centres of more than 4,000 people) justified mapping at a scale of 1:500, and the whole country was represented on smaller-scale, derived maps on the six-inch and one-inch scales. Six inches to one mile initial-scale mapping was used for unpopulated mountain and moorland areas to save the cost of both survey and printing.

The Ordnance Survey 1:2500 maps show public boundaries in great detail, they name roads, and portray not only buildings but also minor features such as ponds in their correct shape. Symbols or abbreviated words indicate paths, trees, lime kilns, smithies, mills, railway signal boxes, milestones, greenhouses, village pounds, direction of stream flow, orchards, allotments, sea walls and other details. Of considerable historical interest is the fact that the names of houses and gardens, farmsteads and other features are given (as far as is practicable). Maps published on the scale of 1:2500 before 1879 were also accompanied by separate Books of Reference in which the acreage of each land parcel and a description of its use is detailed, a practice discontinued after 1879. The

7.7 Ordnance Survey six-inch (1:10,560) map, Devon sheet 103 NW, 1906. This shows the centre of the sea-side town of Exmouth. Building shapes are generalised at this scale. Each sheet 42 × 59 cm; area covered is about 700 × 400 metres (765 × 437 yards).
Reproduced by permission of Department of Geography, University of Exeter.

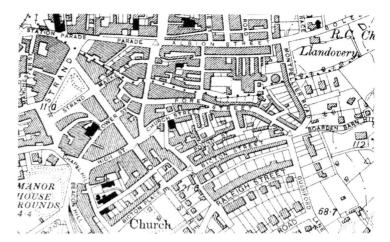

7.8 Improvised mapping: an extract from GSGS 3906 sheet 56/12 NW, 1940. In order to provide complete 1:25,000 cover of Britain for use as a military map in case of invasion during the Second World War, six-inch maps (1:10,560) were reduced by photography to 1:25,000; contours were added, photo-enlarged, from 1:63,360 maps. Each 40 × 60 cm (excluding margins); area covered is about 2.0 × 1.5 km (1.2 × 0.9 miles).
Reproduced by permission of Department of Geography, University of Exeter.

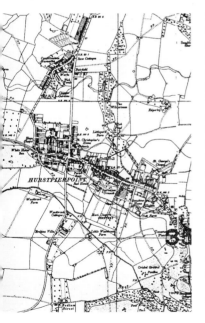

Ordnance Survey made its first edition 1:2500 maps available to the public either plain or hand-coloured to show stone and brick buildings in red, water in blue and metalled roads in brown.[9]

Maps derived from the 1:2500 standard scale

For a short period, from 1841 to 1853, the scale of six inches to one mile was standard for rural areas. After 1853, this continued to be used as the largest scale for mountain and moorland areas. Elsewhere, maps at this scale were derived from those published at the scale of 1:2500 (Fig. 7.7). From 1969, as part of a programme of metrication, the six-inch scale was replaced by the 1:10,000.

Maps published on the scale of six inches to one mile were redrawn, simplified, and photographically reduced to the yet smaller scale of 1:25,000 (two and a half inches to one mile). This second intermediate series began life during the First World War as a map on the scale of 1:20,000. At this time, it covered only small parts of the country, mainly those of military interest. In the 1930s the scale was changed to 1:25,000 and during the Second World War government departments had access to coverage for the whole country produced by direct photo-reduction from six-inch maps (Fig. 7.8). The scale of 1:25,000 proved to be particularly useful for strategic town and country planning, and the Ordnance Survey was authorised to experiment with a civilian version. The aim was to fill the gap between six-inch and one-inch maps and to provide a map for walkers and for educational purposes. Publication began in 1945. The modern 1:25,000 maps are known as the Pathfinder, Explorer and Outdoor Leisure maps, names which indicate their potential market. In the early 1980s it was debated whether the 1:25,000 scale should be discontinued on grounds of cost but pressure from educationalists and ramblers' associations has ensured its survival and the completion of national coverage.

Before 1897, all Ordnance Survey one inch to one mile maps were published in black and white, but from that date coloured versions were gradually introduced and have since became the most popular format. By the First World War, the use of colour had evolved to the extent that on a few sheets twelve separate printings were required. After the war, this was replaced by a greatly simplified map – the Popular Edition – which introduced an eleven-fold road classification but which relied wholly on contours for relief (Plate 20). As a result, landform information is intentionally less visible on these maps while the cultural content of the map is dominant. Hachures made a brief reappearance in the 1930s on the Fifth Relief Edition one-inch map in conjunction with contours, a map considered by many to be the high point of Ordnance Survey landform representation, but the style did not prove popular with the purchasing public and hachures were quickly dropped on the unshaded edition of this map.

After the Second World War, the one inch to one mile map was continued in a New Popular Edition (Plate 21). Although published after the war, the main revision of this particular sheet had been carried out between 1913 and 1937, and only building development was updated to 1939. The New Popular Edition was the first published one-inch map to carry the metric National Grid for locating places by coordinates. In 1947, a full post-war revision was authorised and this was published as the Seventh Series, the last imperial-scale map.[10] Finally, in 1974–6, the Seventh Series was replaced by a metric scale 1:50,000 map, of which two series have been published to date: while the First Series has a metric horizontal scale, the contours are still spaced at 50 feet intervals although they are labelled in metres. The Second Series 1:50,000 is a fully metricated map.

CHARTING COASTAL WATERS: THE HYDROGRAPHIC OFFICE OF THE ADMIRALTY

The needs of national defence, and the awakening of maritime enterprise in England's Age of Discoveries, had stimulated the production of coastal, estuarine and river charts in the sixteenth century, particularly along the English Channel coastline. In the seventeenth century, the Royal Society (founded 1662) encouraged the formulation of a sound, theoretical basis for hydrographic surveying, and the Society's Curator of Experiments, Robert Hooke, himself invented an instrument for measuring water depth to replace the lead line traditionally used for soundings.[11] In the eighteenth century, surveyors such as Murdoch Mackenzie junior, Maritime Surveyor at the Admiralty, and Graeme Spence did much to combine theory with increasingly sound practice. Part of their chart of Poole Harbour on the south coast of England at six inches to one mile is illustrated in Figure 7.9.[12] In total, the eighteenth century saw the production of many new charts of coastal waters, some made by Royal Navy officers like Mackenzie, others made by Trinity House pilots and military engineers, by officials of the Custom House, and by private individuals, the last usually in

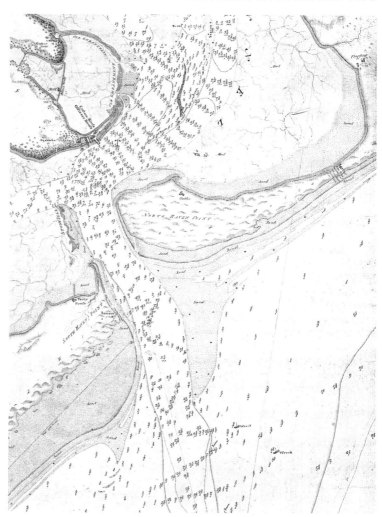

7.9   Chart of Poole Harbour by Murdoch Mackenzie junior, 1786. In the eighteenth century, surveyors such as Mackenzie, Maritime Surveyor at the Admiralty, did much to improve the accuracy of maritime charting. 6 inches to 1 mile (1:10,560); this extract covers an area of approximately 2.5 × 1.5 km (1.5 × 0.9 miles). Hydrographic Data Centre, Taunton, D927.
Reproduced by permission of United Kingdom Hydrographic Office.

association with harbour works.[13] Whatever their provenance, most of the charts made in the eighteenth century either remained as manuscripts or were printed by private, mostly London-based, map-sellers.[14]

A prime reason for the foundation of the Hydrographic Office of the Admiralty in 1795 was to coordinate this fragmented but voluminous output at a time of critical military importance occasioned by the war with France. The Hydrographic Office was set up just four years after the foundation of the Trigonometrical Survey of the Board of Ordnance and in the same year that topographical mapping for the one-inch maps of England and Wales was authorised. Land and marine cartography were twin priorities during the Napoleonic Wars. By 1795, the Trigonometrical Survey had triangulated the whole of southern England, from Kent to Lands End. The geodetic framework for the topographic mapping of southern England in the first decades of the nineteenth century was also used to calculate latitudes and longitudes for the principal triangulation stations, providing valuable data to the Admiralty for correcting its marine charts. In August 1795, the Lords Commissioners of the Admiralty stated that 'on examination of charts in office, we find a mass of information requiring digest, which might be utilised, but owing to the want of an establishment for this duty, His Majesty's Officers are deprived of the advantages of these valuable communications'.[15] In the same year, Alexander Dalrymple, Hydrographer to the East India Company, was appointed Hydrographer to the Admiralty. His brief was limited: to sort and classify the accumulated charts and records, to use their data to update existing charts, and to compile new charts by aggregating data from existing local charts.[16] It was not envisaged at this time that the Hydrographer would be involved in commissioning new surveys; this continued to be the responsibility of the naval authorities themselves. Thus Dalrymple was in charge of what was in essence a chart compilation office, not a survey office. Moreover, he had neither the budget nor authority to publish charts. The only press under his control was a proofing press and he was given a paper allowance for proofs only. Dalrymple carried out an evaluation of pre-existing charts for all United Kingdom coasts and for overseas waters, but naturally was reluctant to give an *imprimatur* to charts whose authority and accuracy were unclear.[17]

Dalrymple set such high standards for the work that many extant charts were rejected and remained unused, a situation which irked some in the Admiralty who believed that more could be made of the accumulation of eighteenth-century work. In 1808 he was dismissed and was succeeded by Thomas Hurd.[18] By the end of his period of office, in 1823, Hurd had advanced the institutionalisation of chart-making on two fronts. He saw the establishment of an independent surveying branch within the Royal Navy and had also persuaded the Board of the Admiralty to make its charts available to the civilian merchant marine. This not only serviced the needs of the increasing coastal traffic generated in rapidly industrialising England, but also raised much-needed revenue to finance further survey.[19] Close, two-way collaboration was also established with the Ordnance Survey which continued to supply triangulation data, and in return, naval surveyors provided coastal detail for Ordnance maps.[20] Just two months after his appointment as Hydrographer, Hurd wrote to the Admiralty about 'the necessity of an application being made to the Board of Ordnance that this office be allowed to have a copy of such parts of Colonel Mudge's military survey of England as respects the sea coast thereof together with all remarkable objects in the vicinity, as may be judged useful to navigation'.[21] By 1811, Hurd was able to publish his *Charts of the English Channel*, an atlas of 31 charts, which included some new compilations by the Hydrographic Office and some which he reprinted from privately published plates. All these developments

7.10   Gravesend Reach by
Capt. F. Bullock R.N., 1833.
This chart forms part of what is
known as the 'Grand Survey of
the British Isles', a recharting of
almost all of the British coast
and inshore waters which
occupied the Hydrographic
Service of the Admiralty for
much of the mid-nineteenth
century. This extract covers
an area of approximately
2.0 × 1.5 km (1.2 × 0.9 miles).
Hydrographic Data Centre,
Taunton, L8880.
Reproduced by permission of United
Kingdom Hydrographic Office.

provided a springboard for the ambitious re-survey of the coastal
waters of Britain known as the 'Grand Survey of the British Isles'
which was to occupy the Hydrographic Office for much of the mid-
nineteenth century.

The Grand Survey recharted almost all of the coast and inshore
waters, including the main navigable rivers.[22] It comprised some
highly detailed and accurate charts, like one of the River Thames
by Frank Bullock (Fig. 7.10). Francis Beaufort, appointed
Hydrographer in 1829, considered that the 'survey of the River
Thames will be an important epoch in its maritime history; for
besides the practical value of a correct plan of its present limits,
depths, currents, etc., we shall be able to watch the smallest indica-

tions of those changes which nature is always producing . . .'.[23] Bullock began his survey in 1830 on scales of between five and 22 inches to the mile (1:12,672 to 1:2880). He quickly found that 'accuracy had to be bought with time'; accordingly, it was not until 1836 that he reached the Nore off Southend.[24] In the 27 years that Beaufort was Hydrographer, the previous total of 44 Admiralty charts of British waters was increased to 255, with the result that there were modern charts for most of the United Kingdom, including virtually all the coast of England and Wales.[25] Only Cornwall around the Lizard and the waters of the Isles of Scilly remained to be completely re-surveyed. Initiated and largely completed under Beaufort's administration, the Grand Survey was the foundation for later revisions necessitated by, for example, the changing physical state of coasts and estuaries and by mariners' need of larger- and larger-scale charts.[26]

Carto-bibliographical studies of the marine charts of home waters reveal that their physical characteristics vary considerably and that their revision and printing history is complex. For example, a wide range of scales was employed for nineteenth-century Admiralty charts. Harbour charts were usually drawn at between ten and twenty inches to one mile (1:6336 to 1:3168), while estuaries, bays and approaches might be represented at scales as small as one inch to one mile (1:63,360).[27] Early printings of charts often predate the Ordnance Survey's large-scale plans of the area and, once published, were subject to almost continuous revision throughout the nineteenth century and up to the present day. The official Admiralty *Catalogues of Charts* series were commenced in 1825. They contain lists of charts by standard number, with details of scale, notes of corrections and revisions.[28] Early catalogues give a true impression of the patchy nature of coastal coverage. Catalogues were issued every few years until the middle of the century, after which they became an annual publication, reflecting the great strides Beaufort had made in improving the quality and comprehensiveness of chart coverage.

MAPS IN MODERN SCIENTIFIC INQUIRY: A NEW GENRE OF THEMATIC MAPS

In one sense, all maps are thematic, since no single map can present and communicate all the potentially mappable information for the area that it covers. All maps are selective, but the degree of specificity varies from, say, topographical maps at one end of the spectrum (representing as broad a range of visible features, boundaries and names as can be mapped at a particular scale), to those highly specific maps which portray particular subjects or phenomena, such as geology or population density. It is the latter category of maps which cartographers refer to by the term 'thematic map'. Thematic maps focus on the differences from place to place of one class of feature, that class being the subject (hence theme, thematic) of that map. Heuristic motives explain why many thematic maps are compiled; they can serve as instruments to reveal the spatial structure of a set of phenomena so that the 'geography' of one mapped distribution may be correlated with that of other variables in a search for causal relationships. In his seminal work on the history of thematic maps, Arthur Robinson contends that although the roots of thematic cartography can be traced to the last half of the seventeenth century, the critical period of most rapid growth and maturation occurred between approximately 1800 and 1860. In his view, 'as an event in the history of cartography this period ranks with that which occurred in the fifteenth century when Ptolemaic ideas were resurrected'.[29]

Most data sets used by nineteenth-century thematic mappers consisted of summary statistics related to areas such as census enumeration districts, parishes or counties. The problem of mapping statistics, that is numbers, was new. Pioneer thematic mappers were faced with a conceptual problem, still as real today as in the last century: that, other things being equal, anything counted in large areas will total more than the same object counted in small areas. Since the primary objective of thematic mapping is the examination and comparison of spatial differences, it is first necessary to remove the effect of variation among the sizes of areas within which data are collected. Mapping relative values rather than absolute numbers was, and still is, a preferred solution. An example is mapping the density of population in numbers of persons per square kilometre.[30]

The temporal context of the development of the thematic map and the concurrent emergence of the modern discipline of geography are products of the social, economic and scientific changes which occurred during the industrial revolution and of the new problems which society faced as a result of those changes, most notably the health problems that accompanied mass urbanisation as industrial activity became concentrated in factories located in towns. Epidemics of disease were commonplace, and mapping individual outbreaks helped to discover causes and established a role for thematic maps which in turn helped develop an applied geography.[31] The continuing processes of nineteenth-century industrial change and the problems they bequeathed also encouraged the collection of comprehensive statistics by central government; for example, the first of the decennial censuses of population in England and Wales was undertaken in 1801. These data sets provided increasing bodies of data for thematic mapping.

The use of thematic maps as heuristic devices in the early nineteenth century characterised what is known as the moral statistics movement, a name coined by André-Michel Guerry, a French social reformer in the title of his *Essai sur la statistique morale de la France*.[32] This book focussed on the collection and analysis of data on what might be termed the downside of the industrial revolution: problems such as crime, delinquency, education, pauperism, disease and mortality.

There had been, however, forerunners of thematic maps in England more than two centuries earlier. Günter Schilder has

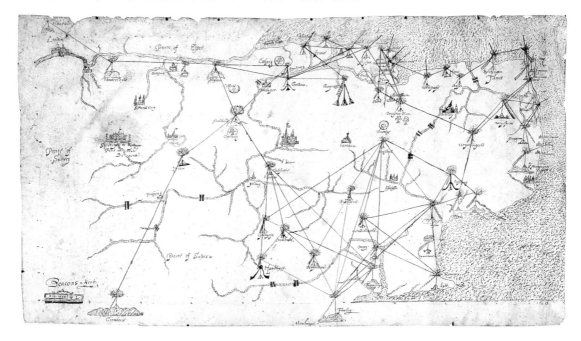

drawn attention to a late sixteenth-century map made by Jodocus Hondius, possibly while he was working as an engraver in London. Hondius's map portrays the number of parishes in the counties of England and Wales, 1589–90, and was probably intended as a book illustration.[33] In 1595, William Lambarde included a printed map based on one by Philip Symonson (now lost), of the network of beacons in Kent in the second edition of his *Perambulation of Kent* (Fig. 7.11). He explained that he wanted the inhabitants of Kent to appreciate the significance of the beacons and to know where they were in case of need.[34] Later on, in the seventeenth century, Edmond Halley, Astronomer Royal in England and celebrated for his observations of comets, was an important early innovator in scientific mapping. His 1686 map of trade winds and monsoons reflects his understanding that the earth's atmospherics could be 'better expressed in the Mapp hereto annexed, than it can well be in words' (Fig. 7.12).[35] Halley's words neatly vindicate the value of cartographical methods in the communication of scientific data, a method which he himself was to use again in his charts of magnetic variation in the north and south Atlantic (1700) and of tides in the English Channel (1700–1).[36] The fascination that Halley's mapping of trade winds held for his contemporaries is reflected in the way that these features were included in maps of the world destined for the popular market.[37]

Today, an almost infinite variety of thematic maps is possible. Arthur Robinson offers the following three-fold classification, based on hierarchical levels of thematic structure.[38]

7.11   Manuscript draft of William Lambarde's copy of Philip Symonson's map of the beacons of Kent, which Lambarde said was the best map of Kent he had ever seen and which he published, greatly reduced, in the 1596 edition of his *Perambulation of Kent*. 13.5 × 22 cm. British Library, Add. MS 62935.
Reproduced by permission of the British Library Board.

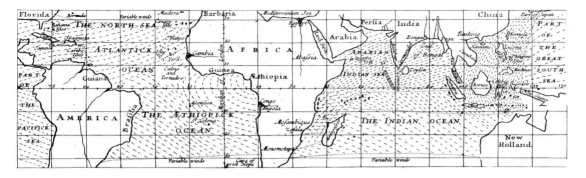

7.12 Edmond Halley, map of trade winds and monsoons, 1686. In Halley's opinion, the earth's atmospherics could be 'better expressed in the Mapp hereto annexed, than it can be in words'. 8 × 25 cm; printed in *Philosophical Transactions*, Vol. 2, 1686.
Reproduced by permission of the Norman Lockyer Observatory, Sidmouth.

Maps of the physical world: the atmosphere, the oceans, geomagnetism, geology, the land surface, vegetation and animal life;

Maps of people and their activities: population, characteristics of peoples, economic activities, movements of goods and peoples;

Maps of the social environment: moral statistics, medical maps, living conditions.

His typology is useful in that it helps focus on the particular subjects of special-purpose mapping, but it is immediately evident (as noted in the introduction to this section) that most of the maps discussed in this book could also be considered thematic! Land cadastres, road and route maps, and street atlases are three such thematic genres already discussed. While there is an undoubted continuum of mapping from the general to the specific, thematic maps do constitute a distinctive genre. Examples of geological mapping, land-use mapping and social, demographic and medical mapping are introduced in this chapter to reveal their common characteristics.

Geological maps of England and Wales

In his study of early geological mapping, Martin Rudwick contends that the development of the subject itself in the early nineteenth century on the one hand, and the proliferation of maps, geological sections and diagrams during the same decades on the other, were not simply coincidental but were related directly to the intellectual goals of the nascent discipline. Associated with the new concepts and methods was a comparable emergence of what Rudwick calls a 'visual language for the science'.[39] He argues that while the 'traveller-naturalists' or proto-geologists of the late eighteenth century 'had a well developed visual awareness of the topographical phenomena that they studied, they generally communicated what they had seen primarily in written descriptions and only subordinately, if at all, in visual terms'.[40]

The making of geological maps had been discussed at the Royal Society as early as the late seventeenth century by John Aubrey, the

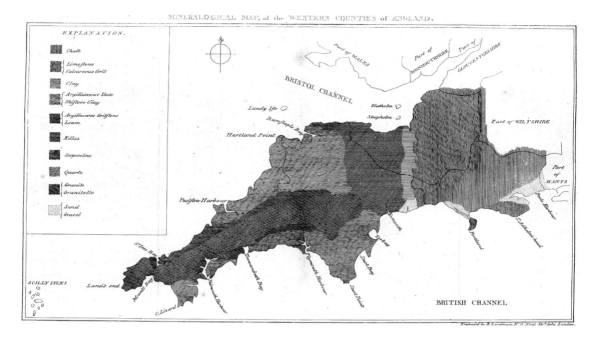

antiquarian, and Martin Lister, a physician. Aubrey wrote about geological mapping in his unpublished *The Natural History of Wiltshire* (1685): 'I have oftentimes wished for a mappe of England coloured according to the colour of the earth with marks of fossils and minerals'.[41] However, the first map known to have been compiled with some geological content is an eighteenth-century map of a part of south-east England.[42] This was Christopher Packe's *New Philosophico-Chorographical Chart of East Kent*, 1743, which describes the valley and drainage pattern and also distinguishes, albeit in general terms, four geologically based regions: Upper East Kent, the Vale of Ashford, the Weald of Kent, and Romney and Walland Marshes.[43] In 1797 William Maton published his *Mineralogical Map of the Western Counties of England*, in which he delimited the outcrops of ten groups of strata and which is claimed to be the first attempt in England to construct a true geological map (Fig. 7.13).[44]

The first geological map of the whole of England and Wales is William Smith's classic of geological cartography, *Delineation of the Strata of England and Wales with part of Scotland*, published by John Cary in 1815 on a specially prepared base map (Plate 22). William Smith's map consists of fifteen sheets at a scale of five miles to one inch (1:316,800) and an index sheet. Smith had begun work in 1805 and employed a complicated and expensive system of hand colouring, using, for example, different intensities of tone of the same colour to distinguish scarp and dip slopes on the same geological outcrop. This contrasted with the methods of the French geological map-makers Georges Cuvier and Alexandre Brongniart

7.13   William Maton, *Mineralogical Map of the Western Counties of England*, 1797. Maton maps the outcrops of ten groups of rock strata on what is claimed to be the first attempt in England to construct a true geological map. The map covers an area of about 200 × 360 km (124 × 223 miles). Reproduced by permission of the Devon and Exeter Institution.

for their *Carte géognostique*, published in their *Essai sur la géographie minéralogique des environs de Paris* (1810–11). Their technique of using plain colours to indicate outcrops and with key boxes to these colours arranged in the true temporal order of the strata was quickly adopted as the international method of geological mapping.[45] Although Smith's variable colour tone technique of geological representation was not destined to become the standard method for representing geological outcrops on maps, he did receive the premium of 50 guineas for a geological map of the nation offered since 1802 by the Society of Arts and, in 1831, eight years before his death, he became the first recipient of the Geological Society of London's Wollaston Medal. Rapid progress in geological mapping was made after 1815, and Smith's map was superseded in 1820 by George Bellas Greenough's more detailed map. The second edition of Greenough's map, published in 1839, 'illustrates the high level of sophistication to which geological cartography had risen in Europe'.[46]

Official geological mapping in England and Wales developed in the 1830s from early experiments in the counties of Devon and Cornwall by Sir Henry De la Beche, founder of the British Geological Survey. In 1832 De la Beche made an offer to the Board of Ordnance to apply geological colour to the eight published one-inch maps of Devon in return for a payment of £300.[47] He completed this work in two years, and his maps were well received by the geological establishment which pressed for an extension of similar detailed geological mapping elsewhere. It was argued that this would serve many practical purposes: the search for minerals, road and building stones, marls and calcareous rocks for agricultural improvement, and for planning water-supply works and canal and railway construction.

The official foundation date for the Geological Survey is 1835, when De la Beche was authorised to continue his survey of Devon first into Cornwall and then into South Wales. In 1845, the Geological Survey was separated from the Ordnance Survey, and in 1890 it completed publication of the geological mapping of England and Wales at the one-inch scale, with some coalfields also published at six inches to one mile (Plate 23).[48] The geological mapping of Scotland at one inch to one mile remains unfinished today.[49] As official geological mapping proceeded only slowly through the nineteenth century, there was plenty of opportunity for Victorian private enterprise to contribute to mapped geological knowledge. For example, the first detailed map of the geology of the London area came not from the Geological Survey but was compiled by the engineer Robert Mylne in 1851: his *Topographical Map of London and Its Environs*.[50] Similar information was also being compiled by the mid nineteenth century, together with other physical data, by men such as Alexander Keith Johnston for his *Physical Atlas of Natural Phenomena* (1848).[51]

At the end of the nineteenth century, the Wharton Committee in its review of the work of the Geological Survey, looked back to what had been accomplished in the six decades of official mapping.

The conclusion was that 'with regard to the practical uses of the Survey ... there is no doubt that, apart from the scientific and educational aspects, it has been of great practical service to the country. It has been shown to us that great benefit has been found to be obtainable from the results of the Survey in the matter of mining, agriculture, water-supply and sanitation'.[52] In this area of cartographical endeavour, aims and achievements are as one; thematic geological mapping had proved its worth.

Maps of agricultural land use

The mapping of agricultural and industrial activity was stimulated by the need to monitor the rapid economic change experienced by most European countries during industrialisation. The changing contributions to gross national products of agriculture and industry caused widespread government concern for the food supply of towns. It was thus of the first importance that governments obtained information about agricultural land use. In England, the first detailed, national agricultural inquiry was made parish-by-parish in 1801 during the emergency of the Napoleonic Wars. Further experiments were undertaken in some counties in 1854, and from 1866 the government required all occupiers of units of land of more than one acre in extent to return statistics on crops and animals in June each year. However, none of these surveys was put on to a map. Yet, at a local level, information on land use had been mapped for centuries by estate surveyors. Land use of fields on estate maps was portrayed by letter codes or by colour, different shades indicating arable, sown grass on leys, permanent pasture, and woodland, for instance. (Fig. 7.14). The depiction of land use was not, however, the prime purpose of an estate map, but just one of the categories of information relating to private property carried on some maps.

The first English example of a special-purpose land-use map of a large area of country was the map made in 1800 by Thomas Milne of some 673 km^2 around London at a scale of two inches to one mile (1:31,815) (Plate 24).[53] Milne was an estate surveyor and county map-maker. For his land-use survey and for his published land-use map he made use of the Ordnance Survey's one inch to one mile maps. Milne's land-use classification was far more elaborate than anything previously attempted. While John Rocque had indicated (or purported to indicate) in general terms different categories of land use on his large-scale county maps, Rocque's objective was the compilation of a general topographical map, whereas Milne's prime focus was the thematic portrayal of land use. His map highlights clearly the metropolitan influence on farming. It shows the extent of market gardens producing vegetables and fruit for the London market, the parks and gardens, and the meadows for grazing cows for London's milk supply. No land-use map of comparable extent followed Milne's map of London until the inter-war years of the twentieth century, when the First Land Utilisation Survey of Great Britain was conducted under the direc-

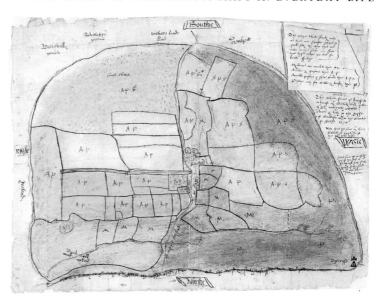

7.14 Land use on an estate map: Kersford in Bridestowe parish, Devon, 1574. A legend to the map explains the way that land use is represented by letters placed in the fields: 'Meadows are marked thus m / Arrable and pasture thus Ap / Arrable and pasture & shepe gronde thus Ap S / Pasture only for cattle and beasts thus p'. No scale; 34 × 42 cm. Devon Record office, 189 Madd 3/E4. Reproduced by permission of Devon Record Office.

tion of the geographer L. Dudley Stamp.[54] This was an official survey of the whole country made for government to aid land-use planning as the threat of war again came to Europe (Plate 25).

Field work for the First Land Utilisation Survey of Great Britain was conducted by school pupils and their teachers. Seven categories of land-use were plotted, field by field, on to Ordnance Survey maps at a scale of 1:10,560 for subsequent reduction and publication at one-inch to one mile (1:63,360). On New Year's Day 1933 the first one-inch maps were published, by 1935 the bulk of the field work had been accomplished, and in 1946 the last of the county reports accompanying the 170 map sheets of the Survey was published. The results of the field survey constitute a record of rural land-use prior to the great changes induced by the ploughing up of grassland during the Second World War. Although not by any means a precisely synchronous cross-section, the maps of the First Land Utilisation Survey approached this ideal more closely than others until remote sensing digital data became available from the LANDSAT satellite thematic mapper.[55] The Land Utilisation Survey pictures the results of the unfettered, inter-war urban, and especially suburban, sprawl around London, Birmingham, Manchester and other large towns and cities and 'unproductive' land in the older, run-down industrial areas of the country. After the Second World War, the information recorded in the land-use maps was employed in a number of ways by the nascent town and country planning movement: to regulate land-use in general and, more specifically, to promote activities such as the establishment of green belts around the major conurbations, to locate new towns, and to establish priorities in the reclamation of derelict land.[56]

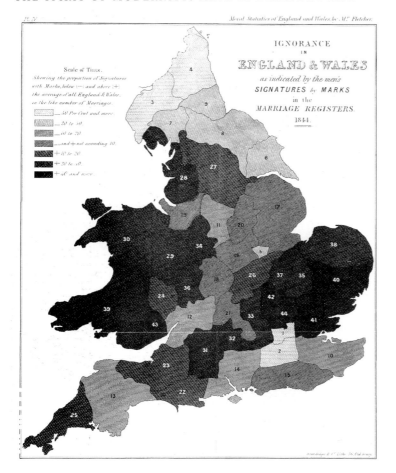

7.15 Joseph Fletcher's map of 'ignorance', 1844 (published 1849). To obtain data to measure literacy, Fletcher sampled marriage registers and counted the number of people who signed their name and those who made marks. 24 × 19 cm. *Journal of the Statistical Society of London*, 12 (1849), Plate 1.
Reproduced by permission of University of Exeter Library.

Social, demographic and medical mapping

The first mapping in France in what can be termed the field of moral statistics was actually earlier than Guerry's celebrated *Essai sur la statistique morale de la France*, for in 1827 Charles Dupin had already published a map of educational provision in France and in 1829 had collaborated on maps comparing the incidence of crimes against persons and property with educational provision.[57] In all fields of statistics and especially in the collection of social data, post-Revolution France was ahead of England.[58] Two Englishmen had also mapped social data in Ireland before any comparable maps appeared in England. Henry Harness mapped Irish population density in 1837 and Thomas Larcom's census maps of Irish literacy rates and other social statistics were published in 1843.[59] In England, anybody needing data on the provision of education, for example, had to obtain them from a variety of surrogate measures. Joseph Fletcher used the record of marks made by illiterate persons to 'sign' the marriage register for

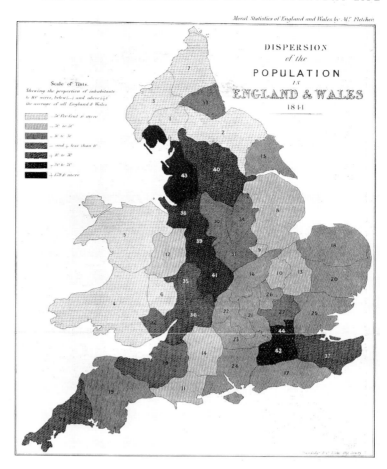

7.16 Joseph Fletcher's map of population density in England and Wales from the 1841 census, published in 1849. 24 × 19 cm. *Journal of the Statistical Society of London*, 12 (1849), Plate 4.
Reproduced by permission of University of Exeter Library.

his map of 'ignorance', published in 1849 (Fig. 7.15).[60] These marks are now widely used as the basis for historical study of literacy, even if their validity as indicators is disputed.

From the first count, taken in 1801, the population censuses of England and Wales gradually provided an increasing and eventually massive data-base for the investigation of population characteristics and distribution. Figure 7.16 shows part of an early population map of England and Wales compiled by Joseph Fletcher from the 1841 census and published in 1849.[61] The German geographer, August Petermann, also mapped English population data and in 1852 he was appointed 'Physical Geographer to the Queen', a post which underscored the English government's recognition of the value of a geographical and, especially, thematic map approach to the analysis of government statistics.[62]

Thematic maps made significant contributions to understanding patterns of disease in the rapidly-growing industrial cities where people lived at high densities.[63] Some of the earliest maps concern the distribution of the incidence of cholera, a disease which became

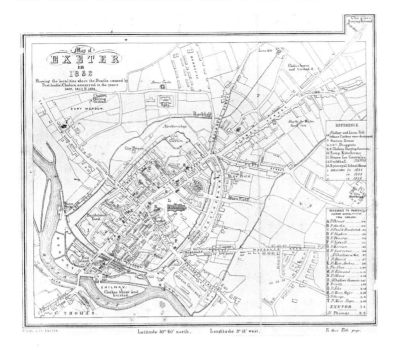

7.17 Thomas Shapter, *Map of Exeter in 1832 Shewing the localities where Deaths caused by Pestilential Cholera occurred in the years 1832, 1833 & 1834.* Shapter's map is one of the earliest cartographical analyses of an infectious disease in England. 20 × 21cm; the map covers area of about 2.2 × 1.8 km (1.3 × 1.1 miles). Reproduced by permission of University of Exeter Library.

pandemic in India in 1817 and which by 1832 had spread over all of Eurasia and North America. The early assumption by proto-medical geographers was that cholera had a physical cause and that somehow (although they knew not why) geographical variation in incidence was a product of the cause. Among the earliest distributional analyses is a map of the 1832–4 epidemics in Exeter compiled by the physician Thomas Shapter (Fig. 7.17).[64] Probably the most celebrated medical map, however, is that prepared in 1855 by Dr John Snow as part of his enquiry into the spread of cholera in London (Fig. 7.18). Snow plotted the location of each house in which death from cholera had occurred. Despite his ignorance of the cause of cholera, the pattern of deaths on the maps convinced him that water from a particular well was the local source of the illness. Snow persuaded the authorities to close the public water pump in Broad Street, and new cases of cholera immediately declined.[65] Later in the same year, Snow went on to use thematic cartography to demonstrate that it really was polluted water that was the carrier of cholera. Prior to 1852, the Southwark and Lambeth water-supply companies had both provided drinking water to south Londoners by drawing supplies from the River Thames. In 1852, after the Lambeth company changed to a source free of sewage pollution, there was a massive contrast between the rate of cholera deaths in the area with polluted water (which averaged 71 per thousand) and in the area supplied with the cleaner water (5 per thousand).[66]

Living conditions, another potential source of health problems, were also surveyed and mapped in the nineteenth century. A pioneer map in this field was W. R. Wilde's *Sanitary Map of Dublin* (1841) on which streets are colour coded from high-class residence (blue) through to slum streets (red and brown).[67] In England, Edwin Chadwick, Secretary of the English Poor Law Commissioners, a champion of social reform and architect of the first Public Health Act in England and Wales, recognised the merit of maps in helping to identify the areas of squalor to which, in his view, the 'labouring classes' were consigned by the capitalist system. One such map was Robert Baker's *Sanitary Map of the Town of Leeds*, compiled in 1833.[68] Another socialist, Henry Mayhew, was also at the forefront of the mid nineteenth-century social-reform movement in England. Mayhew's influential study, *London Labour and the London Poor*, affected both public and government opinion and is a milestone in the study of the disadvantaged elements in the population. Mayhew and/or his collaborators compiled a number of maps for the 1862 edition of his book as a way of helping set London in its national context.[69]

The successor to Chadwick and Mayhew at the end of the nineteenth century was Charles Booth, whose work, *Life and Labour of the People in London*, had by 1902–3 run to seventeen volumes and includes his *Descriptive Map of London Poverty*. Booth's objective was to make a practical contribution to understanding London's social problems by making reliable information available.[70] In David Reeder's words, Booth wanted to 'lift the curtain' and to provide a more informed basis for discussion of contemporary problems than that engendered by sensationalised 'people of the abyss' depictions.[71] Booth obtained his data from School Board visitors who had to carry out house-by-house surveys in their districts to locate and enrol children for schooling. On the basis of their information, Booth and his assistants classified streets on a seven-degree scale ranging from poverty to comfort in order to reveal the greatly contrasting conditions of life in London. His *Poverty Map* emerged as an eloquent expression of the geographical components of the class system imposed by nineteenth-century *laissez-faire* capitalism, a point made more forceful by the choice of colours on the map (Plate 26). The worst streets are either black (the 'lowest grade occupied by occasional labourers and semi-criminals – the elements of disorder') or dark blue (the streets of the 'very poor occupied by casual labourers and others living from hand to mouth').[72] For London as a whole, Booth found that about 30 per cent of the population was living in poverty and 70 per cent in comfort, a serious situation but not as bad as had been suggested by some of the more impressionistic comments on the capital's social geography. His map confirmed the already well-known distinction between east and west London, but it also revealed a much more heterogeneous London than suspected, one made up of 'distinct and separated residential areas, segregated district by district so as to produce striking variations in the social conditions of different parts of the metropolis'.[73] Booth's work was social mapping unprecedented in the detail of inquiry and extent of area surveyed.

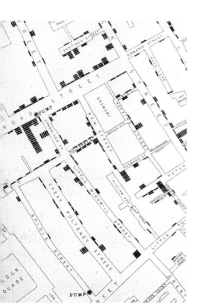

7.18 John Snow's map of deaths from cholera in the Broad Street area of London, published 1855. Probably the most celebrated medical map of all time, Snow's cartographical analysis convinced him that the Broad Street drinking water pump was the source of the disease, a deduction confirmed by the immediate decline in new cases after the pump was closed. Published in *On the Mode of Communication of Cholera* (London, 1855), pp. 44–45. Whole map 39 × 46 cm, portion shown represents about 270 × 200 m on the ground (295 × 218 yards).
Reproduced by permission of the British Library Board.

Maps of the social environment continued, and continue, to be a regular part of socio-demographic analyses. They are offered as supporting evidence in written reports and stand alone in social atlases.[74] Their origin reflected the wave of reform which swept across western Europe and, especially, France during the nineteenth century. They were also important in the development of human geography, for they helped establish the idea that the spatial characteristics and spatial behaviour of human beings ought to be as predictable and as subject to laws as were the physical aspects of the environment, which were being studied as physical geography. With hindsight, we can now see that this concern for physical explanations led human geography into the *cul-de-sac* of environmental determinism, in which explanations for spatial variations were sought exclusively by correlating human attributes and activities with environmental characteristics. However, a more positive outcome was that this work firmly established a two-fold role for thematic maps: first, for the communication of spatial variations and as a method of enabling analysis of regional variation, and second, in the search for explanation through the correlation of different patterns. The thematic map has proved itself to be a powerful heuristic device.

The Lottery of Map Survival

The maps discussed in this book are not all the maps there once were. For all their variety, they are only the survivors. As at the end of the twentieth century, when so much ephemera contains maps, most of which will never be seen again once their moment on television has passed and once the newspapers which contained them are pulped, so in the past: the gap between the number of maps created and the numbers which survive has always been considerable. We may catch, or think we catch, fleeting glimpses of the essential continuity of maps in English culture but we find it difficult sometimes to demonstrate the connections because so much of the contextual material has been lost. In this brief closing chapter we have two aims. One is to underline the haphazard nature of map survival, with its implications for a partial history, and the other is to point to the relationships between map history and book history as one way of gaining some idea of what might be missing from map history.

WHAT HAVE WE LOST?

Maps have a low survival rate for various reasons.[1] They are fragile in every format, even in books. They tend to be disposed of once superseded by new information. Collected together in one place such as a library, they are particularly vulnerable to destruction through fire and flood, revolution and political upheaval. English libraries have indeed suffered from periods of wholesale destruction as a result of the collapse of law and order, first after the departure of the Roman authorities in 410, then following the Danish raids of the ninth century, and then again in the first half of the sixteenth century.[2] It is difficult to imagine what may have been lost in this last period alone when, first, the great monastic collections were raided, neglected or misused under Henry VIII's policy of dissolution in the late 1530s and when, second, the remaining secular libraries were cleansed as a direct result of Edward VI's purge of all that was 'tainted' by the old Catholic faith in the 1550s. At Oxford, the university library, and in London the City's Guildhall library (from which some 900 volumes were carried away in carts) had to be closed.[3] In Cambridge, the 500 to 600 books that had been in the university library in 1530 had been reduced by 1574 to 174 titles (of which 120 related to manuscripts).[4] In the school library at Eton, 60 to 70 manuscript books survived out of some 500.[5] Particularly susceptible, it can be assumed, were such overtly Catholic maps as the large *mappae-mundi* with their highly visible 'idolatrous' religious imagery. Pamphlets, news-sheets and church sheet music, already naturally fragile, were liable to be found offensive and destroyed to the extent that virtually nothing of these whole classes of material has survived.[6] Even if not expressly targeted in all such destruction, what chance had a map? What have we lost from the centuries up to 1600?

By no means all early maps are today found in institutional or large private collections, but for those that are, we should remember that modern map collections reflect a haphazard pattern of document survival, not a steady accretionary process. When a comprehensive history of map collections and map collecting comes to be written it will offer much more than a mere chronicle of acquisitions. It will start with some elementary questions such as: where were the main collections of maps in each period in the past and what other, related, material was also present? Who had access to the collections? Which maps had been there for some time and which were recently acquired? Which were used and which ignored? How were any of the maps we see today judged in their time (we should not assume that only the best work has survived)? Patterns may well emerge, both chronological and geographical. We should begin to see networks of personal connections and be able to identify those centres with long cartographical traditions. If maps were absent from this or that library, we need to know why. For instance, what was the attitude to maps of that great fifteenth-century book collector, Humfrey, Duke of Gloucester, the benefactor of Oxford university's library? Why do we find no maps amongst his books, or even amongst the early donations to Thomas Bodley's reconstructed university library at the start of the seventeenth century? Is it really a case of not a single map reaching the library amongst the bequeathed books? Or is it more to do with contemporary treatment of maps or a result of the unobliging intractability of a sheet map in a library? Unlike maps, books – together with maps bound into atlases – range themselves obediently in harmonious order along the shelves. Or is it simply to do with the way the records of accessions were kept? Bodley's first printed catalogue (1605), which relates to over 1,200 editions in 6,000 volumes (530 in manuscript), includes several of the 'modern' editions of Ptolemy (Ruscelli, 1574; Mercator, 1584; Magini, 1597), Ortelius's *Theatrum orbis terrarum* (1570), and many books which we know contain, or should have contained, maps.[7] But in the entire catalogue, only one sheet map is mentioned, a *mappa totius Angliae*, kept perhaps tellingly in one of the cupboards reserved for especially valuable (or just inconveniently-shaped?) items. It is difficult to know what to read into such lacunae when we know that a wide range of people both within and outside the universities possessed maps. The atlases left in 1589 by Andrew Perne to Peterhouse, Cambridge, are still identifiable by his bookmark, but there is no sign of the many sheet maps from his collection which were also bequeathed to the college library, nor of those which graced the walls of Perne's house at Ely and which Perne left to the Dean of Ely.[8]

Part of the elusiveness of map history lies in the ambivalence of the map. We see this clearly from early domestic inventories. Consider how Sir William More treated the maps he left at Loseley Hall, Guildford, on his death in 1556.[9] His most valuable map, a map of the world, was in his personal closet. All the others – a 'lyttle map of the world' and various maps of France, England, and Scotland – were in his library, where there was also a globe, a copy of Sebastian Münster's Cosmography, Ranulf Higden's *Polychronicon*, and 'a boke of Tholomye'. More, it would seem, either appreciated the commercial value of his map of the world and for this reason kept it in the safest place in the house, or he simply liked to have it handy to view at leisure or in the privacy of his own room. A different attitude to maps is found in the (unidentified) house 'Cockesden' where, in 1610, a map of Flanders hung in the hall along with several pictures, while six maps of various countries and 'the great mappe of London' were to be found in the 'Studdy Room'.[10] At Cockesden, maps had a dual function, as decorative domestic furniture as well as useful works of reference and sources of information. A systematic investigation might well suggest the Cockesden model was probably the commonest in early modern England. Maps used as domestic decoration are even more liable to be discarded once discoloured or torn, or when the family moves, or on the death of the owner, than maps kept in a library.

At the other end of the social scale, cheap maps – like cheap books and prints – were likewise destined for a short life. The 'small godly books' and 'small merry books' of the sixteenth- and seventeenth-century chapbook (that is, *cheap*-book) publisher were scarcely more robust, especially given the wear and tear of domestic use in the humblest of homes, than the other forms of ephemera – pamphlets and political tracts, news-sheets, ballads, sentimental or moral tales, and almanacs – which constituted the licensed pedlar and itinerant hawker's standard stock from the sixteenth century onwards.[11] This door-to-door and market trade was countrywide and rural as well as urban. By 1685 it had its own road-book, the *City and Country Chapman's Almanac*, which contains a month-by-month list of the fairs in England and Wales, county-by-county lists of market towns indicating the day of the week on which the market was held, and a travel aid in the form of a list of the stage towns on the roads out of London with the distances between them.[12] Almanacs – small, slim booklets – had included such information since the sixteenth-century.[13] By the mid seventeenth century a few, like John White's, also contained a map of the counties of England and Wales, number-keyed to a table detailing not only the distance and the compass direction in which the county or shire lies from London but also the total number of cities, market towns, castles, parishes, rivers, parks, and bridges that each county contained.[14] Broadsheets produced for general household use were regularly illustrated, many with

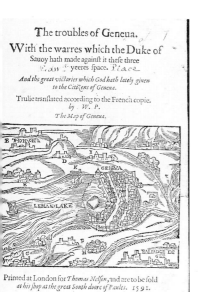

8.1 Maps in early newsletters and pamphlets rarely survive, yet they were much used in the reporting of current affairs. This woodcut map of Geneva was printed both on the title page of *The Troubles of Geneva*, a 16-page pamphlet, and in the text. English readers would have sympathised with Protestant Geneva, home of John Calvin's edition of the Bible, rather than the Catholic Duke of Savoy. 9 × 11.5 cm. 1591. British Library, C.55.d.8 (1).
Reproduced by permission of the British Library Board.

instructive diagrams, some even using the map analogy as a guide to improving reflection and prayer.[15] Such sheets, together with pictures printed on paper or painted on cloth, were not only read avidly but often pinned about the house, especially above the bedheads, in humble homes.[16] Most illustrated biblical themes but there were also royal portraits and genealogies.

The place of maps in the motley of publications which filled the chapman's baskets has not been studied, and the identity of individual maps is likely to be difficult to establish. There are leads, though. One may be through the print trade where printers like William and Charles Dicey (from the 1720s onwards) were better known for their output of cheap popular books and ephemera than for their maps.[17] Wills and probate inventories also reveal not only the value of the maps but the social standing of their owners, as well as the place in the home given to the maps. Approaching the problem through existing bibliographies has already shown how the Geneva bible edition (the most widely distributed English-language bible from 1560 onwards) contained maps, how maps were used to underscore the point in political tracts, and how maps were essential adjuncts in the dissemination of news. Pamphlets and single or folded news-sheets covered a range of topics of current concern, from the military situation in Ireland and wars on the continent to the 'sinking of the earth' which took place between the 18 and 29 December 1596 at Westerham, Kent.[18] Maps helped English readers follow events in the Spanish War in the Low Countries, as did: *A mapp and true situacion of the toune of Grave* (*c.*1586), or *The scituation of the lands of Cleve and Munster, where the Spanish forces now are* (1599) and *A briefe relation of what is hapned since the last of August 1598. by* [the] *comming of the Spanish campe into the dukedom of Cleve* (1559).[19] Rumours about 'troubles' in Geneva were elucidated in a sixteen-page pamphlet of 1591 in which the woodcut map of the city featured

8.2 In one form or another, maps of the world have always been popular as wall decoration. By the late seventeenth century, cheap double-hemispheric maps like Robert Marshall's were being produced – to judge from the wear on the copper plates – in huge numbers. 66 × 98 cm. 1785. British Library, Maps M.T.11.g.1 (9).
Reproduced by permission of the British Library Board.

twice, once on the title-page (Fig. 8.1) and again in the text. Given the physical fragility of such pamphlets and news-sheets and the fleeting relevance of their contents, the fact that any have survived at all from four centuries ago is a reflection of their former quantity and widespread circulation. Possibly more surprising, though, is the implication that, after maps in bibles, maps in the chapman's wares may have accounted for the largest single category of maps in sixteenth-century English printed material.[20]

Rather less can be said about maps printed on separate sheets expressly for the cheap end of the market. Early inventories often give sufficient detail for the maps in the deceased's possessions to be identified, but later inventory-takers tended to note the existence of a map without describing it.[21] We are aware of only one systematic attempt to recognise cheap maps, the British Library's collection of eighteenth-century double-hemispheric maps of the world dating from the 1680s to about 1800.[22] Aimed at an educationally modest and financially constrained market, and printed on two sheets to make wall maps of roughly the same size (on average, some 145 × 60 cms), the maps carry simple, boldly printed titles along the lines of 'A New Map of the World' or 'A New and Correct Map of the World'.[23] They tend to be informative rather than just decorative, with the space between the two hemispheres and the rectangular margins packed with circular insets. Henry Overton's map (1738) has 24 such insets, depicting subjects such as 'The Face of the Moon', the northern and the southern polar hemispheres, and planets, notably Saturn and Jupiter.[24] There is also an inset map of the counties of England and Wales, pictured on a drape. Additional information was more or less topical, like the tracks of William Dampier's voyages on Hermann Moll's map of 1709, and those of James Cook's three voyages on Richard Marshall's map of 1785 (Fig. 8.2).[25] Halley's winds are also marked on most of these maps. In other respects, the maps were grossly out of date; Overton's map shows a very late rendition of California as an island.[26] Coarse in their cartography and plain in their appearance (very few were coloured), the key characteristic of these maps is that they were printed from obviously well-worn – that is, heavily-used – plates. Marshall's two copper plates, first engraved in 1715, were so worn by 1785 that the rivet marks show up clearly on the imprint (Fig. 8.3).[27] We will probably never know how many thousands of such maps were printed. What we do know is that only a very few survive – a telling indicator of the loss of maps over time.

Other factors contributed to the loss of maps. Maps are poor travellers, yet, like books, they travelled widely across Europe at all times. Books were used as a form of currency (as in the case of the book used to purchase land in the seventh century noted in Chapter 2 above) and maps were used in diplomacy (like the Venetian's gift of Battista Agnese's atlas to Henry VIII, and Jean Rotz's presentation of his atlas of sea charts to the same king).[28] They were brought home as prized acquisitions from duty tours abroad, like the *mappamundi* taken in 1219 to Vercelli by a papal

8.3   Detail from the British Library's copy of Robert Marshall's map of the world (1785) showing how over-use of the copper plates has worn them down to the rivets which held them together. Maps M.T.11.g.1 (9).
Reproduced by permission of the British Library Board.

legate returning home from England.[29] The foundation of a new monastic house called for gifts of books from the parent monastery.[30] Manuscripts travelled to be copied or for personal study. Many of the maps ending up in Sir Robert Cotton's collection had once belonged to William Cecil, or to his son, or had been copied from originals in the Cecils' possession.[31] Some plundered books are still in the countries to which they were transported and some maps must have suffered a similar fate. An English copy of the Gospels, taken in a Viking or Danish raid, is now in Stockholm, while, in July 1278 when Archbishop Robert Kilwardby left Canterbury for Rome to claim his cardinal's hat, not only his own books went with him but also many taken from the cathedral library.[32] Whole libraries were bought and sold, and, once again, we can only speculate that there were at least a few maps in some of them. Charles V of France inherited, in the late thirteenth century, 'quite a few books' from his father (Louis IX), to which he added substantially, leaving, in his turn, a royal library of some 900 manuscripts which were among those eventually purchased, in 1424, by the Duke of Bedford and shipped to England.[33] Exile occasioned the displacement of cartographical books as well as others. When the Catholic exile John Ramidge was murdered on the road between Louvain and Mechelen in 1568, the library he had brought with him from England to Louvain had to be auctioned in Louvain. It included some 450 books, amongst which was a work by Ptolemy, valued at 2 shillings in the sales list.[34]

MAP HISTORY AND BOOK HISTORY

Early map history can be viewed as inextricably related to book history. As so much more tends to be known about the history of books, especially for the medieval and early modern periods, the light that book history can shed on map history is potentially invaluable. For a start, book history offers the map historian easy access to medieval and later book lists, already transcribed and edited, which can be combed for whatever map-related hints may be discovered from them.[35] Most of the medieval manuscript maps we know about were in one way or another associated with books, and it can be anticipated that future additions to the census of early maps (printed and manuscript) will also come from books as new generations of map historians continue the search in contexts as yet not fully explored.[36]

Maps are physically related to books in a number of ways. A separate sheet map may be accompanied by a booklet explaining the use of the maps or containing a gazetteer of the places shown on the map. And atlases are books – more than mere aggregates of estate, county, regional or world maps between two covers – and entities in their own right, where the sum is much greater than that of the individual parts. While the place of an atlas in a library is usually on shelves alongside other books, the history of the atlas is as much a part of the wider history of ideas as it is of the narrower history of each map. Books, too, have sometimes sheltered maps, not because the map has anything to do with the text but simply because the solid covers of a bound volume provided protection. And, of course, books contain maps as illustrations, sometimes closely integrated into the text, sometimes seemingly employed without regard to the text, having been selected as a vaguely appropriate visual enhancement.

Maps also relate to books conceptually. In the first place, many maps could not have been drawn without access to information contained in books. The extraordinary variety of material depicted on the medieval *mappaemundi* was drawn from a wide range of ancient authorities whose texts must have been available to the map-makers from the monastic library. Matthew Paris was clearly reading a copy of Geoffrey of Monmouth's History of the Kings of Britain when he penned the sketch of the ancient roads exactly as described by Geoffrey (see Fig. 2.8, p. 17). His sketch of the Anglo-Saxon heptarchy likewise presupposes familiarity with at least some of the writings on that period and we have already commented on the sources used in the making of his maps of Britain and of others like the fourteenth-century Gough map.[37] In the second place, many maps are drawn to replace words, or as a supplement to words. As a means of clarifying or reiterating a point, the phrase 'as this figure shows' is a vital, albeit often missing guide (especially in early books, printed and manuscript) to the fact that the map derives from the author and is not a later addition. In didactic treatises and in textbooks on many subjects, maps help express instructions. In literature, from Thomas More's *Utopia* and Michael Drayton's *Poly-Olbion* onwards, maps have been part of English culture. The teasing ambiguity of maps used as allegory or as political satire comes from the way the cartography was always intended to mean more than it appears to show.[38]

The history of English maps takes us far beyond the cartographical artefact alone. The contexts in which the maps were produced and used, from the earliest centuries of the Middle Ages onwards, underpin our review of map history. They reveal the degree to which the map, manuscript or printed, and in one form or another – from the most lavish and ornate, or the most scientific, to the simplest and most diagrammatic of sketches, or the most abstract expression – is deeply embedded in English society. One conclusion is clear: irrespective of graphic style, and of the difference from age to age in the ways of seeing the world around us, the use of maps as a form of representation and communication has never been absent from English culture. In this sense, modern ideas are deeply-rooted in the past and old maps are thoroughly modern.

Notes and References

Chapter 1

1 The liberalisation of the definition of a 'map' was formally initiated by J. B. Harley and David Woodward in *The History of Cartography*. Vol. 1, *Cartography in Prehistoric, Ancient, and Medieval Europe and the Mediterranean* (Chicago and London, University of Chicago Press, 1987), p. xvi. It has turned out to be a continuing process, as the three subsequent books of the *History*, with their penetrating explorations of non-western forms of mapping, have shown; see the discussion in Volume 2, Book 3, *Cartography in the Traditional African, American, Arctic, Australian, and Pacific Societies*, edited by David Woodward and G. Malcolm Lewis (Chicago and London, University of Chicago Press, 1998), Introduction, especially pp. 1–3. The relevant point to the approach in our book is Woodward's and Lewis's reminder (p. 6) that 'The "mapness" of an artefact depends in great degree on the social or functional context in which it is operating'.

2 Peter Barber, 'Old encounters new: the Aslake world map', in Monique Pelletier (ed.), *Géographie du monde au Moyen Age et à la Renaissance* (Paris, Editions du Comité des Travaux Historiques et Scientifiques, 1989), pp. 69–88, quotation from p. 87.

3 Maps drawn in the margins by glossators should be distinguished from those sketched by readers on the one hand, and those produced as part of the illustration of the original text, on the other. On the latter see Kathleen L. Scott, 'Design, decoration and illustration', in J. Griffiths and D. Pearsall (eds), *Book Production and Publishing in Britain, 1375–1475* (Cambridge, Cambridge University Press, 1989), pp. 31–64, especially p. 43.

4 J. B. Harley, *Maps for the Local Historian: A Guide to the British Sources* (London, National Council of Social Service for the Standing Conference for Local History, 1972); David Smith, *Maps and Plans for the Local Historian and Collector* (London, Batsford, 1988); Paul Hindle, *Maps for Historians* (Chichester, Phillimore, 1998). This last supersedes Hindle's earlier *Maps for Local History* (London, Batsford, 1988).

5 We do not say anything (after Chapter 2) about English mapping overseas or about English maps of foreign countries. We say little about the influence of the colonial experience on the cartography of England and Wales, despite the major contributions of, for example, Andrew Cook and Matthew Edney on English cartographical activities in the East Indies and in India respectively, or of Jeffrey Stone on Africa. Nor do we review the immense literature on the English mapping of North America. The cartography of Empire still awaits its historian. And, while we may touch on Wales, subdued by the English in the thirteenth century, we exclude mapping in Ireland and in Scotland except for an occasional mention where English map-makers were engaged; in colonial Ireland under the Tudors, for example, or in Scotland after the Jacobite rebellion. Some aspects of maps are also missing: we omit the technicalities of surveying, map construction and drawing, and map printing, we do not dwell on map semiotics, and we do not discuss globes or celestial maps.

6 For prehistoric maps in England, see Catherine Delano-Smith, 'Cartography in the prehistoric period in the Old World: Europe, the Middle East, and North Africa', in J. B. Harley and David Woodward (eds), *The History of Cartography. Volume 1: Cartography in Prehistoric, Ancient, and Medieval Europe and the Mediterranean* (Chicago and London, University of Chicago Press, 1987), pp. 54–101.

Chapter 2

1 The roles of oral, aural, and silent reading are discussed in Joyce Coleman, *Public Reading and the Reading Public in Late Medieval England and France* (Cambridge, Cambridge University Press, 1996). Coleman's summary (p. 88) of reading as professional (scholarly, literary), religious (clerical, lay), pragmatic (public, practical) or recreational (public, private) provides us not only with some terms but also with a persuasive underpinning for our own interpretation of the range of contexts of medieval map use. On medieval literacy, and especially changes and advances in the later fourteenth century, see also M. T. Clanchy, *From Memory to Written Record: England 1066–1307* (London, Edward Arnold, 1979; Oxford and Cambridge, Mass., Blackwell, 1993) and Janet Coleman, *English Literature in History: 1350–1400, Medieval Readers and Writers* (London, Hutchinson, 1981).

2 The general point is made by Mary Carruthers, *The Book of Memory. A Study of Memory in Medieval Culture* (Cambridge, Cambridge University Press, 1990), especially pp. 229–57. For a brief discussion of verbal topographical description, see Antonia Gransden, 'Realistic observation in twelfth-century England', *Speculum*, 47 (1972), pp. 29–51, reprinted in Antonia Gransden, *Legends, Traditions and History in Medieval England* (London and Rio Grande, The Hambledon Press, 1992), pp. 175–98, and other of her essays in the same book. For an example of a reader studying a map, see the annota-

tions in a thirteenth-century Hebrew Bible commentary in Corpus Christi College, Oxford (MS 6) where some place-names on a map have been translated into Latin by a Christian reader (eg. fol. 59). Cartographical marginalia in medieval texts is not uncommon. For a glossator's maps, see Fig. 2.12 and text, p. 18.

3 On Gervase, see William Stubbs (ed.), *Gervasii Cantuariensis, Opera Historicae*, 2 vols (Chronicles and Memorials of Great Britain and Ireland during the Middle Ages, 1879). Other examples of verbal *mappaemundi* include, for instance, the 'mappamundi' accompanying a fourteenth-century copy of Bede's *De locis sanctis* (British Library, Add. MS 22,635, fol. 55v). Filling a single large folio page, this follows the traditional format of Classical geographies, describing the three parts of the world and their constituent countries, and terminates with the phrase 'explicit mappa mundi'.

4 For further examples, see Evelyn Edson, *Mapping Time and Space. How Medieval Mapmakers Viewed their World* (London, The British Library, 1997). For scientific drawing in general, see John E. Murdoch, *Album of Science. Antiquity and the Middle Ages* (New York, Charles Scribner's Sons, 1984) and Brian J. Ford, *Images of Science. A History of Scientific Illustration* (London, The British Library; New York, Oxford University Press, 1992).

5 Richard H. Britnell, *The Commercialisation of English Society 1000–1500*, 2nd ed. (Manchester and New York, Manchester University Press, 1996), p. 80.

6 Bede, *Vita sanctorum abbatum monasterii* [Lives of the Abbots], para. 15. Eight hides represented land sufficient for eight families to live from. See also Ernest Savage, *Old English Libraries. The Making, Collection, and Use of Books during the Middle Ages* (London, Methuen, 1911), p. 246.

7 Jonathan J. G. Alexander, *A Survey of Manuscripts Illuminated in the British Isles, vol. 1, Insular Manuscripts from the Sixth to the Ninth Century* (London, Harvey Miller, 1978), pp. 33–34, Plate 23; R. S. L. Bruce-Mitford, 'The art of the Codex Amiatinus', *Journal of the British Archaeological Association*, 32 (1969), pp. 1–32, esp. Plate D; Pierre Courcelle, *Late Latin Writers and their Greek Sources* (Cambridge, Mass., Harvard University Press, 1969), esp. pp. 376–7; James W. Thompson, *The Medieval Library* (Chicago, University of Chicago Press, 1939), pp. 109–10.

8 Fols. IIv–III. The *Codex* is now in the Bibliotheca Laurentiana, Florence (Ms. Laur. Amiatino 1). The plan had certainly been copied by 716, the year Abbot Ceolfrid died at Langres on his way with it to the Pope. The manuscript was kept at Monte Amiata before its transfer to Florence: Courcelle, *Late Latin Writers* (see note 7), p. 376.

9 The Kirkham plan of the Tabernacle is British Library, Add. MS 38,817, fol. 2. Bede's architectural

texts are *De tabernaculo* (after 721); *De templo Solomnis* (between 729 and 731). Modern editions, with introductions, are: Arthur G. Holder, *Bede: On the Tabernacle* (Liverpool, Liverpool University Press, 1994) and Seán Connolly and Jennifer O'Reilly, *Bede: On the Temple* (Liverpool, Liverpool University Press, 1995). While Bede clearly appreciated the importance of graphic representation as an aid to understanding difficult concepts, both spatial and abstract, he himself does not seem to have drawn anything, although he clearly approved of drawings as aids to interpretation. The exceptions are his versions of Arculf's plans, which he included in his own discussion of the holy places of Palestine, *De locis sanctis* (see note 13, below). See also, Faith E. Wallis, *Bede: The Reckoning of Time* (Liverpool, Liverpool University Press, 1999), pp. 345–6.

10 Thomas O'Loughlin, 'The view from Iona; Adomnán's mental maps', *Peritia. Journal of the Medieval Academy of Ireland*, 10 (1996), pp. 98–123, esp. p. 99 and note 10; Denis Meehan, *Adamnan's De locis sanctis* (Dublin, Dublin Institute for Advanced Studies, Scriptores Latini Hiberniae 3, 1958), pp. 4–5 and 9.

11 Adomnán's plans survive in four ninth-century copies in Vienna, Paris, Zurich and Karlsruhe. For a comparative discussion and reproductions, and for a translation of the text, see John Wilkinson, *Jerusalem Pilgrims before the Crusades* (Warminster, Aris & Phillips, 1977), pp. 192–7, and pp. 92–116, respectively. Bede's plans are extant in six manuscripts (ninth to eleventh centuries) and are discussed by Wilkinson. For Bede's version of the buildings on Golgotha (including the Round Church) see also P. D. A. Harvey, 'Local and Regional Cartography in Medieval Europe', in J. B. Harley and David Woodward (eds), *History of Cartography*, vol. 1, *Cartography in Prehistoric, Ancient, and Medieval Europe and the Mediterranean* (Chicago and London, University of Chicago Press, 1987), p. 467, and P. D. A. Harvey, *Medieval Maps* (London, The British Library; Toronto, University of Toronto Press, 1991), p. 13. See also Meehan, *Adamnan's De locis sanctis* (see note 10) and Thomas O'Loughlin, 'Adomnán and Arculf: the case of an expert witness', *Journal of Medieval Latin*, 7 (1997), pp. 127–146.

12 Wilkinson, *Jerusalem Pilgrims* (see note 11), p. 96.

13 Compare Adomnán's version of the buildings on Golgotha (Fig. 2.2 here) with Bede's, reproduced in Harvey, *Medieval Maps* (see note 11). See also W. Trent Foley and A. G. Holder, *Bede: A Biblical Miscellany* (Liverpool, Liverpool University Press, 1999) for a recent commentary and translation of the *De locis sanctis*.

14 John Williams, *The Illustrated Beatus. A Corpus of the Illustrations of the Commentary on the Apocalypse* (London, Harvey Miller, 1994–);

Williams, 'Isidore, Orosius and the Beatus map', *Imago Mundi*, 49 (1997), pp. 7–32.

15 British Library, Add. MS 42,555, fol. 79v (made in Abingdon, Oxfordshire before 1262) and Bodleian Library, Douce 180, fol. 92r (probably made at Westminster between 1254 and 1274). Both are reproduced in a useful survey of early illustrations of the Heavenly City: M. L. Gatti Perer (ed.), *La Gerusalem celeste. Catalogo della mostra, Milano, Università Cattolica del S. Cuore, 20 maggio – 5 guigno 1983* (Milan, Università Cattolica, 1983).

16 Trinity College, Cambridge, MS R.16.2 (C.M.A. 524), fol. 25r.

17 For the cartographical aspects of computus maps see Evelyn Edson, 'World maps and Easter tables: computus maps in context', *Imago Mundi*, 48 (1996), pp. 25–42; also Leonid Chekin, 'Easter tables and the pseudo-Isidorean Vatican map', *Imago Mundi*, 51 (1999), pp. 13–23. For a well-grounded if popular account of the calendraic problem, see David Ewing Duncan, *The Calendar* (London, Fourth Estate, 1998).

18 The editors of *Giraldus Cambrensis* (Rolls Series, VII, pp. 118 and 165) interpret the entry *Mappa Mundi* in the list of the 42 of Hugh's books already in the cathedral library by 1150 as a map rather than a text. The item is also recorded in the mid twelfth-century catalogue of the library: R. M. Woolley, *Catalogue of the Manuscripts of Lincoln Cathedral Chapter Library* (Oxford, Clarendon Press, 1927), no. 33 in the list on p. vi, see also p. ix.

19 P. D. A Harvey, 'The Sawley world map and other world maps in twelfth-century England', *Imago Mundi*, 49 (1997), pp. 33–42, especially p. 36.

20 The Sawley map is in Corpus Christi College, Cambridge (MS 66); see Harvey, 'The Sawley world map', (see note 19). The Vercelli *mappamundi* is described in full in Carlo Capello, *Il mappamondo medioevale di Vercelli (1191–1218?)* (Turin, Università di Torino, Memorie e Studi Geografici 10, 1976). See also David Woodward, 'Medieval *mappaemundi*' in J. B. Harley and David Woodward (eds), *History of Cartography*, vol. 1 (see note 11), pp. 286–370, especially pp. 306–8 and Fig. 18.17, and P. D. A. Harvey, *Mappa Mundi. The Hereford World Map* (London, Hereford Cathedral and The British Library; Toronto, University of Toronto Press, 1996), pp. 30–3 with illustration.

21 The Thorney manuscript (St John's College, Oxford, MS 17) is discussed by Edson, *Mapping Time and Space* (see note 4), pp. 86–92, and Fig. 5.4, who also refers to Faith E. Wallis's doctoral dissertation for the University of Toronto, 'MS Oxford, St John's 17: a Medieval Manuscript in Context' (1985).

22 Many, however, are illustrated in R. A. Skelton and P. D. A. Harvey (eds) *Local Maps and Plans from Medieval England* (Oxford, Clarendon Press, 1986); P. D. A. Harvey, *Topographical Maps. Symbols, Pictures and Surveys* (London, Thames and Hudson,

1980); or P. D. A. Harvey, *Medieval Maps* (see note 11). Other references are given below as appropriate.

23 Lambeth Palace, MS Court of Arches Ff. 291, fol. 58v. We are grateful to Stephen Upex for bringing this map to our attention.

24 P. D. A Harvey, 'Wormley, Hertfordshire, 1220 × 1230', in Skelton and Harvey, *Local Maps and Plans* (see note 22), pp. 59–70 and Plate 2. See also P. D. A. Harvey, 'A 13th-century plan from Waltham Abbey, Essex', *Imago Mundi*, 22 (1968), pp. 10–12; Harvey, *Medieval Maps* (see note 11), Fig. 76.

25 H. E. Hallam, 'Wildmore Fen, Lincolnshire, 1224 × 1249', in Skelton and Harvey, *Local Maps and Plans* (see note 22), pp. 71–81; Harvey, *Topographical Maps* (see note 22), Fig. 95 and Plate 3.

26 Quotations from Richard Vaughan, *Matthew Paris* (Cambridge, Cambridge University Press, 1958), p. 247. The surviving portion of the *mappamundi*, perhaps a quarter of the original, is on the verso of the last folio (fol. 284) of the second part of the *Chronica Maiora* in Corpus Christi College, Cambridge, MS 26. It shows Europe from the Danube westwards, with a disproportionately large Italy, the Apennines, the Alps and a large number of rivers. The peninsula of Iberia is not indicated at all although *Gades Herculis* straddles the straits between Spain and Africa. The map is reproduced, with transcriptions, in Konrad Miller, *Mappaemundi. Die ältesten Weltkarten*, vol. 3, *Die kleineren Welkarten* (Stuttgart, J. Roth, 1895), pp. 71–4. For the Westminster *mappamundi*, see Paul Binski, *The Painted Chamber at Westminster* (London, Society of Antiquaries, Occasional Paper, n.s., 9, 1986), pp. 16 and 44.

27 Thomas O'Loughlin, 'The map of Europe (N.L.I. 700): a window into the world of Giraldus Cambrensis', *Imago Mundi*, 51 (1999), pp. 24–39 (quotation from p. 27) and Plate 2. O'Loughlin transcribes the names of 88 places and geographical features. See also Robert Bartlett, *Gerald of Wales, 1146–1223* (Oxford, Clarendon Press, 1982). The smaller drawings, virtually identical from manuscript to manuscript, are in manuscripts of the third recension (British Library, Add. MS 33991, fol. 26r, and MS Arundel 14, fol. 27v), although this observation needs to be followed up by comparison with other manuscripts. Another diagram relating to the North and South Poles is found in Gerald's *De descriptione mundi* (British Library, Cotton MS Cleopatra D.V, fol. 136).

28 For the map of Palestine which accompanies the London-Rome itinerary and the map of Britain in various volumes of the *Chronica majora*, see note 74 in Chapter 5.

29 David Woodward, 'Roger Bacon's terrestrial co-ordinate system', *Annals of the Association of American Geographers*, 80 (1990), pp. 109–21; David Woodward with Herbert M. Howe, 'Roger

Bacon on geography and cartography', in Jeremiah Hackett (ed.), *Roger Bacon and the Sciences. Commemorative Essays* (Leiden, New York and Cologne, Brill, 1997), pp. 198–222, esp. p. 210.

30 The date of 1290s or *c*.1300 (instead of *c*.1289) has been suggested by Nigel Morgan and M. B. Parkes on the basis of their recent examination of script and artwork on the *mappamundi*: 'The Hereford Mappa Mundi: art historical aspects' and 'The Mappa Mundi at Hereford. Report on the handwriting and copying of the text', respectively, papers presented to the Mappa Mundi Conference, Hereford, 27 June – 1 July 1999, to be published in the forthcoming Proceedings. We are most grateful to Professor Morgan, Dr Parkes, and the Dean and Chapter of Hereford Cathedral for permission to use this date. For a well-illustrated account of the Hereford map, see Harvey, *Mappa Mundi* (see note 20). See also Peter Barber, 'Old encounters new: the Aslake world map' in Monique Pelletier (ed.), *Géographie du` monde au Moyen Age et à la Renaissance* (Paris, Editions du Comité des Travaux Historiques et Scientifiques, 1989), pp. 69–88.

31 N. J. Morgan, *Early Gothic Manuscripts 1250–1285. A Survey of Manuscripts Illustrated in the British Isles* IV, vol. 2 (London, Harvey Miller, 1988), no. 188, pp. 197–8; Harvey, *Medieval Maps* (see note 11), p. 25 and Fig. 20; Peter Barber and Michelle P. Brown 'The Aslake world map', *Imago Mundi*, 44 (1992), pp. 24–44, esp. p. 32 and n. 55; Edson, *Mapping Time and Space* (see note 4), pp. 135–7 and Plate VI.

32 The bible (Bodleian Library, MS BNC 5) was probably made in Oxford about 1230 and glossed before mid-century. Textual and cartographical glosses (fols. 32 and 281) are in the same hand, to which also belong the enlarged versions, and the Temple plan, at the end. See H. H. Glunz, *History of the Vulgate in England, from Alcuin to Roger Bacon* (Cambridge, Cambridge University Press, 1933), pp. 270–72.

33 Richard of St Victor provided his commentary on Ezekiel with 15 architectural plans and sections, each introduced by an appropriate phrase such as *Horum itaque quae diximus tale exemplar formamus* [We make the following drawing of the things we have said]. Not all manuscripts contain the map, however. It is found in Bodleian Library, MS Bodley 494 (dated *c*.1160–75), fol. 166v. J. P. Migne, who used this manuscript amongst others for the Patrologia Cursus Completus edition of *In visionem Ezekielis* (tom. 196, 1855), omits the map, as did the early printers of Richard's works (for example, *Opera omnia*, printed in Venice in 1592).

34 On the maps in Ranulf Higden's *Polychronicon* see Peter Barber, 'The Evesham world map: a late medieval English view of God and the world', *Imago Mundi*, 47 (1995), pp. 13–33 and John Taylor, *The Universal Chronicle of Ranulf Higden* (Oxford,

Clarendon Press, 1966), esp. pp 50–71 and the Appendix of extant manuscripts.

35 F. Hull, 'Isle of Thanet, Kent. Late 14th century × 1414', in Skelton and Harvey, *Local Maps and Plans* (see note 22), pp. 119–126. See also Antonia Gransden, 'Antiquarian studies in fifteenth century England', *Antiquaries Journal*, 60 (1980), pp. 75–97, especially pp. 84–8.

36 Dorothy M. Owen, 'Clenchwarton, Norfolk. Late 14th or early 15th century', in Skelton and Harvey, *Local Maps and Plans* (see note 22), pp. 127–30; Harvey, *Medieval Maps* (see note 11), Fig. 75.

37 M. W. Barley, 'Sherwood Forest, Nottinghamshire. Late 14th or early 15th century', in Skelton and Harvey, *Local Maps and Plans* (see note 22), pp. 131–9 and Plate 10. The map is at Belvoir Castle, Rutland, and difficult to access.

38 K. D. Hartzell, 'Diagams for liturgical ceremonies, late 14th century', in Skelton and Harvey, *Local Maps and Plans* (see note 22), pp. 339–43. Hartzell discusses the ten printed editions (1502–1558) rather than the earlier manuscript, now British Library Add. MS 57,534.

39 James E. Thorold Rogers, *History of Agricultural Prices* 7 vols (Oxford, Clarendon Press, 1866–1902), vol. 4, p. 600; cited by Savage, *Old English Libraries* (see note 5), p. 253.

40 Peter Barber, 'The Evesham world map' (see note 34).

41 British Library, Harley MS 1808, fols. 9v (*Totius Britanniae*) and 45v (York), dated *c*.1400. G. R. Crone, *Early Maps of the British Isles, A.D. 1000–A.D. 1579* (London, Royal Geographical Society, Reproductions of Early Maps 7, 1961), p. 19. The map measures 22 × 15 cm. See also John H. Harvey, 'Symbolic plan of a city. Early 15th century'. in Skelton and Harvey, *Local Maps and Plans* (see note 22), pp. 342–3 (unillustrated).

42 *Historia regum Britanniae* (1130s). British Library, Royal MS 13.A.iii. A numberof the drawings have been found to be accurate in specific architectural details and imply first-hand observation on the spot. York (fol. 6v), for instance, was represented as it would have been before the window in the west front of the Minster had been built. London is on fol. 28v. Carlisle, Canterbury and Bath, amongst others, are represented. The pencilled work is of a high quality, though possibly in places unfinished. The manuscript is addressed to Robert, Earl of Gloucester.

43 P. D. A. Harvey, 'Boarstall, Buckinghamshire. 1444 × 1446', in Skelton and Harvey, *Local Maps and Plans* (see note 22), pp. 211–19 and Plate 18.

44 The Cornish manuscript is in the Bodleian Library, Oxford, MS Bodleian 791. The plans are on folios 27, 56, and 83. They are reproduced in Edwin Norris, *The Ancient Cornish Drama* (Oxford, Oxford University Press, 1859), vol. 1, pp. 219 and 479; vol. 2, p. 201 and F. J. Furnivall and A. W. Pollard

(eds), *The Macro Plays*, 2nd ed. (London, Early English Text Society, Extra Series, No. 91, 1924). See also David Mills, 'Diagrams for staging plays. Early or mid-15th century', in Skelton and Harvey, *Local Maps and Plans* (see note 22), pp. 344–5. William Borlase, *Observations on the Antiquities Historical and Monumental, of the County of Cornwall* (Oxford, W. Jackson, 1754), pp. 195–6 and Plate XVI, fig. 2, and *Natural History of Cornwall* (Oxford, W. Jackson for the author, 1758), pp. 295–8, discusses the origin and performance of the plays and describes two surviving Rounds, at St Just and at Perranzabuloe.

45 The manuscript of 'The Castle of Perseverance' is in Washington (Folger Library, MS V.a. 354, fol. 191v). It is reproduced in M. Eccles (ed.), *The Macro Plays* (London, Early English Text Society, Extra Series No. 262, 1969), frontispiece, where there is also a full discussion of the date, language and staging of the plays. The plan is described briefly by Mills, 'Diagrams for staging plays' (see note 44).

46 British Library, Lansdowne MS 204, fols 226v–227r. The map survives in several versions. See Harvey, *Medieval Maps* (see note 11), p. 73 and Fig. 55.

47 M. G. Snape and B. K. Roberts, 'Tursdale Beck, County Durham. *Circa* 1430 × *circa* 1442' in Skelton and Harvey, *Local Maps and Plans* (see note 22), pp. 171–87.

48 J. V. Somers Cocks, 'Dartmoor, Devonshire. Late 15th or early 16th century', in Skelton and Harvey *Local Maps and Plans* (see note 22), pp. 293–302 and Plate 26. On the special moorland rights (venville rights) and a later map, see also E. M. Yates, 'Dark Age and medieval settlement on the edge of wastes and forests', *Field Studies*, 2 (1965), pp. 133–53, especially pp. 143–8.

49 M. G. Snape, 'Witton Gilbert, County Durham. Mid-15th century' and Susan Reynolds, 'Chertsey, Surrey, and Laleham, Middlesex, mid- or late 15th century', in Skelton and Harvey, *Local Maps and Plans* (see note 22), pp. 229–35, 237–43.

50 Susan Reynolds, 'Staines, Middlesex. 1469 × *circa* 1477', in Skelton and Harvey, *Local Maps and Plans* (see note 22), pp. 245–50 and Plates 3 and 4.

51 Rose Mitchell and David Crook, 'The Pinchbeck Fen map: a fifteenth-century map of the Lincolnshire fenland', *Imago Mundi*, 51 (1999), pp. 40–50 and plates 3 and 4.

52 M. W. Beresford, 'Inclesmoor, West Riding of Yorkshire. *Circa* 1407', in Skelton and Harvey, *Local Maps and Plans* (see note 22), pp. 147–61; Harvey, *Medieval Maps* (see note 11), Fig. 68 and Plate 12.

53 P. D. A. Harvey, 'Shouldham, Norfolk. 1440 × 1441', and B. K. Roberts, 'North-west Warwickshire; Tanworth in Arden. 1497 × 1519', in Skelton and Harvey, *Local Maps and Plans* (see note 22), pp. 195–201 and 317–29 respectively. See also B. K.

Roberts, 'An early Tudor sketch map', *History Studies*, 1 (1968), pp. 33–8.

54 M. D. Knowles, 'Clerkenwell and Islington, Middlesex. Mid-15th century', in Skelton and Harvey, *Local Maps and Plans* (see note 22), pp. 221–8 and Plate 19. Up to the early twentieth century, an early copy of the map also existed.

55 Philip E. Jones, 'Deptford, Kent and Surrey; Lambeth, Surrey; London. 1470–1478', in Skelton and Harvey, *Local Maps and Plans* (see note 22), pp. 251–62.

56 H. S. A. Fox, 'Exeter, Devon. *Circa* 1420' in Skelton and Harvey, *Local Maps and Plans* (see note 22), pp. 163–9.

57 L. D. Reynolds and N. G. Wilson, *Scribes and Scholars: A Guide to the Transmission of Greek and Latin Literature* 3rd ed. (Oxford University Press 1968, Oxford, Clarendon Press, 1991).

58 Richard Talbert, 'Rome's empire and beyond: the spatial aspect', in *Gouvernants et gouvernés dans l'Imperium Romanum. Cahiers des Etudes Anciennes XXVI* (Montreal Université Laval, Département d'Histoire, 1990), pp. 215–23; Kai Broderson, *Terra Cognita. Studien zur römischen Raumerfassung*, Spudasmata 59 (Hildesheim, G. Olms, 1995).

59 See Chapters 4 and 6.

60 See page 114.

61 On the late-fourteenth century spread of literacy, see Coleman, *English Literature* (see note 1), p. 24. For a critique of the orality-to-literacy model, see Coleman, *Public Reading* (see note 1), especially Chapter 1 'On beyond Ong'.

62 The visual simplicity of a map in diagrammatic style (using straight or near-straight lines and smoothed curves and including only the minimum of content consistent with the purpose of the drawing) is a form of generalisation, articulating the essence of an idea or observation. This means that a map in diagrammatic style can be regarded as intellectually more challenging than a mimetic map. The argument will be discussed in the final series of Lectures on the History of Cartography, at the Institut Cartogràfic de Catalunya, Barcelona, February 2000, to be published as *Approaches and Challenges in a Worldwide History of Cartography* [provisional title], edited by David Woodward.

63 Edson, *Mapping Time and Space* (see note 4), pp. 5–6, calls this type of map a 'list map'. For the maps cited in this paragraph, see Skelton and Harvey, *Local Maps and Plans* (see note 22), pp. 131–9 and 71–81.

64 On the Chertsey map, pictorial representation is used for ten landscape features altogether, alone in four cases (rivers, vegetation, field boundaries, bridges) and with words confirming the identity of the pictorial features in six cases: the lane (*venella*), the track (*calcetum*), the monastery of Chertsey, its mill, its barn, and the village of Laleham. Three features are represented by words alone (meadow, common

pasture, arable). On the Fineshead map, straight lines indicate the boundaries of the meadows, numerals give the acreage (in perches and feet) and written names identify the plot holder. On later maps of rural property, ownership was not uncommonly indicated by abstract symbols.

65 On the date of use of field measurement, we are grateful to B. K. Roberts (personal communication). On medieval planned villages in England and Wales in general, see his *The Making of the English Village* (Harlow, Longman, 1987). For towns, the best general text is still Maurice W. Beresford, *New Towns of the Middle Ages: Town Plantation in England, Wales and Gascony* (London, Lutterworth Press, 1967). For a recent discussion of urban design in post-Norman colonisation and consolidation, see Keith Lilley, 'Urban landscapes and the cultural politics of territorial control in Anglo-Norman England' *Planning History*, 20 (1998), pp. 7–15.

66 Cited by Richard Vaughan, *Matthew Paris* (Cambridge, Cambridge University Press, 1958), pp. 239–40 and note 1.

67 O'Loughlin's analysis of Adomnán's mental maps makes a useful contribution in this respect: 'The view from Iona' (see note 10).

68 The full inscription (top right of the map) starts *Mappa dicitur forma* and goes on to relate how Julius Caesar sent out for information on '... *regiones, provincias, insulas, civitates, syrtes, paludes, equora, montes, flumina quasi sub unius pagine visione coadunavit; que scilicet non parvam prestat legentibus utilitatem, viantibus directionem rerumque viarum gratissime speculationis dir*[or *l*]*ectionem*': Konrad Miller, *Mappaemundi. Die ältesten Weltkarten* (Stuttgart, J. Roth, 1895–8) vol. V, p. 8, §24. Note that Youssouf Kamal, *Monumenta cartographica Africae et Aegypti* (Cairo, 1926–51), tom. 4, fascicule 1, fol. 1118 v, renders the last word as *dilectionem* without querying it. For the Ebstorf map (about 1239), see text, p. 38.

69 Wilkinson, *Jerusalem Pilgrims* (see note 11), p. 96.

70 *De locis sancti*, Book VI, paragraph 2, translation from Trent Foley and Holder (see note 13), p. 14.

71 Discussed in D. F. McKenzie, *Bibliography and the Sociology of Texts. The Panizzi Lectures 1985* (London, The British Library, 1986), p. 34.

72 The number of maps that can be closely associated with a legal function remains small: ten examples from the fifteenth-century alone as against five from all previous medieval centuries together. On the role of maps in legal proceedings, see Peter Eden, 'Three Elizabethan estate surveyors: Peter Kempe, Thomas Clerke and Thomas Langdon', in Sarah Tyacke (ed.), *English Map-Making 1500–1650. Historical Essays* (London, The British Library, 1983), pp. 68–84, especially p. 77.

73 Jacques Signot's map of Italy (1495–8); *Code Signot*, Bibliothèque Nationale de France Ge. D.

7,687, was the outcome of reconnoitre requested by Charles VIII of France who was planning an invasion of Italy: David Buisseret, 'Monarchs, ministers, and maps in France before the Accession of Louis XIV', in Buisseret (ed.), *Monarchs, Ministers and Maps. The Emergence of Cartography as Tool of Government in Early Modern Europe* (Chicago, Chicago University Press, 1992), pp. 99–123 and especially p. 101.

74 The notion of text as an 'act of construction' is discussed by D. F. McKenzie, *Bibliography and the Sociology of Texts* (see note 71) who on p. 34 specifically includes maps as 'text'.

75 John H. Harvey, 'Winchester, Hampshire. *Circa* 1390', in Skelton and Harvey, *Local Maps and Plans* (see note 22), p. 145.

76 For a fuller description, see William Urry, 'Canterbury, Kent. *Circa* 1153 × 1161', in Skelton and Harvey, *Local Maps and Plans* (see note 22), pp. 43–58.

77 New College, Oxford MS C.288; currently on deposit in the Bodleian Library. On the function of the four topographical drawings, see Kathleen L. Scott, *Later Gothic Manuscripts, 1390–1490. A Survey of Manuscripts Illuminated in the British Isles*, vol. 6, 2 vols (London, Harvey Miller, 1996, 2 vols), vol. 1, Fig. 426, Vol. 2, pp. 310–11.

78 Harvey, *Medieval Maps* (see note 11), pp. 21–26, and G. R. Crone, *Early Maps of the British Isles* (see note 41) p. 13. On its relationship to the computus manuscript, see Edson, 'World maps' (see note 17), pp. 32–3.

79 Harvey, *Mappa Mundi* (see note 20), p. 29 caption. Harvey has made similar remarks in his *Medieval Maps* (see note 11), pp. 24–5.

80 Edson, 'World maps' (see note 17) pp. 32–3.

81 Edson, *Mapping Time and Space* (see note 4), p. 75.

82 Edson, *Mapping Time and Space* (see note 4), p. 76.

83 Agrippa's map, in turn, may have been based on the survey ordered by Julius Caesar in 44 BC: Harvey, *Medieval Maps* (see note 11), p. 21. See also O. A. W. Dilke, 'Maps in the service of the state: Roman cartography to the end of the Augustan era', in Harley and Woodward, *History of Cartography* (see note 11), pp. 207–9.

84 Harvey, 'The Sawley map' (see note 19). The map is reproduced in colour in Danielle Lecoq, 'La "mappemonde" du *De arca Noe mystica* de Hughes de Saint-Victor (1128–1129)', in Monique Pelletier (ed.), *Géographie du monde au Moyen Age* (see note 30), pp. 9–31, Fig. 5. See also Valerie I. Flint, *Honorius Augustodunensis of Regensburg* (Aldershot, Variorum, Authors of the Middle Ages, 6, 1995).

85 On *mappaemundi* in general, see David Woodward, 'Medieval *mappaemundi*', in Harley and Woodward, *History of Cartography* (see note 11), vol. 1, pp. 286–370 and 'Reality, symbolism, time and space in medieval world maps', *Annals of the*

Association of American Geographers, 75 (1985), pp. 510–21, quotation from p. 519. See also: Barber, 'Old encounters new' (see note 30), pp. 69–88; Barber. 'The Evesham world map' (see note 34); Barber and Brown, 'The Aslake world map' (see note 31); Edson, *Mapping Time and Space* (see note 4); Harvey, *Medieval Maps*, (see note 11), pp. 28–32; and Harvey, *Mappa Mundi* (see note 20). For the link between English *mappaemundi* and northern France, see Marcia Kupfer, 'The lost *mappamundi* at Chalivoy-Milon', *Speculum*, 66 (1991), pp. 540–71.

86 Isidore, *Etymologies*, book 14, chapter 2. The confusion between these secular diagrams and the Christianised maps derived from them has been compounded by misattributions. Miller, for instance, thought that the Christianised world diagram drawn a century later on the front cover of a ninth-century codex from St Gallen (Stiftsbibliothek, MS Sang. 237) which contains, amongst other material, Isidore's *Etymologies*, actually illustrated that work and called it an 'Isidorean' *mappamundi* (Konrad Miller, *Mappaemundi. Die ältesten Weltkarten* 6 vols (Stuttgart, J. Roth, 1895–98) vol. 6, p. 56). On the dating of MS Sang. 237, which ranges from the seventh to the ninth century, see John Williams, 'Isidore, Orosius and the Beatus map', *Imago Mundi*, 49 (1997), pp. 7–32, note 67. The correct diagram for the text, which we reproduce in Fig. 2.24, comes further on in the codex, at the end of the chapter to which it applies (chapter 2, book 14).

87 Woodward, 'Reality, symbolism, time and space' (see note 85).

88 Barber, 'The Evesham world map' (see note 34), p. 32, note 73.

89 Harvey, *Mappa Mundi* (see note 20) p. 7. Valerie I. Flint, 'The Hereford map; its author, two scenes and a border', *Transactions of the Royal Historical Society*, 6th series, 8 (1998), pp. 19–44, continues the debate on the identity of Richard of Haldingham. A slim but perceptive guide to the map is: Gabriel Alington and Dominic Harbour, *The Hereford Mappamundi. A Medieval View of the World* (Leominster, Fowler Wright Books, 1996). For the date of the map, see note 30 above.

90 Peter Wiseman, 'Julius Caesar and the Hereford world map', *History Today*, 37 (1987), pp. 53–7.

91 Barber, 'Old encounters new' (see note 30); Barber and Brown, 'The Aslake world map', (see note 31), pp. 24–44.

92 Graham Haslam, 'The Duchy of Cornwall map fragment', in Pelletier, *Géographie du monde au Moyen Age* (see note 30), pp. 33–44. The original date of *c.*1260 is now thought to be too early: Peter Barber (personal communication).

93 Harvey, *Mappa Mundi* (see note 20), pp. 29–30, with figure; Edson, *Mapping Time and Space* (see note 4), pp. 135–7, Fig. 7.1 and Plate VI.

94 Barber, 'Old encounters new' (see note 30).

95 Barber and Brown, 'Aslake world map' (see note 31), pp. 33–5.

96 Alessandro Scafi, 'Mapping Eden', in Denis Cosgrove (ed.) *Mappings* (London, Reaktion Books, 1999), pp. 50–70 and pp. 273–6 (notes).

97 O'Loughlin, 'The view from Iona' (see note 10), p. 100.

98 Irrespective of their date of accession, none of the manuscripts of the Geography in the British Library or, as far as our enquiries could ascertain, in Oxford, is older than 'mid-fifteenth century'. The *Cosmographia* included with a group of Latin translations by Antonio Pacini was given to the University of Oxford only in February 1444: see *Duke Humfrey's Library and The Divinity School, 1488–1988*. Catalogue to an exhibition at The Bodleian Library June-August 1988 (Oxford, Bodleian Library, 1988), p. 43. The study of Greek in England is thought to have so declined after the seventh century that, even had a Greek text arrived in the country before the thirteenth century, few would have been capable of reading it. In the eighth century, for instance, there were 'books of cosmography written in peculiar letters which no one could copy' and which might have indicated 'a book of Byzantine Geography': Thompson, *Medieval Library* (see note 7), p. 113.

99 O. A. W. Dilke, 'The culmination of Greek cartography in Ptolemy', in Harley and Woodward, *History of Cartography*, vol. 1 (see note 11), pp. 180 ff; Liba Chaia Taub, *Ptolemy's Universe. The Natural Philosophy and Ethical Foundations of Ptolemy's Astronomy* (Chicago and La Salle, Illinois, Open Court, 1993).

100 Charles Burnett, *The Introduction of Arabic Learning into England. The Panizzi Lectures, 1996* (London, The British Library, 1997), especially p. 68; Barber and Brown, 'Aslake world map' (see note 31).

101 A. W. Crosby, *The Measure of Reality. Quantification and Western Society 1250–1600* (Cambridge, Cambridge University Press, 1997), pp. 18–19 et seq.

102 David Woodward 'Chronogeography and coordinates: European *mappaemundi* and Islamic world maps'. Talk given in the Oxford Seminars in Cartography series, Bodley Library, 25 November 1993, p. 2. Quoted here from circulated notes, with permission. See Woodward with Howe, 'Roger Bacon' (see note 29).

103 Woodward, 'Roger Bacon's terrestrial coordinate system' (see note 29), p. 117.

104 Raymond P. Mercier, 'Geodesy', in J. B. Harley and David Woodward (eds), *The History of Cartography*, vol. 2, book 1, *Cartography in the Traditional Islamic and South Asian Societies* (Chicago, University of Chicago Press, 1992), pp. 175–88. Edward S. Kennedy and Mary Helen Kennedy, *Geographical Coordinates of Localities from Islamic Sources* (Frankfurt, Institut für

Geschichte der Arabisch-Islamischen Wissenschaften, 1987) publish some seventy such tables.

105 For an outline of the implications of Adelard's translation, see J. K. Wright, *Geographical Lore of the Time of the Crusades* (New York, American Geographical Society, Research Series 15, 1925; republished with additions, New York, Dover Publications, 1965), esp. pp. 95–6 and then 86–7.

106 The debate is referred to by Bede (see text, p. 30) and by William of Malmesbury (British Library, Arundel MS 161. fol. 146v). See also Add. MS 35,179, fol. 19r.

107 The *tabulae Londonienses* were referred to in 'Another Treatise on the Sphere' (Paris, Bibliothèque Nationale de France, Latin MS 7272, ff. 60v–67v),'thought to have been written by Analdò di Negro, who lived from *c.*1270 to *c.*1342: Lynn Thorndike, *The Sphere of Sacrobosco and Its Commentaries* (Chicago, Chicago University Press, 1949), p. 36 and Appendix 1. For other sources, see Dana Bennett Durand, *The Vienna-Klosterneuburg Map Corpus of the Fifteenth Century. A Study in the Transition from Medieval to Modern Science* (Leiden, E. J. Brill, 1952), Appendix 3.

108 From British Library, Royal MS 17.A.16, fol. 26r, and Arundel MS 66, item 7. Only the lists surviving in the British Library (of which there are a number) have been consulted; a wider search is likely to add both to the tally of such lists and possibly to the number of places in England for which such co-ordinates had been calculated.

109 Thorndike, *Sphere of Sacrobosco* (see note 107) includes Robertus Angelus's commentary which has a section: 'Telling longitude from eclipses' using the Toledan Tables (pp. 244–5).

110 *Liber de composicione universalis astrolabii,* now Bodleian MS Digby 40: described in *Duke Humfrey's Library* (see note 98), p. 56.

111 On the change in European thought from medieval to early modern, see Timothy J. Reiss, *Knowledge, Discovery and Imagination in early Modern Europe* (Cambridge, Cambridge University Press, 1997) and Crosby, *Measure of Reality* (see note 101).

112 O'Loughlin, 'Map of Europe' (see note 27).

113 C. W. Boase, *Oxford* (London, Longmans, Green & Co., 1887), p. 65.

114 Bartlett, *Gerald of Wales* (see note 27), esp. 127–34.

115 Harvey, 'Local and regional cartography' (see note 11), especially p. 495. See also Harvey, *Medieval Maps* (see note 11), p. 71 and Figs. 57 and 58; P. D. A. Harvey, 'Matthew Paris's maps of Britain', in P. R. Coss and D. S. Lloyd (eds), *Thirteenth Century England IV: Proceedings of the Newcastle upon Tyne Conference, 1992* (Woodbridge, Suffolk, The Boydell Press, 1992), pp. 118–21; and Joan B. Mitchell, 'Early maps of

Great Britain: I. The Matthew Paris maps', *Geographical Journal*, 81 (1933), pp. 27–34.

116 Harvey, 'Matthew Paris's maps' (see note 115). pp. 118–21.

117 E. J. S. Parsons, *The Map of Great Britain circa AD 1360 known as the Gough Map. An Introduction to the Facsimile* (Oxford, Oxford University Press for the Bodleian Library and the Royal Geographical Society, 1958); Brian Paul Hindle, 'The towns and roads of the Gough map (*c.*1360)', *Manchester Geographer*, 1 (1980), pp. 35–49; Harvey, *Medieval Maps* (see note 11), pp. 73–8.

118 Harvey, *Medieval Maps* (see note 11), pp. 73 & 76, comments on the use of a portolan chart. G. R. Crone, in discussing the remarkably realistic portrayal of the Yorkshire Ouse-Trent river network shown on the Hereford map ('Landmarks in British cartography. I. Early cartographic activity in Britain', *Geographical Journal*, 128 (1962), pp. 406–10, and Fig. 1), notes that 'the Gough map . . . develops this representation of the Ouse-Trent river system considerably' (p. 407).

119 Parsons, *Gough Map* (see note 117), pp. 6–7.

120 As by Sir Frank Stenton, 'The roads in the Gough map' in Parsons, *Gough Map* (see note 117), pp. 16–20, and Hindle, 'Towns and roads' (see note 117).

CHAPTER 3

1 See Victor Morgan, 'The cartographic image of "The Country" in early modern England', *Transactions of the Royal Historical Society*, 5th series, 29 (1979), pp. 129–54. Also John Gillies, *Shakespeare and the Geography of Difference* (Cambridge, Cambridge University Press, 1994).

2 For a census of maps in books, including news-sheets and pamphlets, see Ruth S. Luborsky and Elizabeth M. Ingram, *A Guide to English Illustrated Books, 1536–1603*, 2 vols (Tempe, Arizona Center for Medieval and Renaissance Studies, 1998); for those in bibles, see also Catherine Delano-Smith and Elizabeth M. Ingram, *Maps in Bibles 1500–1600. An Illustrated Catalogue* (Geneva, Droz, 1991).

3 Morgan, 'The cartographic image' (see note 1). For maps as symbols of sovereignty, see especially Peter Barber, 'Maps and monarchs in Europe 1550–1800', in Robert Oresko, G. C. Gibbs and H. M. Scott (eds), *Royal and Republican Sovereignty in Early Modern Europe* (Cambridge, Cambridge University Press, 1997), pp. 75–124.

4 For an example of the transformation of 'an aspiring merchant's son into the patriarch of a landed Kentish family', see Retha M. Warnicke, *William Lambarde. Elizabethan Antiquary 1536–1602* (London and Chichester, Phillimore, 1973).

5 Elizabeth M. Ingram, 'Maps as readers' aids: maps and plans in Geneva bibles', *Imago Mundi*, 45 (1993), pp. 29–44. See also Eugene R. Kintgen, *Reading in Tudor England* (Pittsburgh, University of Pittsburgh Press, 1996) on 'Reading in a Religious Setting' (pp. 99–139) and on Thomas Blundeville's justification for acquiring the art of reading in *The true order and Method of wryting and reading Hystories* (1574), pp. 140–9.

6 For the tapestry maps made at Sheldon, Warwickshire, about 1588, see Morgan, 'The cartographic image' (see note 1), pp. 152–3 and n. 71. Between the main tapestries and the fragments, the maps depict the midland counties and show 'hills, rivers and streams, each town and hamlet, with the churches, parks and *in some instances roads* [our italics]; and suggest personal knowledge of the county by the designer of each map, for churches are usually correctly represented, with or without spire': John Humphreys, *Elizabethan Sheldon Tapestries* (London, Oxford University Press, 1929), p. 16. The tapestry maps are generally supposed to be modelled on Saxton's county maps but the inclusion of some roads points to additional information. Humphreys also reports an inventory of goods in Chastleton House, Moreton-in-the-Marsh dated 1633 where there were 'fower large quarter maps' hanging in the gallery, as well as tapestries elsewhere in the house. Whilst still on the walls at Weston Park, Warwickshire, the maps were seen by Richard Gough, *British Topography* (1780), vol. 2, p. 309, who noted 'Three large maps in tapestry 80 foot square by Francis and Richard Hickes'. The manuscript map of the Trent made before 1540 (see Fig. 3.3) could have been a cartoon for a tapestry, if not an illustration of riparian rights or flood control: C. R. Salisbury, 'An early Tudor map of the River Trent in Nottinghamshire', *Transactions of the Thoroton Society of Nottinghamshire*, 87 (1983), pp. 54–9 & folded map (in colour). On the reproduction of John Hooker's map of Exeter on a screen, see William Ravenhill and Margery Rowe, 'A decorated screen map of Exeter', in Todd Gray, Margery Rowe and Audrey Erskine (eds), *Tudor and Stuart Devon. The Common Estate and Government. Essays Presented to Joyce Youings* (Exeter, University of Exeter Press, 1992), pp. 1–12, who suggest, on balance, a late sixteenth-century date for the screen.

7 For the latter see Figure 5.1, Chapter 5.

8 Catherine Delano-Smith, 'Map ownership in sixteenth-century Cambridge: the evidence of probate inventories', *Imago Mundi*, 47 (1995), pp. 67–91.

9 A good idea of the range of local subject matter being mapped from the sixteenth century onwards is to be gained from county inventories such as those by Harold Nichols, *Local Maps of Derbyshire. An Inventory and Introduction* (Matlock, Derbyshire Library Service, 1980) and *Local Maps of Nottinghamshire. An Inventory* (Nottingham,

Nottinghamshire County Council Leisure Services, 1987). The British Library and the Public Record Office have major collections of maps of national importance: see Tony Campbell, *Indexes to Material of Cartographic Interest in the Department of Manuscripts and Manuscript Cartographic Items elsewhere in the British Library*, 3 vols (1992; on open shelves in the Maps Reading Room and the Manuscripts Reading Room at the British Library) and *Maps and Plans in the Public Record Office*, vol. 1 *British Isles c.1410–1860* (London, H.M.S.O., 1967).

10 The choice of map subject tended to reflect their owners' main interests. In sixteenth-century Cambridge, for instance, classical scholars had maps of Greece and Italy while those with reformist inclinations had maps of the Holy Land: Delano-Smith, 'Map ownership' (see note 8).

11 P. D. A. Harvey and Harry Thorpe, *The Printed Maps of Warwickshire 1576–1900* (Warwick, Warwick County Council in association with University of Birmingham, 1959), pp. 1–52. Paul Laxton, 'Introductory notes', *250 Years of Map-Making in the County of Hampshire* (Lympne, Harry Margary, 1976). We are indebted to Paul Laxton for his comments on this chapter and for his invaluable suggestions.

12 The second edition of William Caxton's *Mirrour of the World* (1490) contains a number of simple, woodcut, T-O diagrams: Oliver H. Prior, *Caxton's Mirror of the World* (London, Kegan Paul, Trench Trübner; New York, Humphrey Milford, Oxford University Press, Early English Text Society, Extra Series 110, 1913) pp. vi–vii; Tony Campbell, *The Earliest Printed Maps, 1472–1500* (London, The British Library; Berkeley, University of California Press, 1987), pp. 98–9 and Figs. 1 and 2.

13 The processionals, containing music for the chants, words for the prayers, and plans (which adhere closely to those in the manuscript) showing priests where to stand for major church ceremonies conducted according the Sarum Rite, were printed in England and on the continent from 1501 to the middle of the sixteenth century, after which the Sarum Rite ceased to be observed. For an example, see British Library C.35.f.9, printed in Antwerp in 1525 or 1530. The picture in De Sancto Marco Hieronymous, *Opusculum de universali mundi machina* (London, R. Pynson, 1505) shows the four elements arranged as concentric circles around a central landscape. It is reproduced in Edward Hodnett, *English Woodcuts 1480–1535* (Oxford, Oxford University Press, 1973), no. 1590 & Fig. 166.

14 Peter Barber, 'A glimpse of the earliest map-view of London?' *London Topographical Society* 27, (1993), pp. 91–102.

15 Map, decorated title page, and text were all printed on the continent: Delano-Smith and Ingram, *Maps in Bibles* (see note 2), pp. 26–7, Figs. 16–18;

and Elizabeth M. Ingram, 'The map of the Holy land in the Coverdale Bible: a map by Holbein?' *The Map Collector*, 64 (1993), pp. 26–31.

16 In William Patten, *The Expedicion into Scotla[n]d*, printed in London by Richard Grafton in 1547. It is not clear where the blocks were cut but it is likely to have been in England.

17 The total for copperplate maps excludes the 35 maps comprising Saxton's atlas (34 counties and one map of England and Wales) and that for book maps excludes reuses (which would bring the total to nearer 300). We are indebted to Laurence Worms for the data on separately-printed maps. For book maps see Luborsky and Ingram, *Guide to English Illustrated Books* (see note 2). For English map-production in general up to 1640, see Worms, 'The London map trade to 1640', in David Woodward (ed.), *History of Cartography*, vol. 3, *Cartography in the European Renaissance* (Chicago and London, University of Chicago Press, forthcoming). Worms suggests that as many as 1,500 maps may have been put out under London imprints before 1640, 85 per cent of which date from after 1585, but with a noticeable peak in production during the first decade of the seventeenth century. We are immensely grateful to Laurence Worms for letting us see a draft of his contribution for the *History of Cartography* and for permission to quote these figures in advance of publication.

18 Worms, 'London map trade ' (see note 17); Barber, 'Earliest map-view of London?' (see note 14) mentions the possibility of an earlier copperplate map of the world, now lost, which may have been produced in London in 1549.

19 Robert W. Karrow, *Mapmakers of the Sixteenth Century and Their Maps* (Winnetka, Illinois, Speculum Press for The Newberry Library, 1993), p. 252; Worms, 'London map trade' (see note 17).

20 The backs of the plates were re-used for paintings of the Tower of Babel. This 'lost copperplate' map is discussed more fully in Chapter 6.

21 On the relative merits of wood block and copper plate in cartography, see especially: David Woodward (ed.), *Five Centuries of Map Printing. The Kenneth Nebenzahl, Jr. Lectures in the History of Cartography* (Chicago and London, University of Chicago Press for The Newberry Library, 1975).

22 The role of maps in state affairs is the key theme in David Buisseret (ed.), *Monarchs, Ministers and Maps: The Emergence of Cartography as a Tool of Government in Early Modern Europe* (Chicago, University of Chicago Press, 1992) and also Roger J. P. Kain and Elizabeth Baigent, *The Cadastral Map in the Service of the State: a History of Property Mapping* (Chicago and London, University of Chicago Press, 1992).

23 See Robert Baldwin, 'Colonial cartography and the role of the British chartered companies' and Sarah Tyacke, 'English overseas chartmaking', in

Woodward, *History of Cartography*, vol. 3 (see note 17). Kenneth R. Andrews, *Trade, Plunder and Settlement. Maritime Enterprise and the Genesis of the British Empire, 1480–1630* (Cambridge, Cambridge University Press, 1984), provides a useful summary of the context of trade and exploration.

24 John Blagrave's map of the world (1596) was made for his book although none of the three known extant copies is bound up with the text: Rodney Shirley, *The Mapping of the World. Early Printed World Maps 1472–1700*, 3rd ed. (London, Holland Press, 1984, 1993). Shirley lists only seven maps of the world made by Englishmen and printed in London, the earliest in 1576 (by Humphrey Gilbert), including Blagrave's.

25 Speed's map of the world appeared in his atlas *A Prospect of the most Famous Parts of the World* (London, 1627) but even this map was based almost wholly on published Dutch maps: Shirley, *The Mapping of the World* (see note 24), entry 317.

26 The English-language edition of the Genevan Bible, printed in Geneva in 1560 by Rouland Hall, contains the standard set of five maps. The point of the maps, together with other readers' aids, was to help the new, vernacular, bible-reading public to reach a correct way of reading when they read in the vernacular 'for themselves': Catherine Delano-Smith, 'Maps as art and science: maps in sixteenth century Bibles', *Imago Mundi*, 42 (1990), pp. 65–83, and Delano-Smith and Ingram, *Maps in Bibles* (see note 2), esp. pp. xxv and 99. See also Kintgen, *Reading in Tudor England* (see note 5).

27 Foreigners played an important role in many aspects of English life throughout the century but especially during the first half. The book industry (printing and selling) was largely in the hands of Dutch and Flemish immigrants; court painters included some of the best-known names in Europe; the universities welcomed scholars from all parts of the continent; the masters of military architecture and techniques were largely German and Italian.

28 Books like Thomas Elyot's *The Boke named the Governour* (London, 1531), Machiavelli's *Arte della guerra* (1521), and Castiglione's *Il libro del cortegiano* (1528): see Peter Barber, 'England I: pageantry, defense, and government: maps at court to 1550', in Buisseret, *Monarchs* (see note 22), pp. 26–56. See also J. B. Harley, 'Meaning and ambiguity in Tudor cartography' in Sarah Tyacke (ed.) *English Map-Making 1500–1650* (London, The British Library, 1983), pp. 22–45, especially p. 27.

29 For a comprehensive survey of Henry's use of maps, and further references, see Barber, 'England I' (see note 28), pp. 26–56, and Peter Barber, 'Cartography, topography and history paintings', in David Starkey (ed.), *The Inventory of King Henry VIII*, vol. 2, *Text* (in preparation for the Society of Antiquaries). For a view of the wider context, see also

Jerry Brotton, *Trading Territories: Mapping the Early Modern World* (London, Reaktion, 1997).

30 Edward IV of England was Charles the Bold of Burgundy's brother-in-law. On maps in Henry VIII's pageants see Barber, *Maps and Monarchs* (see note 3), pp. 29–30, and Sydney Anglo, *Spectacle, Pageantry and Early Tudor Policy* (Oxford, Clarendon Press, Warburg Studies, 1996).

31 Scot McKendrick, 'Tapestries from the Low Countries during the fifteenth century', in Caroline Barron and Nigel Saul (eds), *England and the Low Countries in the Late Middle Ages* (Stroud, Alan Sutton; New York, St Martin's Press, 1995), pp. 43–60, especially p. 45.

32 Information from Peter Barber (see note 29).

33 Lisa Jardine, *Worldly Goods. A New History of the Renaissance* (London, Macmillan, 1996), pp. 386–92 and plate 13.

34 Barber, 'England I' (see note 28), p. 27. Barber also comments on the previous page on the way 'the political and administrative use of maps by English monarchs seems to stretch back well into the Middle Ages'; such use however remains to be fully explored.

35 J. H. Andrews, *Plantation Acres. An Historical Study of the Irish Land Surveyor and his Maps* (Belfast, Ulster Historical Foundation, 1985), especially pp. 28 and 32; and *Shapes of Ireland* (Dublin, Geography Publications, 1997), pp. 29, 33–4, 45.

36 Ascham's manuscript, now MS 337 in the Beinecke Library, Yale University, is described in Barbara Shearer (ed.), *Catalogue of Medieval and Renaissance Manuscripts, Beinecke Library, Yale*, vol. 2 (MSS 251–500) (Binghamton, N.Y., Medieval and Renaissance Texts and Studies, vol. 48, 1987). Anthony's date of birth is not known and all his printed publications come from the 1550s. However, an idea of his youth in 1526 may be gained from the fact that his younger brother, Roger, was fifteen years old on going up to St John's College, Cambridge in 1530, and his older brother Thomas had been a fellow at the same college since 1523 (see *Dictionary of National Biography* under Roger Ascham). For an idea of medieval interest in the latitude and longitude of places in England, see documents listed in Campbell, *Indexes to Material of Cartographic Interest* (see note 9). In the Middle Ages, the terms latitude and longitude were often used as synonyms for breadth and length (as on the Fineshead map, see Fig. 2.6).

37 P. D. A. Harvey, *The History of Topographical Maps: Symbols, Pictures and Surveys* (London, Thames and Hudson, 1980), p. 162. See also Sarah Tyacke and John Huddy, *Christopher Saxton and Tudor Map-Making* (London, The British Library, 1980), pp. 19–23. J. H. Andrews, *Plantation Acres* (see note 35), Chapters 1 and 2, discusses English surveying techniques in the context of Ireland.

38 J. A. Bennett, 'Geometry and surveying in early-seventeenth century England', *Annals of Science*, 48 (1991), pp. 345–54, especially p. 346.

39 Sebastian Münster, *Erlerung des newen Instruments der Sunnen* (Oppenheim, 1528); Gemma Frisius, *Libellus de locorum describendorum ratione* (Antwerp, 1533).

40 For a description of both Rathborne's and Leybourne's surveying techniques, see Bennett, 'Geometery and surveying' (see note 38). See also p. 127.

41 See Bennett, 'Geometry and surveying' (see note 38), and Stephen Johnston, 'Mathematical Practitioners and Instruments in Elizabethan England', *Annals of Science*, 48 (1991), pp. 319–44.

42 Illustrated in Tyacke and Huddy, *Christopher Saxton* (see note 37), p. 54.

43 British Library, Cotton MS Augustus I.i.9. G. R. Crone, *Early Maps of the British Isles AD 1000–AD 1579* (London, Royal Geographical Society, Reproductions of Early Maps 7, 1961), pp. 22–3 and reproduced in colour in P. D. A. Harvey, *Maps in Tudor England* (London, The Public Record Office and The British Library; Chicago, Chicago University Press, 1993), Fig. 2. That other maps may have existed can be surmised from contemporary allusions: Barber, 'England I' (see note 28), p. 33, n. 116.

44 On Kratzer's role in promoting maps, see Barber, 'England II: monarchs, ministers, and maps, 1550–1625', in Buisseret, *Monarchs* (see note 22), p. 62. The version of Britain in the Venice edition of Ptolemy, with its eastward distortion of Scotland, was repeated by Martin Waldseemüller (Vienne, 1541) and Jacobo Gastaldi (Venice, 1548): see Rodney Shirley, *Early Printed Maps of the British Isles. A Bibliography. 1477–1650* (London, Holland Press Cartographica 5, 1980).

45 For later versions of Lily's map see Shirley, *Early Printed Maps* (see note 44), pp. 31–2; for a reproduction of the original see Crone, *Early Maps* (note 43), no. 14.

46 Peter Barber, 'Les îles Britanniques', in Marcel Watelet (ed.) *Geraldi Mercatoris. Atlas Europa* (Antwerp, Bibliothèque des amis du Fonds Mercatoris, 1994), pp. 70–1, and 'The British Isles', in Marcel Watelet (ed.) *The Mercator Atlas of Europe circa 1570–1572*, English edition (Pleasant Hill, Oregon, Walking Tree Press, 1998), pp. 43–77 and 90–2 (notes). Quotation from pp. 62 and 60–1 in the French and English editions respectively. Further references are to the English edition only. Barber extends his discussion of Mercator's map of the British Isles to consider all possible candidates for authorship of the prototype.

47 Barber, 'The British Isles' (see note 46), pp. 70–71.

48 David Marcombe, 'Saxton's apprenticeship: John Rudd, a Yorkshire cartographer', *The Yorkshire*

Archaeological Journal, 50 (1978), pp. 171–5, and (shortened) 'John Rudd: a forgotten Tudor map-maker?' *The Map Collector*, 64 (1993), pp. 34–7; Barber, 'England II' (see note 44), pp. 57–98, especially p. 63; Barber, 'The British Isles' (see note 46), p. 69.

49 Reproduced in Marcombe, 'John Rudd' (see note 48), p. 37, and in Barber, 'The British Isles' (see note 46), p. 69.

50 Robert Lythe is also known as Robert Legh or Leight: J. H. Andrews, 'The Irish surveys of Robert Lythe', *Imago Mundi*, 19 (1965), pp. 22–37. Lythe had previously worked on the mapping of the Anglo-French boundary around Calais, presumably as a military engineer. Andrews lists 13 extant manuscript maps as possibly by Lythe and points to their influence on later printed maps. For Lluyd, see Barber, 'The British Isles' (see note 46), p. 68, and Karrow, *Mapmakers* (see note 19), pp. 344–8.

51 On Dee in general see Peter J. French, *John Dee: The World of an Elizabethan Magus* (London, Routledge and Kegan Paul, 1972).

52 For the most recent comments on Nowell, see Barber, 'The British Isles' (see note 46), pp. 68–9.

53 Nowell also left a map of Scotland, reproduced in Harvey, *Maps in Tudor England* (see note 43), Fig. 35, together with his manuscript map of East Yorkshire and Humberside (Fig. 12). See also Barber, 'The British Isles' (see note 46), p. 47. Nowell's letter to William Cecil of 1563 is transcribed and translated in R. A. Skelton, *Saxton's Survey of England and Wales with a facsimile of Saxton's Wall-Map of 1583*, with a Preface by J. B. Harley (Amsterdam, N. Israel, Imago Mundi Supplement No. 6 1974), pp. 15–16.

54 Nowell's map is reproduced in full in colour on the cover of Harvey, *Maps in Tudor England* (see note 43), and in Barber, 'The British Isles' (see note 46) pp. 52–3 with details on pp. 62–3. For the itinerary, see Chapter 5.

55 On William Lambarde, see Warnicke, *William Lambarde* (see note 4). The printed map appears, with Lambarde's map of the Saxon Heptarchy, only in the second edition of Lambarde's *A Perambulation of Kent* (1596), twenty years after the unillustrated first edition of the book. Lambarde's manuscript draft (dated 1586) is now in the British Library, Add. MS 62935. On Saxton's possible secondary sources, see Edward Lynam, *The Mapmaker's Art. Essays in the History of Maps* (London, Batchworth Press, 1953), p. 64.

56 Wolfe's comment about his interest in maps appears in Holinshed's *Chronicle* (1577): Lynam, *Mapmaker's Art* (see note 55), p. 79. See also J. B. Harley, 'Christopher Saxton and the first atlas of England and Wales 1579–1979', *The Map Collector*, 8 (1979), pp. 3–11.

57 J. B. Harley, 'The map collection of William Cecil, First Baron Burghley', *The Map Collector*, 3 (1978), pp. 12–19, reference on p. 12; Barber, 'The Minister

leaves his mark' in Peter Barber and Christopher Board (eds), *Tales from the Map Room: Fact and Fiction about Maps and their Makers* (London, BBC Books, 1993), pp. 88–9; John Andrews, *Ireland in Maps* (Dublin, Geographical Society of Ireland with the Ordnance Survey of Ireland, 1961), p. 5. For Walsingham's involvement with maps, see Barber, 'England II', (see note 44), p. 68.

58 The contemporary phrase is cited by John Andrews, in *Plantation Acres* (see note 35), p. 28, n. 3, from a letter to Burghley from Sir Ralph Lane, 4 August 1597 (*Calendar of State Papers of Ireland, 1596–7*, p. 368).

59 Parts of the Northumberland and the Devon maps are reproduced in Harvey, *Maps in Tudor England* (see note 43), Fig. 36, and Harley, 'The map collection of William Cecil' (see note 57), p. 17, respectively. Burghley's atlas came to the British Library (where it is now Royal MS 18.D.iii) from the royal collection. See also R. A. Skelton and John Summerson, *A Description of the Maps and Architectural Drawings in the Collection Made by William Cecil, First Baron Burghley, Now at Hatfield House* (Oxford, for The Roxburghe Club, 1971). On Burghley's use of a map of Lancashire in recording leading recusant families, see Joseph Gillow, *Lord Burghley's Map Of Lancashire in 1590* (London, Arden Press for the Catholic Society, 1907). The map of Lancashire in William Smith's 'Visitacion of Lancashire' (completed in 1567 but written up in 1598) could also have been used by Burghley, for the Visitation gives the genealogies of Lancastrian families, whose names are marked in red on the accompanying large double-folio map (British Library, Harley MS 6159, fols 3v–4r).

60 On the popularity of Ortelius's maps in Cambridge, see Delano-Smith, 'Map ownership' (see note 8), pp. 76–7.

61 Philip Apian's survey involved triangulation. It was drafted at a scale of 1:50,000 and printed on 24 sheets at a scale of 1:144,000. In methodology and scale it anticipated the national surveys of the eighteenth century. Henry VIII possessed works by Apian's father, Peter, which Burghley would have known: Barber, 'Maps and monarchs' (see note 3), p. 83, n. 18, and personal communication.

62 For a list of Saxton's manuscript maps and surveys, see Ifor M. Evans and Heather Lawrence, *Christopher Saxton. Elizabethan Map-Maker* (Wakefield, Wakefield Historical Publications and Holland Press, 1979), pp. 79–80.

63 Summarised by William Ravenhill, 'Christopher Saxton's surveying: an enigma', in Sarah Tyacke (ed.), *English Map-Making 1500–1650* (London, The British Library, 1983), pp. 112–9.

64 Skelton, *Saxton's Survey* (see note 53), pp. 8–9. On the beacon system in particular, see Ravenhill, 'Christopher Saxton's surveying' (see note 63).

65 Remigius Hogenberg is known to have been in London, *c*.1572–1587: Evans and Lawrence, *Christopher Saxton* (see note 62), p. 35.

66 Some differences are detailed in Evans and Lawrence, *Christopher Saxton* (see note 62), p. 10.

67 Elizabeth Leedham-Green, *Books in Cambridge Inventories: Booklists from Vice-Chancellors' Court Probate Inventories in the Tudor and Stuart Periods*, 2 vols (Cambridge, Cambridge University Press, 1986). Wall maps are notoriously vulnerable and it is not clear from the fact that only two extant examples are known whether Saxton's map was rare in his day. The surviving copies are discussed by Evans and Lawrence, *Christopher Saxton* (see note 62), p. 35.

68 The contents of Burghley's atlas are listed in Evans and Lawrence, *Christopher Saxton* (see note 62), Appendix 6, and of other extant copies in Appendices 10 and 11. See also William Ravenhill, *Christopher Saxton's 16th Century Maps: The Counties of England and Wales* (Shrewsbury, Chatsworth Library and Airlife Publishing, 1992), for a modern facsimile.

69 Evans and Lawrence, *Christopher Saxton* (see note 62), pp. 38–9.

70 Richard Helgerson, 'The land speaks: cartography, chorography, and subversion in Renaissance England', *Representations* 16 (1986), pp. 50–85. Queen Elizabeth asked for the original version of her portrait, which forms the frontispiece to the *Atlas*, to be altered to show her skirts falling more flatteringly. Both states of the portrait are reproduced in Evans and Lawrence, *Christopher Saxton* (see note 62), p. 21 (State I) and frontispiece (State II, in colour). On Seckford, and for the wording of the pass, see Tyacke and Huddy, *Christopher Saxton* (see note 37), pp. 24–5.

71 For the order of production, see Evans and Lawrence, *Christopher Saxton* (see note 62), pp. 18–19.

72 Helgerson, 'The land speaks' (see note 70), p. 53, using information detailed in Evans and Lawrence, *Christopher Saxton* (see note 62), pp. 15–17.

73 We are grateful to John Andrews for pointing this out.

74 The later history of Saxton's atlas is detailed in Evans and Lawrence, *Christopher Saxton* (see note 62), pp. 45–65.

75 The multiple authorship of a map is eloquently demonstrated by David Woodward's diagram of a 'Framework for the study of maps as artefacts', first published in 'The study of the history of cartography: a suggested framework,' *American Cartographer*, 1 (1974), pp. 101–15, and reproduced in M. J. Blakemore and J. B. Harley, 'Concepts in the History of Cartography. A Review and Perspective', *Cartographica*, 17:4 (1980), p. 46, Fig. 12.

76 Graduated non-Ptolemaic maps of Britain had been drawn from at least the mid 1530s, including the anonymous *Angliae Figura* (1534–46) and Lawrence Nowell's little map of the British Isles (*c*.1564). The first English county map with a graduated margin seems to have been Philip Symonson's *A New Description of Kent* (1596), for which see Edward Heawood, *English County Maps in the Collection of the Royal Geographical Society* (London, Royal Geographical Society, 1932, Reproductions of Early Engraved maps II), p. 11 and Sheet 10. In 1680, John Adams was complaining about 'the want of sufficient *information* for the *Placing* in their true *Latitude* and *Longitude*' villages and country houses which had hitherto not appeared on maps: *Index Villaris* (Preface), quoted by William Ravenhill, 'John Adams, his map of England, its projection, and his *Index Villaris* of 1680', *Geographical Journal*, 144 (1978), pp. 424–37.

77 Harley, 'Christopher Saxton' (see note 56), p. 3.

78 Cited in Tyacke and Huddy, *Christopher Saxton* (see note 37), pp. 31–2.

79 For a biographical approach, see Frank Kitchen, 'John Norden (*c*.1547–1625): estate surveyor, topographer, county mapmaker and devotional writer', *Imago Mundi*, 49 (1997), pp. 43–61, and Frank Kitchen, *Cosmo-choro-poly-grapher: An Analytical Account of the Life and Work of John Norden* (unpublished D.Phil. thesis, University of Sussex, 1993) (also deposited in the British Library Map Library).

80 The book concludes with a short note addressed to King James, headed 'Touching your Majesties Mineralls in Cornwall', which expands on the value of copper and its usefulness in 'rais[ing] a greater yearlie profit than the value of Your Majesties Lande Revenues. So riche as the Workes . . . your Majestie may be therefore pleased to cause a further view, and more due search by the skillfull in this misterie [of mining]': John Norden, *Speculi Britanniae Pars Topographicall & Historical Description of Cornwall [sic]*. For a facsimile of the maps, see William Ravenhill, *John Norden's Manuscript Maps of Cornwall and Its Nine Hundreds* (Exeter, University of Exeter, 1972).

81 British Library Add. MS 2596. Reproduced in Henry B. Wheatley and Edmund W. Ashbee (eds), *William Smith, 'The Particular Description of England, 1588, With Views of some of the Chief Towns and Armorial Bearings of Nobles and Bishops'* (London, [by subscription], 1879), p. vi. Smith's authorship of a number of county maps of England was only belatedly recognised, the maps in question being attributed to 'the anonymous map-maker'.

82 Edward Heawood, *English County Maps*, (see note 76) nos. 11–17. For the identification of Smith as the 'anonymous map maker of 1602/3' see R. A. Skelton, 'Four English county maps', *British Museum Quarterly*, 22 (1960), pp. 47–50. Autographed manuscript versions of some of William Smith's county maps, drawn in their own right and not just as drafts for printing, also survive: for example, Lancashire

(British Library, Harley MS 6159) and Cheshire (British Library, Harley MS 1046 fol. 132). For Smith's town maps, and the maps in his chorography of Nuremberg, see Chapter 6 below.

83 J. B. Harley, 'From Saxton to Speed', *Cheshire Round*, 1 (1966), pp. 174–88.

84 Notable exceptions to the lack in the history of cartography of a satisfactory comprehensive treatment of this period are several essays by Peter Barber: 'British cartography', in Robert P. Maccubbin and Martha Hamilton-Phillips (eds), *The Ages of William III and Mary II: Power, Politics and Patronage 1688–1702. A Reference Encyclopaedia and Exhibition Catalogue* (Williamsburg, College of William and Mary in Virginia, 1989), pp. 92–104; 'Necessary and ornamental: map use in England under the later Stuarts, 1660–1714', *Eighteenth Century Life*, 14 (1990), pp. 1–28; and 'Maps and monarchs' (see note 3). See also Helen M. Wallis, ' "Geographie is better than Divinitie": maps, globes and geography in the days of Samuel Pepys', in Norman J. Thrower (ed.), *The Compleat Plattmaker. Essays on Chart, Map and Globe Making in England in the Seventeenth and Eighteenth Centuries* (Berkeley, Los Angeles, University of California Press, 1978), pp. 1–43.

85 On map use in the second half of the century, see Barber, 'Necessary and ornamental' (see note 84).

86 On the Quartermaster's map, see Skelton, *Saxton's Survey* (see note 53), pp. 14–15.

87 Barber, 'Maps and monarchs' (see note 3), p. 105, also p. 95 for a similar lack of enthusiasm from parliament a century later.

88 Barber, 'British cartography' (see note 84), p. 96.

89 Charles II appears to have been willing to support Ogilby in any way except with money: Katherine S. Van Eerde, *John Ogilby and the Taste of his Times* (Folkestone, Dawson, 1976), p. 128 and, for Ogilby's advertisement of his intention to 'make a survey of every county', p. 127. See also J. B. Harley's Introduction to the facsimile edition of *John Ogilby, Britannia (1675)* (Amsterdam, Theatrum Orbis Terrarum, 1970). John Adams, as Richard Gough notes, made an offer to the Royal Society on 27 April 1681 to measure 'the bounding-line [of each county], the distances between places both in the round and the strait lines, and [to take] the latitude and angles of position': *British Topography* (London, 1780), p. 52. Although he was still engaged in 1683 on maps of Cumberland and Gloucestershire in the course of what was intended to be a systematic county survey, this came to nothing, either though Adams's own shortcomings or for lack of official support, or a combination of both: Blake Tyson, 'John Adams's cartographic correspondence to Sir Daniel Flemming of Rydal Hall, Cumbria, 1676–1687', *Geographical Journal*, 151 (1985), pp. 21–39, especially pp. 34–5; E. G. R. Taylor, 'Notes on John Adams and contem-

porary map makers', *Geographical Journal*, 97 (1941), pp. 182–4.

90 Barber, 'Maps and monarchs' (see note 3), pp. 92 and 105.

91 Rodney W. Shirley, *Printed Maps of the British Isles 1650–1750* (Tring, Map Collector Publications; London, The British Library, 1988), pp. 82–3.

92 Edward Lynam, *British Maps and Map-makers* (London, Collins, 1944), p. 31, suggests the selection of the London meridian may have been prompted by the rebuilding of St Paul's after the Great Fire of 1666, despite the fact that Charles II had founded in 1675 the Observatory at Greenwich to 'settle the vexed question of longitude'.

93 In the event, only five other maps were completed: Surrey, Kent, both (like Hertfordshire) at one inch to two miles; Middlesex at one inch to one and half miles; Oxfordshire at one inch to three miles, and Buckinghamshire at one inch to four miles. The atlas exists only as a manuscript mock-up presented to King George III (British Library, Maps 1.Tab.18): R. A. Skelton, *County Atlases of the British Isles 1579 to 1850. A Bibliography. 1579–1703* (London, Carta Press, 1970), pp. 186–7.

94 Laurence Worms, *Some British Mapmakers* (London, Ash Rare Books, Catalogue [1992]), n.p.

95 Tony Campbell, 'Cutting costs', in Barber and Board, *Tales from the Map Room* (see note 57), pp. 34–5.

96 Tony Campbell, 'For those in peril on the sea', in Barber and Board, *Tales from the Map Room* (see note 57), pp. 164–5. The event was recorded in the second edition of the atlas (1722) with a note by the name of the Gilstone Rock: ' Sr Clously lost'.

97 Gough, *British Topography* (see note 6), vol. 1, p. 36.

98 H. C. Darby, 'The age of the improver: 1600–1800', in H. C. Darby (ed.), *The New Historical Geography of England* (Cambridge, Cambridge University Press, 1973), pp. 302–88.

99 As on the map of lead mines around Wirksworth (1632) and the map of Milne Close Grooves, Matlock (1688): PRO, DL 44/1121 and Derbyshire Record Office. D. 239, respectively. In contrast to these two maps of mines from the seventeenth century, a dozen are listed for the eighteenth century for Derbyshire: Harold Nichols, *Local Maps of Derbyshire to 1770. An Inventory and Introduction* (Matlock, Derbyshire Library Service, 1980). See also Harold Nichols, *Local Maps of Nottinghamshire to 1800. An Inventory* (Nottingham, Nottingham County Council Leisure Services, 1987).

100 John Norden, 'An Abstract of divers Manors . . .' (British Library, Add. MS 6027). Stephanos Mastoris and Sue Groves (eds), *Sherwood Forest in 1609: a Crown Survey by Richard Bankes* (Nottingham, Thoroton Society, Thoroton Society Record Series, vol. 40 1997); Sarah Bendall, 'Mapping the English forests: Needwood 1598–1834'

in John Langton (ed.) *Where History and Geography Met: English Forests from the Seventeenth to the Nineteenth Century* (forthcoming).

101 On enclosures on the edge of Needwood Forest see E. M. Yates, 'Dark Age and medieval settlement on the edge of wastes and forests', *Field Studies*, 2 (1965), pp. 133–53, especially 137–43.

102 *The Diary of Samuel Pepys*, Robert Latham and William Matthews (eds), vol. 1 (London, Bell, 1972–83), 20 June 1662. We are grateful to Peter Barber for drawing our attention to this item.

103 Joan Thirsk, *Fenland Farming in the Sixteenth Century* (Leicester, University of Leicester Department of English Local History, Occasional Papers No. 3, 1953).

104 Agas's letter to Lord Burghley is in the British Library (Lansdowne MS 84, n. 32). Only one local Fenland map by John Hexham is known from Burghley's papers at Hatfield House (CPM, supp. 29; reproduced in Skelton and Summerson, *Maps and Architectural Drawings* (see note 59), pp. 53–4 and plate 10). For Hayward's map, see Francis Willmoth, *Sir Jonas Moore* (Woodbridge, Suffolk, Boydell Press, 1993), p. 90, and H. G. Fordham, 'Descriptive list of maps of the Great Level of the Fens', *Studies in Carto-Bibliography* (Oxford, Clarendon Press, 1914), pp. 65–8. Hayward's map served for the next fifty years as the basis of printed representations of the potentially drainable land. For the general history see, besides Willmoth, *op. cit.*, H. C. Darby, *The Draining of the Fens* (Cambridge, Cambridge University Press, 1940) and The Changing Fenland (Cambridge, Cambridge University Press, 1983).

105 Willmoth, *Sir Jonas Moore* (see note 104), p. 104.

106 Hondius copied from the map now British Library, Cotton MS Aug. I.i 78 (*c*.1604): see Skelton *County Maps* (note 93), p. 222. Hondius's map was copied by Joannes Blaeu (*Regiones inundatae*, 1645).

107 Willmoth, *Jonas Moore* (see note 104), p. 90.

108 On the involvement of the Dutch, see Willmoth, *Jonas Moore* (see note 104), pp. 94–5 and 108–10. Vermuyden worked widely in Derbyshire, and elsewhere, on the unwatering of lead mines, and in eastern England on land drainage, notably at Hatfield Chase, Yorkshire. The 'Act for the Draining of the Great Level of the Fens', which gave the Earl of Bedford and his co-Adventurers (financiers) authority to proceed, was passed on 29 May 1649.

109 Cambridge University Library, Atlas 2. 68.1. This is presumably the map referred to by Pitt on the verso of the title page of an anonymous pamphlet, *History or Narrative of the Great Level of the Fenns, called the Bedford Level* (1685, British Library, G.15858), which he describes as 'New Printed and Enlarged' by himself.

110 The only known copy of the first edition of Moore's map is in the Public Record Office, MPC 88; the only known copy of the second version, published by Moses Pitt in 1684, is in the Bodleian Library, Gough Maps, Cambridgeshire 2; two examples of Christopher Browne's edition are in the British Library, Maps 184.1.1 and Maps K.Top. 6 (72) 11 Tab End.

111 Quoted by Willmoth, *Jonas Moore* (see note 104), p. 114.

112 Willmoth, *Jonas Moore* (see note 104), p. 114. On Moore's surveying techniques, see also Francis Willmoth, ' "The Genius of All Arts" and the use of instruments: Jonas Moore (1617–1679) as a mathematician, surveyor, and astronomer', *Annals of Science*, 48 (1991), pp. 355–65.

113 Willmoth, *Jonas Moore* (see note 104), p. 115.

114 J. B. Harley, *William Yates's Map of the County of Lancashire, 1786* (Birkenhead, Historic Society of Lancashire and Cheshire, 1968), p. 7. Paul Laxton used the term 'a cartographic revolution' in his paper on 'The geodetic and topographical evaluation of English county maps, 1740–1840', *Cartographic Journal*, 13:1 (1976), pp. 37–54, especially p. 37, and again in his *250 Years of Map-Making* (see note 11). See also J. B. Harley, 'The re-mapping of England 1750–1800', *Imago Mundi*, 19 (1965), pp. 56–67.

115 The Board of Agriculture and Internal Improvement was founded in 1793. Two series of county reports were produced (1793–6 and 1795–1815). The challenge came from William Marshall in his own regional studies, published between 1788 and 1798. The quotation, from his *Rural Economy of the West of England* (1796), vol. 1, p. 2, is taken from R. A. Butlin, 'Regions in England and Wales *circa* 1500–1914' in R. A. Dodgshon and R. A. Butlin (eds), *An Historical Geography of England and Wales*, 2nd ed. (London and New York, Academic Press, 1990), pp. 223–54. We are grateful to Paul Laxton for drawing our attention to this point.

116 David Spadaforda, *The Idea of Progress in Eighteenth-Century Britain* (New Haven and London, Yale University Press, 1990), p. 48.

117 Tobias Smollett's words, cited by Spadaforda, *Idea of Progress* (see note 116), p. 49.

118 E. Pawson, 'The framework of industrial change 1730–1900', in R. A. Dodgshon and R. A. Butlin (eds), *An Historical Geography of England and Wales*, 1st ed. (London and New York, Academic Press, 1978), pp. 267–89.

119 R. Lawton, 'Population and society, 1730–1914', in Dodgshon and Butlin, *Historical Geography of England and Wales*, 2nd edition, (see note 115), pp. 285–321.

120 A. Moyes, 'Transport 1730–1900', in Dodgshon and Butlin, *Historical Geography of England and Wales*, 1st ed. (see note 118), pp. 401–29.

121 H. G. Lewin, *Early British Railways 1801–44* (London, Locomotive Publishing Company, 1925), p. 37, cited by Moyes, 'Transport 1730–1900' (see note 120), p. 415.

122 Data in Christopher M. Klein, *Maps in Eighteenth-Century British Magazines. A Checklist* (Chicago, The Newberry Library, Herman Dunlap Smith Center, Publication No. 3, 1987).

123 J. B. Harley, 'John Strachey of Somerset: an antiquarian cartographer of the early eighteenth century', *Cartographic Journal*, 3 (1966), pp. 2–5, quotation on p. 5.

124 From P. P. Burdett's 'Proposals for a Map of Lancashire, 1768', cited by Harley, *Yates's Map of the County of Lancashire* (see note 114), p. 7.

125 Gough, *British Topography* (see note 6), p. 50.

126 J. B. Harley, 'The bankruptcy of Thomas Jefferys: an episode in the economic history of eighteenth century map making', *Imago Mundi*, 20 (1967), pp. 27–48. For a brief account of successful and failed local initiatives, see: William Ravenhill, 'The South West in the eighteenth-century re-mapping of England', in Katherine Barker and Roger Kain (eds), *Maps and History in South-West England* (Exeter, Exeter University Press, 1991), pp. 1–27.

127 Detailed in Harley, *Yates's Map of the County of Lancashire* (see note 114), pp. 9–10. See also J. B. Harley 'Uncultivated fields in the history of cartography', *Cartographic Journal*, 4 (1967), pp. 7–11, on the usefulness of the bibliographical approach in uncovering more about the personal and commercial circumstances behind map production.

128 Paul Hindle, *Maps for Historians* (Chichester, Phillimore, 1998), pp. 15–16.

129 J. B. Harley, *The County Maps from William Camden's Britannia 1695 by Robert Morden. A Facsimile* (Exeter, David and Charles, 1970), Introduction; Skelton, *County Atlases* (see note 93), p. 194.

130 J. B. Harley, *Yates's Map of the County of Lancashire* (see note 114), p. 7.

131 William Ravenhill, 'Joel Gascoyne: a pioneer of large-scale county mapping', *Imago Mundi*, 26 (1972), pp. 60–70; W. L. D. Ravenhill and O. J. Padel, *A Map of the County of Cornwall newly Surveyed by Joel Gascoyne. Reprinted in Facsimile with an Introduction* (Exeter, Devon and Cornwall Record Society, n.s. 34, 1991).

132 Laxton, 'Geodetic and topographical evaluation' (see note 114), p. 44.

133 Donald Hodson, *County Atlases of the British Isles. A Bibliography*. vol. 1 *1704–1742* (Tewin, Welwyn, Tewin Press, 1984), p. 167.

134 D. Kingsley, *Printed Maps of Sussex 1575–1900* (Lewes, Sussex Record Society, 1982), pp. 57–63.

135 Donald Hodson, *Printed Maps of Hertfordshire 1577–1900* (Folkestone, Dawson, 1974), p. 227; Harvey and Thorpe, *Printed Maps of Warwickshire* (see note 11), pp. 19–35.

136 John Varley, 'John Rocque. Engraver, surveyor, cartographer and map-seller', *Imago Mundi*, 5 (1948), pp. 83–91. Varley includes a listing of Rocque's work in England and abroad.

137 Paul Laxton, 'Rocque, John', in C. S. Nicholls (ed.), *The Dictionary of National Biography: Missing Persons* (Oxford, Oxford University Press, 1993), pp. 563–4.

138 The assembled map is 2.77 m wide and 1.67 m high. See Paul Laxton, 'John Rocque's survey of Berkshire', *Journal, Durham University Geographical Society*, 8 (1966), pp. 41–9; Paul Laxton, *John Rocque's Map of Berkshire, 1761* (Lympne, Harry Margary, 1973), Introduction; and Paul Laxton and R. I. Hodgson, 'Commonfield and enclosure – two case studies', *Journal, Durham University Geographical Society*, 9 (1967), pp. 41–74.

139 We thank John Andrews for his advice on the nature of Roque's field boundaries.

140 Paul Laxton, 'Geodetic and topographical evaluation' (see note 114), pp. 44 and 47.

141 William L. D. Ravenhill, *Two Hundred and Fifty Years of Map-making in the County of Sussex* (Lympne, Kent, Harry Margary, 1974), Introduction.

142 Ravenhill, *Two Hundred and Fifty Years* (see note 141). Laxton comments on 'serious discrepancies in the depiction of land use, field boundaries and other features' in 'Geodetic and topographical evaluation' (see note 114), p. 47.

143 J. B. Harley, 'The Society of Arts and the surveys of English counties 1759–1809', *Journal of the Royal Society of Arts*, 112 (1963), pp. 43–6, 119–24, 269–75, 538–43; quotation from p. 43.

144 Harley, 'Society of Arts' (see note 143), pp. 44–5.

145 Those receiving awards are identified in Hindle's list of County Maps (after P. Laxton, 1976) in Paul Hindle, *Maps for Local History* (London, Batsford, 1978), pp. 143–6 and in Hindle, *Maps for Historians* (see note 128), pp. 15–16.

146 Not surprisingly, the military were preoccupied with geodetic accuracy and intervisibility, the latter wholly inadequately conveyed by traditional methods of relief depiction (hill-signs and hachures). For their comments on Isaac Taylor's maps, see Laxton, *250 Years of Map Making* (see note 11).

147 J. B. Harley and J. C. Harvey, 'Introduction' to the facsimile edition of *A Survey of the County of Yorkshire by Thomas Jefferys, 1775*, 2 vols (Lympne Castle, Harry Margary, 1973), p. i. For Jefferys's contribution to magazines, see David C. Jolly, *Maps in British Periodicals* (Brookline, Mass., David C. Jolly, 1992).

148 Gough made the comment in connection with his otherwise laudatory comment on Jeffery's map of Yorkshire (1717), *British Topography* (see note 6), vol. 2, p. 478. See also Harley, 'The bankruptcy of Thomas Jefferys (see note 126).

149 Harley and Harvey, 'Introduction', *A Survey of the County of Yorkshire* (see note 147), p. i.

150 Harley and Harvey, 'Introduction', *A Survey of the County of Yorkshire* (see note 147), p. v.

151 W. L. D. Ravenhill, *Benjamin Donn. A Map of the County of Devon, 1765. Facsimile and Introduction* (Exeter, University of Exeter, 1965), pp. 4-5.

152 Donn's published texts include *The Mathematical Essays* (London, 1758) and *The Accountant and Geometrician* (London, 1765).

153 In the event, Taylor received no award from the Society.

154 Ravenhill, *Benjamin Donn* (see note 151), pp. 9-10.

155 J. B. Harley, D. V. Fowkes and J. C. Harvey, *P. P. Burdett's Map of Derbyshire, with an Explanatory Introduction* (Matlock, Derbyshire Archaeological Society, 1975), pp. i-xix. See also Paul Laxton, 'Burdett, Peter Perez' in C. S. Nicholls (ed.) *Dictionary of National Biography: Missing Persons* (see note 137), p. 102.

156 Harley, *Yates's Map of the County of Lancashire* (see note 114), p. 7.

157 Harley, *Yates's Map of the County of Lancashire* (see note 114), p. 12.

158 As Paul Laxton kindly points out, the real figure is 2° 59' 45".

159 Taylor's county maps include: Herefordshire, 1754; Hampshire, 1759; Dorset, 1765; Worcestershire, 1772, and Gloucestershire, 1777. He also drew three town plans: Oxford, 1751; Wolverhampton, 1751, and Hereford, 1757. Probably the bulk of his work was as an estate surveyor: 100 such surveys are known for Dorset alone: Paul Laxton, *250 Years of Map Making* (see note 11). Taylor also mapped estates in Devon; we thank Mary Ravenhill for this last point.

160 John Hutchins wrote to Richard Gough in 1780 to this effect: Gough, *British Topography* (see note 6), vol. 1, p. 407, cited by Laxton, *250 Years of Map Making* (see note 11). When the Society adjudicated in favour of Donn, they set Taylor's aside, resolving to reconsider it later: Ravenhill, *Benjamin Donn* (see note 151), p. 6.

161 The explanation is given by Hutchins: see Laxton, *250 Years of Map Making* (see note 11).

162 The description is Hutchins's: see Laxton, *250 Years of Map Making*, (see note 11).

163 Hindle, *Maps for Historians* (see note 128), p. 21.

164 J. B. Harley, *Christopher Greenwood, County Mapmaker and his Worcestershire Map of 1822* (Worcester, Worcestershire Historical Society, 1962), p. vii.

165 'Proposals for Publishing by Subscription' were announced in the *Leeds Mercury*, January 1815: see Harley, *Christopher Greenwood* (see note 164), p. 2, and pp. 4-5. See also Paul Laxton, 'Christopher Greenwood', *New Dictionary of National Biography* (Oxford, Oxford University Press, forthcoming).

166 Laxton, 'Greenwood' (see note 165). Middlesex was published at a scale of 2 inches to one mile, Yorkshire at three-quarters of an inch.

167 Laxton, 'Greenwood' (see note 165).

168 J. B. Harley, 'A proposed survey of Lancashire by Francis and Netlam Giles', *Transactions of the Lancashire and Cheshire Historic Society*, 116 (1965), pp. 197-206.

169 Harley, *Christopher Greenwood* (see note 164), p. 19.

170 See correspondence quoted in Harley, *Christopher Greenwood* (see note 164), pp. 21-3.

171 When the Greenwoods started in 1818, on what was to be a comprehensive survey of England and Wales, the only maps published by the Ordnance Survey all lay to the south of a line between Bristol and London (except for the map of Essex). By 1834, the Midlands, East Anglia, and South Wales had been covered: Laxton, 'Christopher Greenwood' (see note 165); Harley, *Christopher Greenwood* (see note 164), p. 24.

172 J. B. Harley, 'English county map-making in the early years of the Ordnance Survey: the map of Surrey by Joseph Lindley and William Crossley', *Geographical Journal*, 132 (1966), pp. 372-8; the quotation is from Edward Williams and William Mudge, 'An account of the trigonometrical survey carried on in 1791, 1792, 1793 and 1794', *Philosophical Transactions of the Royal Society*, 85 (1795), pp. 414-591.

173 Harley, 'English county map-making' (see note 172).

174 Harley, *Christopher Greenwood* (see note 164), pp. 28-9.

175 Harley, *Christopher Greenwood* (see note 164), pp. 30-1, quotes a critic who challenged Greenwood on this point in relation to the map of Westmorland, but received no answer.

176 Laxton, 'Geodetic and topographical evaluation' (see note 114), p. 42 and Fig. 7. Laxton finds Bryant's map of Gloucestershire (1824) 'much less praiseworthy' (p. 42 and Fig. 8).

177 Greenwood's secondary sources are discussed by Harley, *Christopher Greenwood* (see note 164), pp. 30-8.

178 See Harley's detailed comments, *Christopher Greenwood* (see note 164), pp. 49-51.

179 Harley, *Christopher Greenwood* (see note 164), pp. 53-4.

180 Blome's *Britannia*, with its county maps, was intended as one part of a three volume 'English Atlas': Skelton, *County Atlases* (see note 93), pp. 139-45. On Ogilby's lotteries and subscriptions, see Van Eerde, *John Ogilby* (see note 89), pp. 85-90.

181 Harley, Fowkes, and Harvey, *Burdett's Map of Derbyshire* (see note 155), p. vi.

182 Reproduced in Harley, Fowkes and Harvey, *Burdett's Map of Derbyshire* (see note 155), p. iii.

183 Peter Barber, 'Finance and Flattery' in Barber and Board, *Tales from the Map Room* (see note 57), pp. 140–1.

184 Moule's maps are reproduced (in colour) in Roderick Barron (ed.), *The County Maps of Old England: Thomas Moule* (London, Studio Editions, 1990). Barron contrasts the pleasantness of Moule's paper countryside with the harsh realities of contemporary events (Swing riots, 1830–1832, New Poor Law, 1834; Chartist riots, 1839). For a detailed account of the history of Moule's map, see Tony Campbell, 'The original monthly numbers of Moule's "English Counties"', *The Map Collector* 31 (1985), pp. 26–39.

185 'Cartographical curiosities 45', *The Map Collector*, 55 (Summer 1991), p. 48. This regular feature in *TMC* contains a wide range of examples of maps put to unusual uses.

186 For the playing cards signed 'W.B. invent. 1590', the several issues of Robert Morden's pack with 'The 52 Counties of England and Wales' (1676) and William Redmayne's version (1676–7), see Skelton *County Atlases* (see note 93) Items 2, 94–7; 'Cartographical curiosities 46', *The Map Collector* 56 (Autumn 1991), p. 50.

187 For the fan map, see 'Cartographical curiosities 25', *The Map Collector*, 33 (December 1985), p. 50; for the glove map, see Ralph Hyde, 'A "handy" map', *The Map Collector*, 35 (June 1986), p. 47.

188 The date of the screen is controversial. Ravenhill and Rowe 'A decorated screen map of Exeter' (see note 6), pp. 1–12, report that of four art history experts, two placed the screen at this late date, seeing the selection of the map as an expression of nostalgia, and two preferred, like both authors, to see it as a late sixteenth-century creation, perhaps associated in some way with Lord Burghley (pp. 8–9).

189 Reported by Tony Campbell, 'Chronicle', *Imago Mundi* 47 (1995), p. 208.

190 Tony Campbell, 'Four-fold map screen', *National Art Collections Fund Review for 1997*, p. 84, and personal communication. The screen was acquired in 1997 by The British Library. For a reproduction of Willdey's advertisement, see Donald Hodson, *County Atlases of the British Isles. A Bibliography*. Vol. 1, p. 142. Dennis Reinhartz, *The Cartographer and the Litterati – Herman Moll and his Intellectual Circle* (Lampeter, Edwin Mellon, 1997), pp. 145–6 reports on another extant screen.

191 Sarah Tyacke, *London Map-Sellers 1660–1720* (Tring, Map Collector Publications, 1978), p. xi.

192 Biographical details are from Tyacke, *London Map Sellers* (see note 191), pp. xii–xv and personal communication.

193 On Faden, see Laurence Worms, 'Faden, William', in C. S. Nicholls (ed.), *The Dictionary of National Biography: Missing Persons* (Oxford, Oxford University Press, 1993), p. 218; also Mary Pedley, 'Maps, war, and commerce: business correspondence with the London map firm of Thomas Jefferys and William Faden', *Imago Mundi*, 48 (1996), pp. 161–73.

194 H. G. Fordham, *John Cary, Engraver, Map, Chart and Print-Seller and Globe-Maker, 1754 to 1853* (Cambridge, Cambridge University Press, 1925; reprinted Folkestone, Dawson, 1976), Introduction.

195 Skelton, *County Atlases* (see note 93), p. v. By 'England' in this context is usually subsumed 'England and Wales'. Both genres, the English county map and the county atlas, are outstandingly well served by carto-bibliographers. For the latter, see not only Skelton's volume but also the three-volumed (to date) continuation by Donald Hodson, *County Atlases of the British Isles*, vol. 1, *Atlases Published 1704 to 1702*; vol. 2, *Atlases Published 1743–1763* (Tewin, Tewin Press, 1984–1989); vol. 3, *Atlases Published 1764–1789* (London, The British Library, 1997). For the former, see the invaluable listing of county bibliographies by Geoffrey Armitage, 'County carto-bibliographies of England and Wales. A selected list', *The Map Collector*, 52 (1990), pp. 16–24 and 'Addendum', *The Map Collector*, 73 (1995), pp. 20–3, to which subsequent publications should be added, such as R. A Carroll, *Printed Maps of Lincolnshire, 1579–1900. A Carto-bibliography. With an Appendix on Road-Books 1675–1900* (Woodbridge, Suffolk, Boydell Press for The Lincolnshire Record Society [vol. 84], 1996), and Eugene Burden's *Printed Maps of Berkshire 1574–1900*, 3 parts (Ascot, Berkshire, the author, 1988–95). Skelton and Hodson also include a good deal of important comment, although this is not always easy to locate in their respective volumes.

196 Abraham Ortelius, *Theatrum orbis terrarum* (Antwerp, 1570); Maurice Bouguereau, *Le Théâtre Françoys* (Tours, 1594); Gerard Mercator, *Atlas sive cosmographicae meditationes de fabrica mundi et fabricati figura* (Duisberg, 1595). On Mercator in general, see Marcel Watelet (ed.), *Gérard Mercator Cosmographe. Le Temps et l'Espace* (Antwerp, Fonds Mercator Paribus, 1995). Giorgio Mangani, 'Abraham Ortelius and the hermetic meaning of the cordiform map', *Imago Mundi*, 50 (1998), pp. 59–83 especially pp. 75–7. Tom Conley, *The Self-Made Map. Cartographic Writing in Early Modern France* (Minneapolis and London, University of Minnesota Press, 1997), especially pp. 202–47.

197 Evans and Lawrence, *Christopher Saxton* (see note 62), pp. 45–65. For all atlas editions mentioned in the following paragraphs in the text, see either Skelton, *County Atlases* (see note 93) or the appropriate volume of Hodson, *County Atlases* (see note 195).

198 J. B. Harley, 'Introduction', *The County Maps from William Camden's Britannia* (see note 129).

199 The world atlas is entitled 'A Prospect of the most Famous Parts of the World . . .'. Speed's 'Civil Wars' map may well have been made as a moral statement: Walter Goffart, personal communication.

200 Skelton, *County Atlases* (see note 93), no. 90, pp. 139–45; E. G. R. Taylor, 'Robert Hooke and the cartographical projects of the late seventeenth century', *Geographical Journal*, 90 (1937), pp. 529–40.

201 A copy of the 1701 octavo edition is British Library, 579.d.28, and the 1704 octavo edition is 577.f.3. The quarto edition, with Moll's maps, is more rare but there are examples in the Bodleian Library (G. A. gen.top. 4°358) and in Cambridge University Library (Atlas 6.70.6).

202 Hodson, *County Atlases* (see note 195), vol. 1. pp. 21–2.

203 Hodson, *County Atlases* (see note 195), vol. 2, p. 97. On Kitchin in general, see Laurence Worms, 'Thomas Kitchin's "Journey of Life": Hydrographer to George III, mapmaker and engraver', The *Map Collector*, 62 and 63 (1993), pp. 2–8 and 14–20, respectively.

204 New details came from maps by Jonas Moore (Cambridgeshire), Joel Gascoyne (Cornwall), John Warburton (County Durham), Isaac Taylor (Herefordshire), William Gordon (Huntingdon), John Strachey (Somerset), Richard Budgen (Sussex), Henry Beighton (Warwickshire), amongst others. Coastal charts were also used, such as those by Joseph Avery (for Poole Harbour on the Dorset map, and the Isle of Wight and the main coastline on the Hampshire map), Charles Labelye (inset on the map of Kent), and Samuel Fearon and John Eyres (parts of the Cheshire and Cumberland coasts): information in Hodson, *County Atlases* (see note 195), vol. 2, pp. 109–221.

205 Hodson, *County Atlases* (see note 195), vol. 2, p. 105. As an example of the fluidity of the printing business in the eighteenth century, Hodson notes that between 1755 and 1761 Tinney worked with at least a dozen different partners, including Thomas Jefferys, John Rocque and Thomas Kitchin.

206 Hodson, *County Atlases* (see note 195), vol. 3, p. 5.

207 Hodson, *County Atlases* (see note 195), vol. 3, p. 28.

208 Fordham, *John Cary* (see note 154), p. 47.

209 R. P. Sprent's phrase: 'Introduction' to Thomas Chubb, *The Printed Maps in the Atlases of Great Britain and Ireland. A Bibliography, 1579–1870* (London, Ed. J. Burrow, 1927; reprinted in facsimile, London, Dawson, 1966), p. xvii. Cary is also well known for his *Actual Survey of the Country Fifteen Miles Round London: On a Scale of One Inch to a Mile* (1786).

210 Laxton, 'Geodetic and topographical evaluation' (see note 114), p. 37.

211 Paul Laxton, personal communication.

212 See David Smith, 'The representation of industry on large-scale county maps of England and Wales 1700–*c*.1840,' *Industrial Archaeology Review*, 12 (1990), pp. 153–77; Paul Laxton, 'Wind and water power' in J. Langton and R. J. Morris (eds), *Atlas of Industrializing Britain* (London, Methuen, 1986, pp. 69–71.

213 Laxton, 'Geodetic and topographical evaluation' (see note 114), p. 43.

214 Harley et al., *Burdett's Map of Derbyshire* (see note 155), p. xi. Burdett omitted not only smaller woods (including one at his own family seat of Foremark) but also some large commons (for instance, 516 acres of Hognaston Winn) and, less accountably given his obligation to subscribers, a small number of major parks (Melbourne and Hopwell) and at least one country mansion (Norbury Hall, home of the Fitzherberts). Burdett was no more (or less) faithful to the realities of the landscape in his later map of Cheshire (1777), which Harley also describes as 'sporadic' in its depiction of many types of industries and details of mining: J. B. Harley, 'Maps of early Georgian Cheshire', *Cheshire Round*, 1 (1966), p. 266; and J. B. Harley and Paul Laxton, *A Survey of the County Palatine of Chester by P. P. Burdett* (Liverpool, Lund Press for the Historic Society of Lancashire and Cheshire, Occasional Series 1, 1974), esp. pp. 25–34. For an overall assessment, see Laxton, 'Geodetic and topographic evaluation' (see note 114).

215 Harley, *Christopher Greenwood* (see note 164), p. 54.

216 Cited in Blake Tyson, 'John Adams's cartographic correspondence to Sir Daniel Flemming of Rydall Hall, Cumbria, 1676–1687', *Geographical Journal*, 151 (1985), pp. 33 and 23 respectively.

217 Paul Laxton cites the 'hopeless meridians on P. P. Burdett's map of Cheshire (1777)' and those of William Yates's map of Warwickshire (1793) 'where the figures for St Mary's Church, Warwick, would give a prime meridian at what is now London's Liverpool Steet Station, at a time when Greenwich had been universally adopted as the prime meridian': Laxton, 'Geodetic and topographical evaluation (see note 114), p. 39.

218 Laxton, 'Geodetic and topographical evaluation' (see note 114), p. 44.

CHAPTER 4

1 We are most grateful to Sarah Bendall and Paul Harvey for their generous and most helpful comments on an early draft of this chapter.

2 Owners or tenants are usually identified by name. In some cases, though, an abstract sign, possibly related to brand marks used for the livestock of the parish, is used instead or in addition, as at East Stoke, Nottinghamshire, in 1561: 'A mappe of Stoke Ground [Nottinghamshire] wc was [. . .] betwixt Mr Markham and John Mullin[eux?] of Thorpe' (1561, Nottingham Archives Office, East Stoke MS 107). The complex signs identifying the plots of each

owner/tenant are keyed into a reference list, also on the map. The map is listed by Harold Nichols, *Local Maps of Nottinghamshire to 1800. An Inventory* (Nottingham, Nottingham County Council Leisure Services, 1987), pp. 41–2.

3 In France, where cadastral mapping goes back to the Napoleonic administration, and in many continental countries, the unit is the *commune*.

4 A point which has been made elsewhere: R. J. P. Kain and E. Baigent, *The Cadastral Map in the Service of the State: A History of Property Mapping* (Chicago and London, University of Chicago Press, 1992), p. xviii.

5 F. G. Emmison, *Catalogue of Maps in the Essex Record Office* (Chelmsford, Essex County Council, 1947), p. v; I. H. Adams, 'Estate plans', *The Local Historian*, 12 (1976), pp. 26–30; quotation from p. 26. See also, David Buisseret, 'Defining the estate map', in David Buisseret (ed.), *Rural Images: Estate Maps in the Old and New Worlds* (Chicago and London, University of Chicago Press, 1996), pp. 1–4.

6 P. D. A. Harvey, 'Shouldham, Norfolk, 1440 × 1441', in R. A. Skelton and P. D. A. Harvey (eds), *Local Maps and Plans from Medieval England* (Oxford, Clarendon Press, 1986), pp. 195–201.

7 On the nature and role of maps for boundary disputes, see P. D. A. Harvey, *Maps in Tudor England* (London, The Public Record Office and The British Library; Chicago, University of Chicago Press, 1993), pp. 102–15. For examples of estate maps serving boundary disputes, see Sarah Bendall, 'Enquire "When the same platte was made and by whome and to what intent": sixteenth-century maps of Romney Marsh', *Imago Mundi*, 47 (1995), pp. 34–48.

8 Sarah Bendall, 'Interpreting maps of the rural landscape: an example from late sixteenth-century Buckinghamshire', *Rural History*, 4 (1993), pp. 107–21.

9 O. A. W. Dilke, *The Roman Land Surveyors: An Introduction to the Agrimensores* (Newton Abbot, David and Charles, 1971), p. 88; O. A. W. Dilke, 'Maps in the service of the state: Roman cartography to the end of the Augustan Era', in J. B. Harley and David Woodward (eds), *The History of Cartography*, vol. 1, *Cartography in Prehistoric, Ancient and Medieval Europe and the Mediterranean* (Chicago and London, University of Chicago Press, 1987), pp. 201–11.

10 Della Hooke, *Worcestershire Anglo-Saxon Charter-Bounds* (Woodbridge, Suffolk, Boydell Press, 1990); Della Hooke, *Pre-Conquest Charter-Bounds of Devon and Cornwall* (Woodbridge, Suffolk, Boydell Press, 1994).

11 Maurice W. Beresford, *History on the Ground: Six Studies in Maps and Landscapes* (London, Lutterworth Press, 1957; revised edition, London, Methuen, 1971), pp. 31–7.

12 H. C. Darby, 'The agrarian contribution to surveying in England', *Geographical Journal*, 82 (1933),

pp. 529–35; Beresford, *History on the Ground* (see note 11), p. 46.

13 F. G. Emmison, 'Estate maps and surveys', *History*, 48 (1963), pp. 34–7.

14 Beresford, *History on the Ground* (see note 11), pp. 25–62; Alain Pottage, 'The measure of land', *Modern Law Review*, 57 (1994), pp. 361–84. Reference to natural features as boundary markers was still common in the seventeenth century, when they were also sometimes included on maps. The survey of Misson Commons (on the Lincolnshire-Nottinghamshire border) made in 1629 defines the area in question as 'From the Parsons Crosse to the grained oake tree, go to the Swans well, then to Finningley Parke corner, and so to Hewing balk, then to the furtherest Stoope, and so along the Stoopes to parsons crosse'. See Nichols, *Local Maps of Nottinghamshire* (see note 2), p. 81.

15 See, for example: E. G. R Taylor, 'The surveyor,' *Economic History Review*, 17 (1947), pp. 121–33; D. J. Price, 'Medieval land surveying and topographical maps', *Geographical Journal*, 121 (1955), pp. 1–10. See also the catalogue of an exhibition associated with a series of lectures on estate mapping delivered in 1988 at the Newberry Library, Chicago: David Buisseret, *Rural Images: The Estate Plan in the Old and New Worlds* (Chicago, The Newberry Library, 1988). The lectures themselves are published as David Buisseret (ed.), *Rural Images: Estate Maps in the Old and New Worlds* (see note 5).

16 Edward Lynam, *British Maps and Map-Makers*, 3rd impression, rev.; London (Collins, 1947), pp. 11–12; P. D. A. Harvey, *The History of Topographical Maps: Symbols, Pictures and Surveys* (London, Thames and Hudson, 1980), p. 102.

17 B. K. Roberts, 'An early Tudor sketch map: its context and implications,' *History Studies* 1 (1968), pp. 33–8.

18 Roberts, 'An early Tudor sketch map', (see note 17), p. 36.

19 Harvey, *Maps in Tudor England* (see note 7), p. 93.

20 Harvey, *Maps in Tudor England* (see note 7), pp. 78–93.

21 Ralph Agas, *A Preparative to the Platting of Landes and Tenements for Surveigh* (1596); John Norden, *The Surveyors Dialogue . . .* (1607).

22 Quotations in this paragraph are from: Andrew McRae, 'To know one's own: estate surveying and the representation of the land in early modern England', *Huntington Library Quarterly*, 56 (1993), pp. 332–57 and see also his recent book, *God Speed the Plough: The Representation of Agrarian England, 1500–1660* (Cambridge, Cambridge University Press, 1996); F. M. L. Thompson, *Chartered Surveyors: The Growth of a Profession* (London, Routledge and Kegan Paul, 1968), p. 2.

23 Agas, *A Preparative to Platting,* (see note 21), pp. 14–15.

24 Norden, *Surveyors Dialogue* (see note 21), pp. 15–16.

25 David H. Fletcher, *Estate Maps of Christ Church, Oxford: The Emergence of Map Consciousness, c.1600–c.1840*, unpublished University of Exeter Ph.D. thesis, 1990; see also his *The Emergence of Estate Maps: Christ Church, Oxford 1600–1840* (Oxford, Oxford University Press, 1995) and 'Map or terrier? The example of Christ Church Oxford estate management 1600–1840', *Transactions of the Institute of British Geographers*, 23 (1998), pp. 221–37.

26 Data taken from Peter Eden (ed.), *Dictionary of Land Surveyors and Local Cartographers of Great Britain and Ireland 1550–1850* (3 vols. and supplement; Folkestone, Dawson, 1975, 1976 and 1979). For similar analyses, see A. Sarah Bendall, *Maps, Land and Society: A History with a Carto-Bibliography of Cambridgeshire Estate Maps c.1600–1836* (Cambridge, Cambridge University Press, 1992), and *Dictionary of Land Surveyors and Local Map-Makers of Great Britain and Ireland 1530–1850*, second edition by Sarah Bendall, 2 vols (London, The British Library, 1997).

27 Joan Thirsk (ed.), *The Agrarian History of England and Wales* (Cambridge, Cambridge University Press, 1967), vol. 4, p. 15.

28 Ian H. Adams, 'Large-scale manuscript plans in Scotland', *Journal of the Society of Archivists*, 3 (1967), pp. 286–90, quotation on p. 286; see also Adams's 'Economic process and the Scottish land surveyor', *Imago Mundi*, 27 (1975), pp. 13–18; and his 'The agents of agricultural change', in M. L. Parry and T. R. Slater (eds), *The Making of the Scottish Countryside* (London, Croom Helm, 1980). On Wales see R. Davies (ed.), *Estate Maps of Wales, 1600–1836* (Aberystwyth, National Library of Wales, 1982), p. 9. The contrasting Irish experience of large-scale property mapping is reconstructed by that country's doyen of map historians, John Andrews, in his *Plantation Acres: an Historical Study of the Irish Land Surveyor* (Belfast, Ulster Historical Foundation, 1985).

29 P. D. A. Harvey, 'An Elizabethan map of manors in north Dorset', *British Museum Quarterly*, 29 (1965), pp. 82–4, quotation on p. 83.

30 P. Eden, 'Three Elizabethan estate surveyors: Peter Kempe, Thomas Clerke and Thomas Langdon', in Sarah Tyacke (ed.), *English Map Making 1500–1650* (London, The British Library, 1983), pp. 68–84, quotation on p. 70; Sarah Tyacke and John Huddy, *Christopher Saxton and Tudor Map-Making* (London, The British Library, 1980), p. 48.

31 Beresford, *History on the Ground* (see note 11), p. 90; P. D. A. Harvey, 'English estate maps: their early history and their use as historical evidence', in Buisseret, *Rural Images* (see note 5): pp. 27–61.

32 Elizabeth M. Elvey, *A Handlist of Buckinghamshire Estate Maps* (Buckingham, Buckinghamshire Record Society, Lists and Indexes No. 2, 1963), p. 56.

33 G. R. Batho, 'Two newly discovered maps by Christopher Saxton', *Geographical Journal*, 125 (1959), pp. 70–74; Ifor M. Evans and Heather Lawrence, *Christopher Saxton, Elizabethan Map-Maker* (Wakefield and London, Wakefield Historical Publications and The Holland Press, 1979), pp. 116–18.

34 G. R. Batho, 'The finances of an Elizabethan nobleman: Henry Percy, Ninth Earl of Northumberland (1564–1632)', *Economic History Review*, 2nd series, 9 (1957), pp. 433–50.

35 Batho, 'Two newly discovered maps' (see note 33) p. 72.

36 Fletcher, *The Emergence of Estate Maps* (see note 25), pp. 120–2.

37 Nottingham Archives Office, DD CW 5a/2; Nichols, *Local Maps of Nottinghamshire* (see note 2), p. 152.

38 Nottingham Archives Office, EP 2 R; Nichols, *Local Maps of Nottinghamshire* (see note 2), p. 47.

39 John Fitzherbert's treatise, *Here begynneth a right frutefull mater: and hath to name the boke of surveying and improveme[n]ts* (1523), which remained the only didactic text on the subject until Benese's in 1537, makes no mention of a map although it is full of advice to the steward on various ways of improving the productivity of the estate.

40 Price indices, 1470–1670. Sources: W. Abel, *Agricultural Fluctuations in Europe from the Thirteenth to the Twentieth Centuries* (London, Methuen, 1980); P. Bowden, 'Agricultural prices, farm profits and rents', in Thirsk, *Agrarian History of England and Wales* (see note 27), vol. 4, pp. 593–695; H. Neveux, J. Jacquart, and E. Le Roy Ladurie, *Histoire de la France rurale*, vol. 2, *L'âge classique des paysans, 1340–1789* (Paris, Seuil, 1975); and N. J. G. Pounds, *An Historical Geography of Europe, 1500–1840* (Cambridge, Cambridge University Press, 1979).

41 B. H. Slicher van Bath, *Agrarian History of Western Europe, 1500–1850* (London, Arnold, 1963).

42 P. Bowden, 'Agricultural prices' (see note 40), quotation on p. 690; and see Eric Kerridge (ed.), *Surveys of the Manor of Philip, First Earl of Pembroke and Montgomery 1631–2* (Devizes, Wiltshire Archaeological and Natural History Society, Records Branch, No. 9, 1953).

43 Darby, 'Agrarian contribution' (see note 12); E. G. R. Taylor, *Tudor Geography, 1485–1583* (London, Methuen, 1930), and Taylor, 'Surveyor' (see note 15); D. Chilton, 'Land measurement in the sixteenth century,' *Transactions of the Newcomen Society*, 31 (1957–9), pp. 111–29.

44 Bendall, *Maps, Land and Society* (see note 26), pp. 168–76 and Sarah Bendall, 'Estate maps of an

English county: Cambridgeshire, 1600–1836', in Buisseret (ed.) *Rural Images* (see note 5), pp. 63–90.

45 A. D. M. Phillips, 'The seventeenth century maps and surveys of William Fowler', *Cartographic Journal*, 17 (1980), pp. 100–10.

46 J. B. Harley and E. A. Stuart, 'George Withiell: a west country surveyor of the late-seventeenth century', *Devon and Cornwall Notes and Queries*, 35 (1982), pp. 45–58, quotation on p. 49.

47 J. B. Harley, 'Meaning and ambiguity in Tudor cartography', in Tyacke (ed.), *English Map-Making* (see note 30), pp. 22–45, quotation on p. 37.

48 Bowden, 'Agricultural prices' (see note 40), p. 674. See Bendall, *Maps, Land and Society* (see note 26), pp. 177–84 on estate maps as status symbols.

49 John V. Beckett, *A History of Laxton. England's Last Open-field Village* (Oxford, Blackwell, 1989), p. 57.

50 Mark Pierce's map and terrier are in the Bodleian Library, Oxford: MS C17:48 (9a) and MS Top. Notts.c.2, respectively. A full-sized photographic copy of the map, in colour, is displayed in the Visitors' Centre at Laxton.

51 Harley, 'Meaning and ambiguity' (see note 47), p. 37.

52 Victor Morgan, 'The cartographic image of "The Country" in early modern England', *Transactions of the Royal Historical Society*, 5th series, 29 (1979), pp. 129–54, who quotes John Dee on the display of maps.

53 William Leybourne, *The Compleat Surveyor* (London, 1653), p. 275. Leybourne tells the mapmaker to draw 'divers little Trees' in the most important places and to use lively colours for them and for the topographical features. He then goes on to say that the coat of arms should be represented in the upper part of the map – correctly coloured – together with a compass rose, scale bar and a picture of the manor house, and provides a diagram to show where these should go. The diagram is reproduced in Harley, 'Meaning and ambiguity' (see note 47), plate 10. Leybourne in the mid seventeenth century was echoing advice given in *An Anonymous Text Book of Surveying c.1600*, British Library Sloane MS 838, ff. 23v–24.

54 Thompson, *Chartered Surveyors* (see note 22), p. 10. See also A. G. R. Smith, *Science and Society in the Sixteenth and Seventeenth Centuries* (London, Thames and Hudson, 1972), on the differences between the mental worlds of 1470 and 1670; Antonia McLean, *Humanism and the Rise of Science in Tudor England* (London, Heinemann, 1972); Alfred W. Crosby, *The Measure of Reality. Quantification and Western Society, 1250–1600* (Cambridge, Cambridge University Press, 1997).

55 John Holwell, *A Sure Guide to the Practical Surveyor, in Two Parts* (London, Christopher Hussey, 1678), preface.

56 For further details see John Chapman, 'The interpretation of enclosure maps and awards', in K. Barker and R. J. P. Kain (eds), *Maps and History in South-West England* (Exeter, Exeter University Press, 1991), pp. 72–88.

57 John D. Chambers and Gordon E. Mingay, *The Agricultural Revolution, 1770–1860* (London, Batsford, 1966), pp. 79–80; Michael Turner, *Enclosures in Britain, 1750–1830* (London, Macmillan, 1984), pp. 36–52; H. C. Darby, 'The age of the improver', in H. C. Darby (ed.), *A New Historical Geography of England after 1600* (Cambridge, Cambridge University Press, 1976), pp. 1–88; J. R. Walton, 'Agriculture and rural society, 1730–1914', in R. A. Dodgshon and R. A. Butlin (eds), *An Historical Geography of England and Wales* 2nd ed. (London, Academic Press, 1990), pp. 323–50; J. A. Yelling, *Common Field and Enclosure in England 1450–1850* (London, Macmillan, 1977).

58 H. C. Darby, 'The agrarian contribution to surveying in England', *Geographical Journal*, 82 (1933), pp. 529–35, especially p. 530; J. B. Harley, 'Maps, knowledge and power', in Denis Cosgrove and Stephen Daniels (eds), *The Iconography of Landscape: Essays on the Symbolic Representation, Design and Use of Past Environments* (Cambridge, Cambridge University Press, 1988), pp. 277–312, especially pp. 284–6.

59 William Marshall, *On the Landed Property of England* (London, G. and W. Nichol, 1804), pp. 29 and 345.

60 William Marshall, *On the Appropriation and Inclosure of Commonable and Intermixed Lands* (London, W. Bulmer, 1801), p. 47.

61 Marshall, *Appropriation and Inclosure* (see note 60), p. 61.

62 41 Geo III c.109 (1801): 4 and 5.

63 Henry S. Homer, *An Essay on the Nature and Method of Ascertaining the Specifick Shares of Proprietors upon the Enclosure of Common Fields* (Oxford, S. Parker, 1766), pp. 43–6.

64 Homer, *An Essay on the Nature and Method* (see note 63) pp. 53–4.

65 Maurice W. Beresford, 'The decree rolls of Chancery as a source for economic history', *Economic History Review*, 2nd series, 32 (1979), pp. 1–10; J. R. Wordie, 'The chronology of English enclosure, 1500–1914', *Economic History Review*, 36 (1983), pp. 483–505; John Chapman, 'The chronology of English enclosure', *Economic History Review*, 37 (1984), pp. 557–9. Agreements do extend beyond these centuries in both directions.

66 John Chapman, 'Some problems in the interpretation of enclosure awards', *Agricultural History Review*, 26 (1978), pp. 108–14.

67 Fletcher, *The Emergence of Estate Maps* (see note 25).

68 John Chapman, 'The extent and nature of Parliamentary enclosure', *Agricultural History Review*, 35 (1987), pp. 25–35, especially p. 27.

69 Northamptonshire County Record Office, Map 561.

70 W. E. Tate, *A Domesday of English Enclosure Acts and Awards*, ed. Michael E. Turner (Reading, University of Reading, 1978). See also Turner's *English Parliamentary Enclosure: Its Historical Geography and Economic History* (Folkestone, Dawson, 1980); 'The landscape of parliamentary enclosure', in Michael Reed (ed.), *Discovering Past Landscapes* (London, Croom Helm, 1984), pp. 132–66; and *Enclosures in Britain 1750–1830* (London, Macmillan, 1984).

71 Turner, 'Landscape of parliamentary enclosure' (see note 70), p. 143; Roger J. P. Kain and Richard R. Oliver, 'Government-sponsored, large-scale mapping of England and Wales before the Ordnance Survey', a research project funded by the Economic and Social Science Research Council. A monograph provisionally titled *The Enclosure Maps of England and Wales* is in preparation for Cambridge University Press.

72 The Cotgrave maps (Nottingham University Library, Ma 2 P 16/1 and Ma 2 P 17) were probably both surveyed by William Calvert: Nichols, *Local Maps of Nottinghamshire* (see note 2), pp. 33–4.

73 Nottingham University Library EY 513; Nichols, *Local Maps of Nottinghamshire* (see note 2), p. 58.

74 F. G. Emmison, *Types of Open Field Parishes in the Midlands* (London, Historical Association, Pamphlet No. 108, 1937); J. B. Harley, *Maps for the Local Historian: a Guide to the British Sources* (London, National Council for Social Service for the Standing Conference for Local History, 1972).

75 Barbara English, *Yorkshire Enclosure Awards* (Hull, University of Hull Studies in Regional and Local History No. 5, 1985), cover and p. 131.

76 Roger J. P. Kain and Hugh C. Prince, *The Tithe Surveys of England and Wales* (Cambridge, Cambridge University Press, 1985); Roger J. P. Kain, *An Atlas and Index of the Tithe Files of Mid-Nineteenth-Century England and Wales* (Cambridge, Cambridge University Press, 1986).

77 Kain and Baigent, *Cadastral Map* (see note 4), pp. 58–145, 225–33, 289–97, 325–8.

78 Kain and Prince, *Tithe Surveys* (see note 76), pp. 6–15.

79 William Cobbett, *Rural Rides* (London, Dent, 1912), vol. 2, p. 124.

80 George F. E. Rudé, 'English rural disturbances on the eve of the first Reform Bill, 1830–1', *Past and Present*, 37 (1967), pp. 87–102; Eric J. Hobsbawm and George F. E. Rudé, *Captain Swing* (London, Lawrence and Wishart, 1969); Andrew Charlesworth, *Social Protest in a Rural Society* (Norwich, Geo Abstracts for Historical Geography Research Group, Research Series No. 1, 1979); Andrew Charlesworth

(ed.), *An Atlas of Rural Protest in Britain, 1548–1900* (London, Croom Helm, 1983).

81 Eric J. Evans, *The Contentious Tithe: The Tithe Problem and English Agriculture, 1750–1850* (London, Routledge and Kegan Paul, 1976); Eric J. Evans, *Tithes: Maps, Apportionments and the 1836 Act* (Chichester, British Association for Local History, 1993).

82 This section on the nature of maps required by the Tithe Commutation Act 1836 is based closely on Kain and Baigent, *Cadastral Map* (see note 4), pp. 246–51.

83 'Copy of Papers Respecting the Proposed Survey of Lands under the Tithe Act', *British Parliamentary Papers (House of Commons, Session 1837)*, vol. 41, pp. 11–16.

84 'Copy of Papers . . . Tithe Act' (see note 83), p. 3.

85 Cited in George H. Whalley, *The Tithe Act and the Tithe Act Amendment Act* (London, Shaw and Sons, 1838), p. 240.

86 'Copy of Papers . . . Tithe Act' (see note 83), p. 10.

87 'Copy of Papers . . . Tithe Act' (see note 83), p. 10.

88 Richard Oliver and Roger Kain, 'Maps and the assessment of parish rates in nineteenth-century England and Wales, *Imago Mundi*, 50 (1998), pp. 156–73.

89 'Copy of Papers . . . Tithe Act' (see note 83), p. 4. See also Thompson, *Chartered Surveyors* (see note 22); Roger J. P. Kain, 'R. K. Dawson's Proposals in 1836 for a Cadastral Survey of England and Wales', *Cartographic Journal*, 12 (1975), pp. 81–8; Kain and Prince, *Tithe Surveys* (see note 76); Geraldine Beech, 'Tithe maps', *The Map Collector*, 3 (1985), pp. 20–5.

90 'Report from the Select Committee on Survey of Parishes (Tithe Commutation Act); with the Minutes of Evidence', *British Parliamentary Papers (House of Commons)*, vol. 6 (1837), pp. 15, 24, 27.

91 Thompson, *Chartered Surveyors* (see note 22), p. 105.

92 Whalley, *Tithe Act* (see note 85), pp. 194–8.

93 [Tithe Commissioners for England and Wales], *Instructions as to Forms of Apportionment and Maps* (London, Tithe Commission, 31 July 1837).

94 Roger J. P. Kain and Richard R. Oliver, *The Tithe Maps of England and Wales: A Cartographic Analysis and County by County Catalogue* (Cambridge, Cambridge University Press, 1995), pp. 708–819.

CHAPTER 5

1 For a fuller discussion of categories of medieval and post-medieval travellers and travelling aids (itineraries and maps), see Catherine Delano-Smith, 'Milieus of mobility: itineraries, road maps, and route maps', in James R. Akerman (ed.) *Maps for Mobility. The Twelfth Nebenzhal Lectures in the History of*

Cartography, October 1996, at The Newberry Library (Chicago and London, University of Chicago Press, forthcoming).

2 For a useful summary of the scales of regular travel in the Middle Ages, see Philip Beale, *A History of the Post in England from the Romans to the Stuarts* (Aldershot, Ashgate, 1998), Chapter 3.

3 The Southampton carriers served Winchester and Oxford; others carried goods from the Cotswolds by road and river to London, from the midland counties to the Stourbridge Fair near Cambridge, and from Winchester and Oxford to York and Newcastle-on-Tyne: Michael Postan, 'The trade of medieval Europe', in M. M. Postan and Edward Miller (eds), *The Cambridge Economic History of Europe*, vol. 2, *Trade and Industry in the Middle Ages*, 2nd ed. (Cambridge, Cambridge University Press, 1987), pp. 191–6. The carriers were not only for trade but also for the transport of people, such as university students going to Oxford or Cambridge and law students going to London: Beale, *The Post in England* (see note 2), Chapter 7. In Hastings, Sussex, in the eighteenth century, there were still 'rippiers' who carried herring and mackerel along the postroads to London: G. Joan Fuller, 'The development of roads in the Surrey-Sussex Weald and coastlands between 1700 and 1900', *Transactions of the Institute of British Geographers*, 19 (1953), pp. 37–49.

4 H. G. Fordham prefaced his account of road-books and itineraries by asserting that 'roads hardly existed [in England] at the time Leland travelled through the country' (1535–43) and then went on to document the contemporary existence of postroads and a Post Master: *Studies in Carto-Bibliography, British and French, and in the Bibliography of Itineraries and Road-Books* (Oxford, Clarendon Press, 1914), pp. 26–7. The second quotation is from J. J. Jusserand, *Les Anglais au Moyen Age: la vie nomade et les routes d'Angleterre au IXe siècle* (Paris, 1884); 4th edition translated by L. T. Smith as *English Wayfaring Life in the Middle Ages* (London, Ernest Benn, 1950), p. 41, who likewise proceeded to fill a still widely-quoted book with descriptions of travellers and their travels in medieval England. The tendency to make sweeping and misleading generalisations about the inadequacy of early roads persists: see, for example, Alan M. MacEachren and Gregory B. Johnson, 'The evolution, application and implications of strip format travel maps', *Cartographic Journal*, 24 (1987), pp. 147–58, especially p. 150. For a more judicious and informed review of travel conditions in and after the Middle Ages, see Sir William Addison, *The Old Roads of England* (London, Batsford, 1980).

5 Charles A. J. Armstrong, 'Some examples of the distribution and speed of news in England at the time of the wars of the Roses', in R. W. Hunt, W. A. Pantin and R. W. Southern (eds), *Studies Presented to Frederick Maurice Powicke* (Oxford, Clarendon

Press, 1948), pp. 429–59; H. G. Fordham, 'Earliest tables of highways of England and Wales, 1541–61', *The Library*, 4th series, 8 (1927–8), pp. 349–54. Two-wheeled carts were in use in Cornwall during Roman times, according to Pliny, and carts and waggons were much used in the Middle Ages for the transport of goods such as grain and wood; see references in R. A. Donkin, 'Changes in the early Middle Ages', in H. C. Darby (ed.), *A New Historical Geography of England* (Cambridge, Cambridge University Press, 1973), p. 119. The real damage to road surfaces was caused by the four-wheeled waggons, hauled by eight to ten horses and carrying up to, or over, 3000 kg, especially when these transgressed into crops to avoid the ruts and potholes of the main track.

6 Beale, *The Post in England* (see note 2), p. 109.

7 The importance of obtaining information *en route* can be verified from the phrasebooks which began to be published on the continent for commercial travellers from the 1530s onwards. In Noël Berlaimont's seven-language edition (*Colloquia et dictionariolum septem linguarum, Belgicae, Anglicae, Teutonicae, Latinae, Italicae, Hispanicae, Gallicae*, Antwerp, 1586), Chapter 4 contains the phrases needed for asking the way. We are indebted to Dr Christopher Wells, St Edmund Hall, Oxford, for this invaluable corroboration of what common sense has always suggested.

8 Sigeric's itinerary (British Library, Cotton MS. Tiberius B.V. f. 23) is reproduced in P. D. A. Harvey, *Medieval Maps* (London, Public Record Office and The British Library, 1991), p. 9. See also Veronica Orteberg, 'Archbishop Sigeric's Journey to Rome in 990', *Anglo-Saxon England* 19 (1990), pp. 197–276, especially pp. 228–44 for a transcription of the route.

9 William Worcestre's itineraries (Corpus Christi College, Cambridge, MS 210) have been edited by John H. Harvey (ed.), *William Worcestre, Itineraries* (Oxford, Clarendon Press, 1969).

10 Zillah Dovey, *An Elizabethan Progress. The Queen's Journey into East Anglia, 1578* (Stroud, Alan Sutton, 1996), pp. 4 and 17–19.

11 John Dee, *Mathematical preface to The Elements of Geometrie* (1570), sig. D.iiii verso. Cited by David W. Waters, *The Art of Navigation in Elizabethan and Early Stuart Times* (London, Hollis and Carter, 1958), p. 3, who also underlines the distinction between coastal navigation (pilotage) and oceanic navigation, the former an ancient skill, the latter new in the sixteenth century. We are grateful to Dr Andrew Cook, British Library, for reading our sections on maritime travel and for his constructive comments.

12 The use of sailing directions can be traced back to the *peripli* of classical Greece. On the distinction between directions and charts, see Tony Campbell, 'Portolan charts from the late thirteenth century to 1500' in J. B. Harley and David Woodward (eds), *The*

History of Cartography Vol. 1, *Cartography in Prehistoric, Ancient, and Medieval Europe and the Mediterranean* (Chicago and London, University of Chicago Press, 1987), pp. 371–463. Itineraries are also both ubiquitous and of considerable antiquity; see the various entries indexed in Harley and Woodward, *History of Cartography*, vol. 1; O. A. W. Dilke, 'Itineraries and geographical maps in the early and late Roman Empires', *ibid*. pp. 234–8.

13 The medieval tide table for London Bridge (British Library, Cotton MS. Julius D. VII, f. 45b) is printed in J. O. Halliwell, *Rara Mathematica. A Collection of Treatises on the Mathematics* (London?, 1834), p. 55. See also E. G. R. Taylor and M. W. Richey, *The Geometrical Seaman. A Book of Early Nautical Instruments* (London, Hollis and Carter for The Institute of Navigation, 1962), pp. 101–2. For sixteenth-century examples of tables and charts in colour of the tides 'about the coasts of France, Flanders, Brittany, Wales, Ireland, and Spain' by John Marshall, written on robust vellum sometime after 1544, see British Library, Royal MS 17.A.2, ff. 9–10, and Add. MS 22721, written between 1540 and 1550.

14 Sailors also liked to know which dates were regarded locally as inauspicious for setting sail etc: Louis Dujardin-Troadec, *Les cartographes bretons du Conquet. La navigation en images, 1543–1650* (Brest, Imprimerie commerciale et administrative, 1966).

15 D. W. Waters, *The Rutter of the Sea. The Sailing Directions of Pierre Garcie. A Study of the First English and French Printed Sailing Directions. With facsimile reproductions* (New Haven and London, Yale University Press, 1967), pp. 9–10.

16 See Francesco Balducci Pegolotti, *La Pratica della mercatura*, edited by Allan Evans (Cambridge, Mass., Medieval Academy of Arts, 24, 1934), pp. 256–8, for a journey made between 1310 and 1340.

17 On rutters in general see: E. G. R. Taylor, *The Haven-Finding Art: A History of Navigation from Odysseus to Captain Cook* (London, Hollis and Carter, 1956), pp. 102–9 and 132–6; Taylor and Richey, *The Geometrical Seaman* (see note 13), pp. 97–9; Waters, *Art of Navigation* (see note 11), pp. 11–16; Waters, *Rutter of the Sea* (see note 15), pp. 7–9.

18 Ebesham's copy, bound with a miscellany of treatises dealing with aspects of ceremony, heraldry and English feats of arms in France, is now British Library, MS Lansdowne 285, formerly ff. 138–42, now 136–40. See Waters, *Art of Navigation* (see note 11), p. 12, n.1 and G. A. Lester, *Sir John Paston's 'Grete Book'. A Descriptive Catalogue, with an Introduction, of British Library MS Lansdowne 285* (Cambridge, D. S. Brewer, 1984) pp. 164–6 for comments on the rutter.

19 Cited in Waters, *Art of Navigation* (see note 11), p. 12.

20 British Library C.21.a.43.

21 There is no extant manuscript of 'Le routier de la mer jusques au fleuve de Jourdain', only a single copy of an anonymous printed rutter from between 1502 and 1510 now in the Bibliothèque Nationale, Paris: Waters, *Rutter of the Sea* (see note 15), pp. 4–5.

22 This went through many editions in the sixteenth century (the British Library copy, C.97.bb.23, is of 1531) and a score of editions were published up to the mid seventeenth century. See Christopher Terrell, 'The seaman's view', *The Map Collector*, 31 (1985), pp. 2–7.

23 The simple style of the profiles, designed for ease of recognition, has tended to be misunderstood. Waters, *Art of Navigation* (see note 11), p. 14, calls them 'crude'.

24 Tides are of little importance in most of the Mediterranean and were scarcely mentioned in directions such as the *Compasso di navigare*, an 'outstanding piece of objective, factual, indeed scientific writing' dated January 1296 but probably compiled nearly fifty years earlier: Campbell, 'Portolan charts' (see note 12), pp. 382–3. The *Compasso* systematically describes sailing conditions along the coasts of the Mediterranean in much the same way as a modern pilot book. Distances (in 'little' sea-miles), landmarks, directions and soundings are given in some detail, and the quality of the port, harbour or landing is commented on. See B. R. Motzo, *Il Compasso di Navigare* (Cagliari, Università di Cagliari, 1947), and Taylor, *Haven-finding Art* (see note 17), pp. 104–8.

25 No copy of the first edition of Copeland's rutter is known. The British Library has a copy dated 1541, to which has been added Richard Proude's *Rutter of the North* (C.21.a.43). There are no illustrations.

26 One of Brouscon's manuscripts, signed 'Fait par G. Brouscon du Conquet' and dated 1540–50, is now in the British Library (Add. MS 22721). It comprises two small volumes, boxed, with individual maps folded to fit the 12 × 8 cm cases.

27 British Library, Royal MS 17.A.2. The map of England has south at the top of the page (f. 10v). See Tony Campbell, 'Finding a safe haven when at sea', in Peter Barber and Christopher Board, *Tales from the Map Room. Fact and Fiction about Maps and their Makers* (London, BBC Books, 1993), pp. 168–9, and Dujardin-Troadec, *Les Cartographes bretons* (see note 14), pp. 49–52.

28 *Volume of Great and Rich Discoveries* (1577): E. G. R. Taylor, 'John Dee and the map of North-East Asia', *Imago Mundi*, 12 (1955), pp. 103–6.

29 *Leeskaart boek van Wisbey*: see Waters, *Art of Navigation* (see note 11), p. 167, n. 2. The British Library's copy of Norman's *Safeguard* (G.7311(3)) is dated 1590. For a facsimile edition (without introductory notes) of the copy in the Bodleian Library (Savile L.18), see *The Safeguard of Sailors, or the Great Rutter. London 1584* (Amsterdam, Theatrum Orbis Terrarum; Norwood N. J., Walter J. Johnson, 1976).

30 Günter Schilder, 'A Dutch manuscript rutter: an unique portrait of the European coasts in the late sixteenth century', *Imago Mundi*, 43 (1991), pp. 59–71. Anthonisz's guide was entitled *Hier beghint die Caerte van der Ooster See*.

31 Cited in Waters, *Art of Navigation* (see note 11), p. 167, n. 2.

32 For Waghenaer's *Spieghel*, see Waters, *Art of Navigation* (see note 11), pp. 168–75; Cornelis Koeman, *The History of Lucas Jansoon Waghenaer and his 'Spieghel der Zeevaerdt'* (Lausanne, Elseviers, 1964); and R. A. Skelton (ed.), *The Mariners Mirrour* (Amsterdam, Theatrum Orbis Terrarum, 1966). See also Schilder, 'A Dutch manuscript rutter' (see note 30), p. 61. Waghenaer published his work in two parts, that for the 'Western Navigation' (22 charts and an outline map) in 1584 and that for the 'Eastern Navigation' in 1585 (23 charts), a division reflected in the English edition.

33 Skelton, *The Mariners Mirrour* (see note 32), p. v, citing Koeman. Other navigational guides for the use of mariners were published in the 1580s and 1590s, such as those by William Borough, *A Discourse of the Variation of the Compasse; or Magneticall Needle* (dated 1581, but printed in 1596 in London). Bound with Robert Norman's *The New Attractive* and Thomas Hood's *The Marriners guide set forth in the forme of a dialogue*, Borough's *Discourse* was added to a re-issue of William Bourne's *A Regiment for the Sea*: British Library, G.7311. Richard Eden published a translation from the Spanish of Martin Cortes's *Arte de Navegar* (1551) as *The Arte of Navigation* (London 1596), generously illustrated with diagrams and mobiles.

34 British Library, Royal MS 12.D.11, f. 81v.

35 British Library, Royal MS 17.C.38.

36 British Library, Sloane MS 683, f. 42r, following a list of the names of the English counties in a volume of miscellaneous documents.

37 British Library, Add. MS. 70,507. ff. 33v–45v and 73–5 respectively. For a transcription of the 29 itineraries in England, see Bruce Dickens, 'Premonstratensian itineraries from Titchfield Abbey MS at Welbeck', *Proceedings of Leeds Philosophical and Literary Society (Literary and Historical Section)*, 4 (1938), pp. 349–61, with a map reproduced also in Brian Paul Hindle, *Medieval Roads* (Princes Risborough, Shire Publications, 1982), p. 16. The journey to Rome was made in 1400–5 in connection with an appeal over the case of Hook Chapel: J T. Munby, *People and Fields in Medieval Portchester* (Winchester, Hampshire County Council, Portsmouth Record Series, forthcoming). We are grateful to Julian Munby for letting us read his text on the itineraries from Titchfield in advance of publication and to P. D. A. Harvey for help with transcribing and translating the Rome itinerary.

38 J. H. Harvey, *William Worcestre* (see note 9).

39 1503: British Library 21.a.10. Reprinted in 1811 as *The Customs of London. Otherwise called Arnold's Chronicle*.

40 See chapter 8 and Margaret Spufford, *Small Books and Pleasant Histories. Popular Fiction and its Readership in Seventeenth-century England* (Cambridge, Cambridge University Press, 1981) and Bernard Capp, *English Almanacs 1500–1800. Astrology and the Popular Press* (Ithaca, N.Y., Harvard University Press, 1979).

41 British Library, Add. MS 62,540, f. 7. We are indebted to Peter Barber for this reference.

42 British Library, Add. MS 61,342. This collection of Blenheim Papers (vol. CCXLII) contains a number of fair copies of marching orders and route orders, mostly from campaigns between 1702 and 1707, together with battle orders and various maps and plans. We are again grateful to Peter Barber for this reference.

43 See Justin Stagl, 'The methodising of travel in the sixteenth century. A tale of three cities', *History and Anthropology*, 4 (1990), pp. 303–38.

44 The social and economic changes of the sixteenth century greatly contributed to the expansion of the 'middling sorts' of people and to an increase in the numbers of people whose fortunes were based not on long-established land holding but on new forms of entrepreneurship, trade and industry. Later, Daniel Defoe, *Tour of Great Britain* (1724–6), spoke disparagingly of those among the wool traders of the Wiltshire Cotswolds ' who now pass for gentry'. John Adams's seventeenth-century acquaintance from the Aberdovey fishery, which led eventually to the creation of Adams's distance line maps (see text below), seems likely to have been one of the new type of entrepreneur.

45 For some introductory comments on road-books, see Fordham, *Studies in Carto-Bibliography* (see note 4), pp. 23–60 ('British and Irish itineraries and road-books') and pp. 120–7 ('An itinerary of the sixteenth-century: *La guide des chemins d'Angleterre*'). Earlier than any mentioned by Fordham are the nine itineraries in *A cronycle of yeres*... (London, W. Middleton, 1544) which seem to represent the standard early modern network.

46 The main articulation of the communications network in England has not altered to any great degree since Roman times. There has always been a major route from London to north-west England (Carlisle and the western Scottish border) and another to the north-east (Newcastle, Berwick on Tweed and the eastern Scottish border), for instance, another into the south-western peninsula, and one from Dover to London. The detailed alignment of each route, as well as of each track on the ground, and the sequence of towns passed through, however, has altered considerably from period to period.

47 British Library, Royal MS 14.C. 7, ff. 2r–4r and Cotton MS Nero D. I, ff. 182v–183r, and Corpus

Christi College, Cambridge, MS 26, ff. ir–iiir and MS 16, fol. iir (incomplete). See Richard Vaughan, *Matthew Paris* (Cambridge, Cambridge University Press, 1958) and Susanne Lewis, *The Art of Matthew Paris in the 'Chronica majora'* (Berkeley, University of California Press, 1987; Aldershot, Scolar Press with Corpus Christi College, Cambridge, 1987). A comparative listing of places (excluding British Library Cotton MS Nero D.I.) is given in Friedrich Ludwig, *Untersuchungen über die Reise- und Marschgeschwindigkeit im XII und XIII Jahrhundert* (Berlin, Ernst Siegfried Mittler & Son, 1897), pp. 126–9. The itinerary was placed, together with the maps of Britain and the Holy Land, at the beginning of each volume of, or relating to, the *Chronica Maiora*.

48 For a discussion of medieval routes to Rome, see Debra J. Birch, *Pilgrimage to Rome in the Middle Ages* (Woodbridge, Suffolk, Boydell and Brewer, 1998), pp. 38–71.

49 Especially since the Lateran Council's meeting in Rome in 1235. In 1252, for example, the Pope offered the crown of Sicily to Richard, Earl of Cornwall, an event alluded to on one of the versions of the map: Vaughan, *Matthew Paris* (see note 47), pp. 239 and 249.

50 Campbell, 'Portolan charts' (see note 12), p. 371.

51 The earliest extant example, the Pisan chart, is thought to have been made after the first known reference (*c*.1270) to any such chart: Campbell, 'Portolan charts' (see note 12), p. 380ff. Their late appearance raises all sorts of questions about the reasons for, and context of, their origin which have yet to be resolved.

52 Peter M. Barber, 'Old encounters new; the Aslake world map', in Monique Pelletier (ed.), *Géographie du monde au Moyen Ages et à la Renaissance* (Paris, Editions du Comité des Travaux Historiques et Scientifiques, 1989), pp. 69–88, esp. pp. 84–6. The Gough map (*c*.1360), which may also owe something to a portolan chart, could have been based on the same source. Andrea Bianco's chart of 1448 is signed as if made in London. However, Bianco, an Italian, was probably merely completing his chart while his ship was docked there: Tony Campbell, 'Portolan charts' (see note 12), p. 433.

53 We have had to exclude discussion of charts for overseas use. For these, see Robert Baldwin, 'Colonial cartography and the role of the British chartered companies' and Sarah Tyacke, 'English overseas chartmaking', in David Woodward (ed.), *The History of Cartography* vol. 3, *Cartography in the European Renaissance* (Chicago and London, Chicago University Press, forthcoming).

54 Thomas Hood's map of the 'King's Chambers', that is, English territorial waters, (1604) is at Hatfield House, CPM. 67. For a reproduction, see R. A. Skelton and John Summerson, *A Description of Maps and Architectural Drawings in the Collection made by William Cecil, First Baron of Burghley, now at*

Hatfield House (Oxford, The Roxburghe Club, 1971), no. 2.

55 British Library, Cotton MS Augustus I.ii.75.

56 The last three maps are British Library Cotton MS Augustus I.i.53, I.i.56, and I.i.35, 36, 38, 39 respectively.

57 In 1542. See Helen M. Wallis (ed.), *The Map and Text of the Boke of Idrography Presented by Jean Rotz to Henry VIII now in the British Library* (Oxford, The Roxburghe Club, 1981).

58 G. de Boer and R. A. Skelton, 'The earliest English chart with soundings', *Imago Mundi*, 23 (1969), pp. 9–16. The chart in question is in Burghley's atlas, British Library, Royal MS 18.D.III. Later charts are listed in Appendix III, p. 16.

59 British Library, Cotton MS Augustus I.ii.5.

60 Hatfield House, CPM II. 37a; British Library Cotton MS Augustus I.i.44 and Augustus I.i.16; Hatfield House CPM II. 33 respectively.

61 Public Record Office, SP. Dom. Eliz. 213/57. Cited by Arthur H. Robinson, *Marine Cartography in Britain. A History of the Sea Chart to 1855* (Leicester, Leicester University Press, 1962), pp. 29–31 and p. 153, from Borough's letter to Walsingham, dated 28 July 1588, transcribed in full by J. K. Laughton (ed.), *State Papers Relating to the Defeat of the Spanish Armada, Anno 1588*, 2 vols (London, Navy Records Society, 1894), vol. 1, pp. 336–8.

62 British Library, Royal MS 18.D.III, fol. 123.

63 In Robert Hitchcock, *A pollitique platt for the honour of the prince . . .* (1580). British Library C.27.f.3. For a summary of Anglo-Dutch trade at this time, see E. G. R. Taylor, *Late Tudor and Early Stuart Geography* (London, Methuen, 1934), pp. 100–4. For aspects of earlier relations across the North Sea, see Caroline Barron and Nigel Saul (eds), *England and the Low Countries in the Late Middle Ages* (Stroud, Alan Sutton; New York, St Martin's Press, 1995).

64 See E. G. R. Taylor, *A Regiment for the Sea and Other Writings on Navigation by William Bourne of Gravesend, a gunner (c.1535–1582)* (Cambridge, Cambridge University Press for the Hakluyt Society, 1963). Blundeville's tome ran to eight editions, the last published in 1638. See also Waters, *Art of Navigation* (see note 11), pp. 212–15.

65 The first of the Royal Society's several awards was the Copley Award, started in 1709. One recipient was John Harrison (1741) for his sea clock which at last permitted accurate timekeeping.

66 Coolie Verner, 'John Seller and the chart trade in seventeenth-century England', in Norman J. W. Thrower (ed.), *The Compleat Plattmaker. Essays on Chart, Map, and Globe Making in England in the Seventeenth and Eighteenth Centuries* (Berkeley, Los Angeles and London, University of California Press, 1978), pp. 127–57, especially pp. 156–7.

67 The earliest known surviving signed and dated examples of the school's work are John Daniel's

charts of the southern Atlantic (1614) and of the Mediterranean (also 1614): Tony Campbell, 'The Drapers' Company and its school of seventeenth-century chart-makers', in Helen Wallis and Sarah Tyacke (eds), *My Head is a Map. Essays and Memoirs in Honour of R. V. Tooley* (London, Francis Edwards and Carta Press, 1973), pp. 81–106; Thomas R. Smith, 'Manuscript and printed sea-charts in seventeenth century London: the case of the Thames School', in Thrower, *Compleat Plattmaker* (see note 66), pp. 45–100. We are grateful to Sarah Tyacke for information about the Mediterranean chart.

68 For example, all fifteen charts made by one person, discovered at Chatsworth House and listed by H. M. Wallis and W. P. Cumming, 'Charts by John Friend preserved at Chatsworth House, Derbyshire, England', *Imago Mundi*, 25 (1971), p. 81, concern coastal regions in Asia. See also Thomas Smith, 'Manuscript and printed sea-charts' (see note 67), Appendix 1.

69 Desmaretz's charts include those of the River Medway (1724), Harwich harbour (1732), Portsmouth (1750), Shoreham (1753) and Ramsgate (1755): Robinson, *Marine Cartography* (see note 61), pp. 92–3. For further details on these chart-makers, see also *Dictionary of National Biography*, E. G. R. Taylor, *The Mathematical Practitioners of Tudor and Stuart England* (Cambridge, Cambridge University Press, 1954), and A. Sarah Bendall, *Dictionary of Land Surveyors and Local Map-Makers of Great Britain and Ireland 1530–1850*, 2 vols (London, The British Library, 1997), vol. 2, p. 141 who reveals Desmaretz to have been far more than just a chart-maker.

70 For example, John Adair along the eastern Scottish coasts (from 1688), Lewis Morris along the Welsh coasts (1737), and two generations of the Mackenzie family (from 1750) on the far north of Scotland, Ireland, Wales and only in 1777 the Thames estuary and parts of the south coast of England: summarised in David Smith, *Maps and Plans for the Local Historian and Collector* (London, Batsford, 1988), esp. pp. 136–7.

71 Cabot was paid 20 shillings in May 1512 for 'Making a carde of Gascoigne and Guyon': British Library, Add. MS 21,481, f. 92r; transcribed in J. S. Brewer (ed.), *Letters and Papers of Henry VIII* (London, Longman, Green, *et al.*, 1864), vol. 2, part 2, p. 1456.

72 For an illuminating account of how an army travelled, see Geoffrey Parker, *The Army of Flanders and the Spanish Road 1567–1659* (Cambridge, Cambridge University Press, 1972).

73 For a reproduction, see David Buisseret, 'Monarchs, ministers, and maps in France before the accession of Louis XIV', in David Buisseret (ed.), *Monarchs, Ministers and Maps. The Emergence of Cartography as a Tool of Government in Early Modern Europe* (Chicago and London, University of Chicago Press, 1992), pp. 99–123, Fig. 4.2 (p. 101).

74 Now in the St Albans Bible, Corpus Christi College, Oxford, MS. 2 ff. 2v–1r. The quotation is from Evelyn Edson, 'Matthew Paris's "other" map of Palestine', *The Map Collector*, 66 (1994), pp. 18–22. She bases her comment on attributes such as the map's restrained (that is, strictly functional) style, the estimates of travel time written along the coast and other relevant remarks such as warnings of lions in the forest near Arsuf, a comment as to the best road to Jerusalem, and the indication of gradients on the Jericho road.

75 Brigitte Englisch, 'Erhard Etzlaub's projection and method of mapping', *Imago Mundi*, 48 (1996), pp. 103–23, especially p. 106.

76 By measuring with dividers the spaces between the dots which indicated each route and reading the total against one of the scale bars on the map, the intending traveller would be able to estimate the total distance to be travelled, and thus work out the number of overnight stops needed. By referring to the graduated scale of length of daylight along the side of the map, he was warned roughly how many hours were available each day for travel according to latitude. Finally, by using the map in conjunction with a sun compass, he could also work out the general direction in which he should be heading.

77 Lajos Stegena, *Lazarus Secretarius: The First Hungarian Mapmaker and His Work* (Budapest, Akadémiai Kiadó, 1982), pp. 24–5, p. 103. Stegena reproduces in facsimile the 1528 map and four later versions.

78 '*Totius Hungariae Chorographia, itinerariaq[ue]*'. Stegena, *Lazarus Secretarius* (see note 77), pp. 24–5, misleadingly renders *itinerarium* as 'road map'.

79 The teaching of mathematics had only recently been introduced into the universities and was scarcely touched on in most schools: Keith Thomas, 'Numeracy in early modern England', *Transactions of the Royal Historical Society*, 5th series, 37 (1987), pp. 103–32; Lesley B. Cormack, *Charting an Empire. Geography at the English Universities, 1580–1620* (Chicago and London, University of Chicago Press, 1997), especially Chapter 3 'Mathematical Geography: Theory and Practice'.

80 Adams's map *Angliae Totius Tabula Cum Distantibus Notoribus in Itinerantium Usum Accommodata* is described by R. W. Shirley, *Printed Maps of the British Isles 1650–1750* (Tring, Map Collector Publications; London, The British Library, 1988), pp. 18–22. See also William Ravenhill, 'John Adams, his map of England, its projection, and his *Index Villaris* of 1680', *Geographical Journal* 144 (1978), pp. 427–37, and Blake Tyson, 'John Adams's cartographic correspondence to Sir Daniel Flemming of Rydal Hall, Cumbria, 1676–1687', *Geographical Journal*, 151 (1985), pp. 21–39.

81 Blake Tyson, 'John Adams's cartographic correspondence' (see note 80), p. 27.

82 Tyson, 'John Adams's cartographic correspondence' (see note 80), p. 27. See also Beale, *The Post in England* (see note 2), p. 60.

83 The whereabouts of the map prior to 1774, when it was purchased by Richard Gough, are unknown. Two topographical maps with similar lines had been published in Germany and at least three in Italy before 1600. The map of the Holy Land in the Coverdale Bible (1535) has lines radiating from Jerusalem to a number of other European and Near Eastern cities with a note of the distance.

84 William Berry (1671 etc.), Robert Morden (1673 etc.), Robert Walton (1679), John Overton (1685 etc.), Herman Moll (1673 etc.), Thomas Bowles (*c*.1710 etc.), George Willdey (1713 etc.). See 'Adams-type distance maps' in the subject index to Shirley, *Printed Maps 1650–1750* (see note 80).

85 The critical matter was to reach a safe place before dusk and to find good lodgings in the centre of town before the drawbridge was raised, as underlined in Berlaimont's phrasebook for commercial travellers (see note 7).

86 Richard Gough, *British Topography*, (London, 1780), vol. I. p. 52; Letter to Flemming dated 29 June 1679. Cited by Tyson, 'John Adams's cartographic correspondence' (see note 80), p. 29.

87 For some medieval examples, see Delano-Smith, 'Milieus of mobility' (see note 1). See also Margaret Aston, *The Fifteenth Century. The Prospect of Europe* (London, Thames and Hudson, 1979), pp. 49–84; Beale, *The Post in England*, (see note 2), pp. 12–18.

88 Howard Robinson, *The British Post Office. A History* (Princeton, NJ, Princeton University Press, 1948), pp. 7–9; Beale, *The Post in England*, (see note 2), Chapter 8.

89 In comparison with the 'Little Ice Age' of *c*.1530–1700, the Middle Ages had enjoyed relatively mild and dry conditions. See Gordon Manley, *Climate and the British Scene* (London, Collins, 1952); H. H. Lamb, *Climate, History and the Modern World* (London and New York, Methuen, 1982); M. L. Parry, *Climatic Change, Agriculture and Settlement* (Folkestone, Dawson, 1978).

90 Philip Harrison and Mark Brayshay, 'Post-horse routes, royal progresses and government communications in the reign of James I', *Journal of Transport History*, n.s. 18 (1997), pp. 116–33, Fig. 1. 'Riding in post' referred to those making a journey by means of a relay of hired horses, usually but not inevitably on official business (p. 118).

91 Robinson, *British Post Office* (see note 88), p. 29.

92 The unpopularity of road users who ruined the roads for others was sometimes acknowledged by those responsible, such as the Sussex gun founder John Fuller who wrote in a letter dated 26 February 1743 that he had 'gotten 20 9 pounders of nine feet to

Lewes' for shipment by sea to Woolwich, but that 'These 20 have torn the roads so that nothing can follow them and the Country curse us heartily': quoted by Herbert Blackman, 'Gun founding at Heathfield in the XVIII century', *Sussex Archaeological Collections*, 67 (1926), pp. 25–59. Such road users were often obliged to maintain the roads their loads had damaged.

93 On English turnpikes in general, see Eric Pawson, *Transport and Economy. The Turnpike Roads of Eighteenth Century Britain* (London, Academic Press, 1977), who makes no mention of maps, however.

94 The publisher of the *London Magazine*, embarking on a series of road maps in September 1765, asked for 'any information relative to the perfection of these maps': David C. Jolly, *Maps in British Periodicals* (Brookline, Mass., David C. Jolly, 1992), vol.1, p. 92.

95 John Cary's *Traveller's Companion or a Delineation of the Turnpike Roads of England and Wales* (1790) offered comprehensive coverage in road-book format.

96 Shirley, *Printed Maps 1650–1750* (see note 80), pp. 40–1.

97 Hollar's map was sold in London by William Place: Shirley, *Printed Maps 1650–1750* (see note 80), pp. 115–7.

98 Shirley, *Printed Maps 1650–1750* (see note 80), pp. 147–8.

99 Described but not illustrated in Shirley, *Printed Maps 1650–1750* (see note 80), pp. 104–5. On *Britannia* in general, see J. B. Harley's Introduction in *J. Ogilby, Britannia (1675)*. Reproduced in facsimile (Amsterdam, Theatrum Orbis Terrarum, 1970), pp. v–xxxi.

100 Ken Garland, *Mr Beck's Underground Map* (Harrow Weald, Capitol Transport Publishing, 1994).

101 The apt term 'diagrammatic map' is R. W. Shirley's: Shirley, *Printed Maps 1650–1750* (see note 80), p. 16 and elsewhere. The maps cited are described on pp. 29 and 126–7 respectively.

102 Shirley, *Printed Maps 1650–1750* (see note 80), pp. 48 and 88–9 respectively.

103 Shirley, *Printed Maps 1650–1750* (see note 80), pp. 150–2. Willdey also used some distance lines.

104 Both examples came to light only in 1998 and are now in the possession of a private collector. See Jonathan Potter Ltd., *Choice Items from Stock*, Catalogue 13, 1998, Items 30 and 31.

105 Carr's map originated from drafts made by a Post Office official, James Hicks, possibly as early as 1665: see Shirley, *Printed Maps 1650–1750* (see note 80), pp. 40–1.

106 See Berlaimont, *Colloquia* (see note 7).

107 For just one of many early examples of travelling with locally-hired drivers, see Fynes Morison, *An Itinerary written by Fynes Moryson first in the Latin Toungue containing his ten yeeres travell through the twelve dominions of Germany, Bohemialand,*

Sweitzerland, Netherland, Denmarke, Poland, Italy, Turkey, France, England, Scotland and Ireland (London, 1617).

108 The exact weight of the Royal copy, British Library 192.f.1, including binding, is 6.941 kg. Our thanks go to John Goldfinch for his willing help in this practical exercise!

109 A new perspective on Ogilby and his monumental projects, including *Britannia*, is being prepared by Garrett Sullivan, who suggests that the road-book, just one element in one of Ogilby's typically grandiose schemes, was created for library and drawing room use: pure arm-chair geography, in other words. See the maps of Ogilby's open and enclosed roads in Harley, *J. Ogilby* (see note 99) where Harley also discusses (p. xi) Ogilby's sources. Harley's Introduction is still the best analysis of Ogilby's achievement. On Ogilby more generally, see Katherine S. Van Eerde, *John Ogilby and the Taste of his Times* (Folkestone, Dawson, 1976).

110 Peutinger's map was published in 1598 by a relative of the family, Markus Welser. The standard edition of the Peutinger map is a reissue of Konrad Miller (ed.), *Die Peutingersche Tafel* (Stuttgart, F. A. Brockhaus, 1962). See also O. A. W. Dilke, 'Itineraries and geographical maps in the early and late Roman Empires', in Harley and Woodward, *History of Cartography*, vol. 1 (see note 12), pp. 234–42.

111 Matthew Parker (ed.), *Matthew Paris Chronica Maiora* (Zurich, Froshauer, 1571). William Wats's four seventeenth-century editions are 1640, 1641, 1644 and (after the publication of *Britannia*), 1684.

112 Preface to the Reader, f. 2v. We are grateful to Alessandro Scafi for assistance with the translation.

113 Six copies of the catalogue are known: see Colin G. C. Tite, *The Manuscript Library of Sir Robert Cotton: The Panizzi Lectures 1993* (London, The British Library, 1994), p. 30, to whom we are especially grateful for enlightenment on possible links between Ogilby and Cotton MS Claudius D.VI. Ogilby is not listed amongst the users of the Cotton Library but then neither are others known to have frequented the library. Later in the century, a copy was made of Paris's maps of Britain, now British Library, Maps. K.Top. 5.27.1. We are greatly indebted to Peter Barber for drawing this map to our attention and for suggesting a date (about 1700).

114 Van Eerde, *John Ogilby* (see note 109), p. 121.

115 Harley, *J. Ogilby* (see note 99), p. xvi. Hooke himself makes no such reference but for his regular and intimate contacts with Ogilby see Henry W. Robinson and Walter Adams (eds) *The Diary of Robert Hooke 1672–1680*, (London, Taylor and Francis, 1935).

116 Quoted by Harley, *J. Ogilby* (see note 99), p. v.

117 The suggested interpretation is Peter Barber's.

118 Harley, *J. Ogilby*, (see note 99), p. v.

119 The book was advertised as open to subscription but no list of subscribers is known: Harley, *J. Ogilby* (see note 99), p. xiv.

120 For a listing of Ogilby-type road maps, see Fordham, *Carto-Bibliography* (note 4) and sections on road-books in county-map histories such as D. Kingsley, *Printed Maps of Sussex 1575–1900* (Lewes, Sussex Record Society, 1982); and R. A. Carroll, *Printed Maps of Lincolnshire, 1576–1900: A Carto-Bibliography. With an Appendix on Road-Books* (Woodbridge, Suffolk; Boydell Press for the Lincoln Record Society, vol. 84, 1996). Some road-books are described in general carto-bibliographies like Donald Hodson, *County Atlases of the British Isles*, vol. I, *Atlases Published 1704 to 1742*, vol. 2, *Atlases Published 1743–1763* (Tewin, Tewin Press, 1984–1989), vol. 3, *Atlases Published 1764–1789* (London, The British Library, 1997); and Shirley, *Printed Maps 1650–1750* (see note 80). Further references are given in David Smith, *Maps and Plans* (see note 70), pp. 122–3.

121 Geoffrey of Monmouth, *The History of the Kings of Britain*, translated by Lewis Thorpe (London, Folio Society, 1969), p. 74.

122 Cited by Tyson, 'John Adams's cartographic correspondence' (see note 80), p. 23.

123 Christopher Taylor, *Roads and Tracks of Roman Britain*, 2nd ed. (London, Orion Books, 1994), p. 121, fig. 57.

124 See Beale, *The Post in England* (see note 2), especially Chapters 7 and 10.

125 According to an Act of Parliament of 1691, cartways leading to a market town were to be 'eight feet wide at the least' and bridlepaths ('horse causeys') were to be not less than three feet wide: Fuller, 'The development of roads' (see note 3), pp. 37–49. Under the terms of enclosure by Act of Parliament, the commissioners were responsible for the layout of carriage roads, bridleways and footpaths.

126 Cited by Donald Hodson, *County Atlases 1764–1789* (see note 120), pp. 173–5. We are indebted to Donald Hodson for drawing this note to our attention and for all his helpful comments on our interpretation of road-books. Smith, *Maps and Plans* (see note 70), p. 118, Fig. 54, mistakes Cary's *Actual Survey of the Country Fifteen Miles Round London . . .* for a 'small road map' and notes that 'much non-road information' was included.

127 British Library, Maps C.27.b.45. Cary's titles are given in full in H. G. Fordham, *John Cary, Engraver, Map, Chart and Print-Seller and Globe-Maker* (London, Cambridge University Press, 1925; reprinted Folkestone, Dawson, 1976).

128 British Library, Maps 1.Tab.21.

129 British Library, Maps 15.bb.10.

130 More precisely, Arrowsmith's map comprises fifteen full sheets and three half-sheets. The assembled, roller-mounted copy in the British Library (Maps S.T.E. (3).) measures almost 3.5×3 m.

131 Fordham, *John Cary* (see note 127), p. 40.
132 See Timothy R. Nicholson, *Wheels on the Road. Maps of Britain for the Cyclist and Motorist 1870–1940* (Norwich, Geo-Books, 1983). Nicholson points out that in the early years of motoring, publishers of maps for cyclists tended to change only the cover of the map to allude to motorists (for example, A. and A. K. Johnston's maps of about 1902, *Cycling and Automobile map of the English Lake District,* and *Cycling and Automobile Map of Middlesborough District,* both published in Edinburgh and London at a scale of one inch to three miles are amongst the first such maps). Maps at scales larger than half or quarter of an inch to one mile rapidly proved useless for the long-distance motorist, whose route was only partially contained on one sheet, and demand from motorists in the first decade of the twentieth century led to huge sales of the Ordnance Survey's smaller-scale maps.

CHAPTER 6

1 For example, the relics of the marble *Forma Urbis Romae* and of the cadastral map of the colony of Orange, Gaul are well known to map historians: see O. A. W. Dilke, 'Roman large-scale mapping in the early Empire', in J. B. Harley and David Woodward (eds), *History of Cartography,* vol. 1, *Cartography in Prehistoric, Ancient, and Medieval Europe and the Mediterranean* (Chicago and London, University of Chicago Press, 1987), pp. 222–30, as are a number of mosaics showing views of towns. In March 1998, the discovery of a fresco buried below the Baths of Trajan, Rome, was reported in the world's press. This gives a bird's-eye view of part of a walled town (possibly Rome itself) with amphitheatre, colonnades, and various brick-built buildings. In Early Christian mural and documentary art, the New Jerusalem was frequently portrayed from an oblique viewpoint as well as from ground level.
2 See, for example, S. S. Frere and J. K. St Joseph, *Roman Britain from the Air* (Cambridge, Cambridge University Press, 1983), Chapter 9 (Urban Sites), especially pp. 147–81 (Silchester, Wroxeter, Mildenhall).
3 O. A. W. Dilke, 'Maps in the treatises of Roman land surveyors', *Geographical Journal,* 127 (1967), pp. 417–26; O. A. W. Dilke, 'Illustrations from Roman surveyors' manuals', *Imago Mundi,* 21 (1967), pp. 9–29, especially pp. 20–2, on 'plans of towns and surrounding areas'; O. A. W. Dilke, 'Roman large-scale mapping' (see note 1), pp. 212–33; Roger J. P. Kain and Elizabeth Baigent, *The Cadastral Map in the Service of the State: A History of Property Mapping* (Chicago and London, University of Chicago Press, 1992), pp. 1–3.
4 There is a rich literature on Vitruvius and the history of city planning. See, for example, Gerald Burke,

Towns in the Making (London, Edward Arnold, 1983 edition); A. E. J. Morris, *History of Urban Form before the Industrial Revolutions* 3rd ed. (London, Longman, 1994). See also: O. A. W. Dilke, 'Ground survey and measurement in Roman towns', in Francis Grew and Brian Hobley (eds), *Roman Urban Topography in Britain and the Western Empire* (London, The Council for British Archaeology, Research Report No. 59, 1975), pp. 6–13, especially p. 10. For a modern translation of Vitruvius see F. Granger, *On Architecture* (2 vols., London, Heinemann, 1931–34). Vitruvius's treatise was rediscovered in 1412 and thereafter became a source book of classical urban aesthetics for Renaissance town-planning theoreticians.
5 Liba Taub, 'The historical function of the *Forma urbis Romae*', *Imago Mundi,* 45 (1993), pp. 9–19; Dilke, 'Roman large-scale mapping' (see note 1), p. 226.
6 Antonia Gransden, 'Realistic observation in twelfth-century England', *Speculum,* 47 (1972), pp. 29–51; republished in Antonia Gransden, *Legends, Traditions and History in Medieval England* (London, Hambledon Press, 1992), pp. 175–97.
7 On the town signs of the Gough map, see B. P. Hindle, 'The towns and roads of the Gough map (*c.*1360)', *Manchester Geographer,* 1 (1980), pp. 35–49.
8 Extant documents date from the ninth century onwards, although the earliest surviving English copy (Eton College, MS 177, f. 56r) dates only from the thirteenth century.
9 British Library, Royal MS 13.A.111. The drawings are mostly in pencil, sometimes with banners in red ink, and may be unfinished.
10 Folio 28v. Billingsgate (*Belmysgate*) and the Tower (*Turris edificata London*) are named.
11 John Harvey, 'Symbolic plan of a city, early 15th century', in R. A. Skelton and P. D. A. Harvey (eds), *Local Maps and Plans from Medieval England* (Oxford, Clarendon Press, 1986), pp. 342–3.
12 John Chartres, 'Lincoln's inns', *Historian,* 7 (1983), pp. 16–19. See also John H. Harvey, 'Symbolic plan of a city, early 15th century', in Skelton and Harvey, *Local Maps and Plans* (see note 11), pp. 342–3, especially p. 342.
13 Bodley Rolls 5. See Jonathan J. G. Alexander and Otto Pächt, *Illuminated Manuscripts in the Bodleian Library, Oxford.* vol. 3 *British, Irish and Icelandic Schools* (Oxford, Clarendon Press, 1973), Plate CV, no. 1121b. Some of the architectural details, such as a tall turret surmounted by a dome, seem distinctly un-English. Both the Percy chronicle (the Geneaology of the Kings of England to Richard III) and Peter of Icham's chronicle are in the Bodleian Library, Bodley Rolls 5 and Laud. Misc. 730, respectively. See folios 9 in the former and 10 in the latter, for example.
14 Eton College MS 177, fol. 55v (thought to have been made at Westminster) and, as in Fig. 2.3, Trinity

College, Cambridge, MS R.16.2 (C.M.A. 524), fol. 25r.

15 The version from British Library, MS Royal 14. C.VII is reproduced as a line drawing in Konrad Miller, *Die ältesten Weltkarten* volume III *Die Kleineren Weltkarten* (Stuttgart, J. Roth, 1895), p. 91; as a monochrome photograph in P. D. A. Harvey, *The History of Topographical Maps. Symbols, Pictures and Surveys* (London, Thames and Hudson, 1980), pp. 56–7, Fig. 28; and in colour in Evelyn Edson, 'Matthew Paris' "other" map of Palestine', *The Map Collector*, 66 (1994), pp. 18–22, especially p. 20.

16 H. S. A. Fox, 'Exeter, Devonshire, *circa* 1420', in Skelton and Harvey, *Local Maps and Plans* (see note 11), pp. 163–79.

17 Respectively – M. G. Snape, 'Durham 1439 × *circa* 1420' pp. 163–9, especially pp. 189–94; M. D. Knowles, 'Clerkenwell and Islington, Middlesex, mid–15th century', pp. 221–8; Philip E. Jones, 'Deptford, Kent and Surrey; Lambeth, Surrey; London, 1470–1478', pp. 251–62; Fox, 'Exeter' (see note 16) , pp. 329–36 – in Skelton and Harvey, *Local Maps and Plans* (see note 11).

18 The reference comes from Ferdinand's library catalogue: Tony Campbell, *The Earliest Printed Maps, 1472–1500* (London, The British Library, 1987), p. 219, no. A4; Peter Barber, 'A glimpse of the earliest map-view of London?', *London Topographical Record*, 27 (1995), pp. 91–102.

19 Thomas Frangenberg, 'Chorographies of Florence. The use of city views and city plans in the sixteenth century', *Imago Mundi*, 46 (1994), pp. 41–64.

20 P. D. A. Harvey, *History of Topographical Maps* (see note 15), pp. 75–8. For a facsimile of the *Très riches heures du duc de Berry, Musée Condé, Chantilly*, see Thames and Hudson's 2nd edition (London, 1989). The Venice Ptolemy is Biblioteca Apostolica Vaticana Vat.Urb.Lat.227.

21 The relationship between the spatial complexity of a town and its mapping is touched on in Catherine Delano-Smith and Roger J. P. Kain, *La Cartografia Anglesa* (Barcelona, Institut Cartogràfic de Catalunya, 1997), pp. 196–7 and is echoed in David Buisseret (ed.), *Envisioning the City: Six Studies in Urban Cartography* (Chicago and London, University of Chicago Press, 1998), pp. ix–x.

22 The British Academy is sponsoring the compilation of a catalogue of English town maps, an essential preliminary to such a study.

23 New College, MS 288, f. 4v. The description comes from the list of contents (f. 2 verso). See Kathleen L. Scott, *Later Gothic Manuscripts, 1390–1490* (London, Harvey Miller, 1996), vol. 1, Fig. 426 and vol. 2, Cat. 114, pp. 310–11.

24 Elizabeth Ralph, 'Bristol, *circa* 1480' in Skelton and Harvey, *Local Maps and Plans* (see note 11), pp. 309–16, especially p. 314. On contemporary Bristol, see J. H. Bettey, 'Late-medieval Bristol: from town to city,' *The Local Historian*, 28 (1998), pp. 3–15.

25 Ruth S. Luborsky and Elizabeth M. Ingram, *A Guide to English Illustrated Books, 1536–1603*, 2 vols (Tempe, Arizona Center for Medieval and Renaissance Studies, 1998).

26 J. R. Bennett, *The Divided Circle*, London, Phaidon, 1987, p. 40.

27 J. Willis Clark and Arthur Gray, *Old Plans of Cambridge, 1574–1798, Reproduced in Facsimile with Descriptive Text*, 2 vols (Cambridge, Bowes and Bowes, 1921); Edward Lynam, 'English maps and map-makers of the sixteenth century', *Geographical Journal*, 116 (1950), pp. 7–28.

28 For Harrison's project, see R. A. Skelton, 'Tudor town plans in John Speed's *Theatre*', *Archaeological Journal*, 108 (1951), pp. 109–20. For the view of Edinburgh, later used by G. Braun and R. Hogenberg for their *Civitates orbis terrarum*, see Luborsky and Ingram, *Guide to English Illustrated Books* (see note 25).

29 The title page of the manuscript (British Library, Sloane MS 2596) bears the date 1588 but there is reason to believe, from internal evidence, that Smith continued to add to the book at least up to 1603 and the reign of James I. A facsimile edition, edited by Henry B. Wheatly and Edmund W. Ashbee, was published in London in 1879.

30 Lambeth Palace Library, MS 508; Nuremberg, Stadtbibliothek, Nor. H 1142. See William Roach, 'William Smith "A Description of the Cittie of Nuremberg" (Beschreibung der Reichsstadt Nürnburg)'. *Mitteilungen des Vereins für Geschichte der Stadt Nürnberg*, 48 (1958), pp. 194–245, especially Plates VI and VII. The Stadtbibliothek's copy, thought to be the original of several versions Smith made for presentation, reached Nuremberg only in 1954.

31 The manuscript of the '*Speculum Northam[p]toniae*' is in Paris (Bibliothèque Nationale, MS 706 Anglais, Acq. nouv. 58) but a photographic copy of the text, with reduced county and town map, is in the British Library, Maps C.7. b.20.

32 The point was first made by Edward Lynam, *The Mapmaker's Art: Essays on the History of Maps* (London, Batchworth, 1953), p. 67. The map is in the British Library, Add. MS 38,065. Norden's exact words (on p. 48, the last page of his preface) are 'I have set down this former plott of Higham Ferrers'. For a reproduction of the map, see Maurice Beresford, *History on the Ground. Six Studies in Maps and Landscapes* (London, Methuen, 1957, 2nd ed. 1971), Chapter 6, Plate 13.

33 Frank Kitchen, *Cosmo-choro-poly-grapher: An Analytical Account of the Life and Work of John Norden* (unpublished University of Sussex D.Phil. thesis, 1992), p. 37.

34 Kitchen, *Cosmo-choro-poly-grapher* (see note 33), pp. 43–223. See also Frank Kitchen, 'John Norden (*c*.1547–1625): estate surveyor, topographer, county mapmaker and devotional writer', *Imago Mundi*, 49 (1997), pp. 43–61.

35 David Smith, 'The enduring image of early British townscapes', *Cartographic Journal*, 28 (1991), pp. 163–75.

36 Ralph Hyde, 'Introductory Notes', in *The A to Z of Restoration London (The City of London, 1676)* (London, London Topographical Society, Publication No. 145, 1992). See also M. D. Lobel, 'The value of early maps as evidence for the topography of English towns', *Imago Mundi*, 22 (1968), pp. 50–61.

37 For an assessment, see Hyde, 'Introductory Notes' (see note 36), p. xi. Parts of the plan are reproduced in Ida Darlington and James Howgego, *Printed Maps of London, circa 1553–1850* (London, George Philip, 1964), Entry 28, Plate 6 (2nd edition, 1978, lacks the plate), and in Philippa Glanville, *London in Maps* (London, The Connoisseur, 1972), Plates 14 and 15.

38 See David Smith, 'The demise of a nineteenth-century cartographic project: the publication of Lysons' *Magna Britannia*', *Journal of the International Map Collectors' Society*, 53 (1993), pp. 5–13, and his 'The preparation of the town plans for Lysons' *Magna Britannia*', *Cartographic Journal*, 32 (1995), pp. 11–17.

39 *The Beauties of England and Wales* (1810) was a more successful topographical, historical and descriptive survey combining antiquarian and contemporary material and was later reissued. For a modern photo-lithographic reproduction of just the maps, see Brian Stevens, *Plans of English Towns by G. Cole and J. Roper, 1810* (Monmouth, Historic Prints, 1970).

40 The fundamental study of de' Barbari's map (which assembled measures some 1.3×2.8 metres) is by Jürgen Schultz, 'Jacopo de' Barbari's View of Venice: map-making, city views and moralized geography before the year 1500', *Art Bulletin*, 60 (1978), pp. 425–74. Rosselli's views, created before his death in 1525, are known only from derivatives: Harvey, *Topographical Maps* (see note 15), pp. 75–8.

41 For example, Stephen Marks, in a lecture at the London Museum, May 1998, on the occasion of the first exhibition of the third plate, recently discovered in Dessau, Germany: Stephen Marks 'The copperplate map – a lecture and an exhibition', *London Topographical Society Newsletter*, 47 (1998), pp. 5–6. See also Stephen Powys Marks, *The Map of Mid Sixteenth Century London: an Investigation into the Relationship between a Copper Engraved Map and its Derivatives* (London, London Topographical Society, Publication 100, 1964); M. A. Holmes, *Moorfields in 1559. An Engraved Copper Plate from the Earliest Known map of London* (London, H.M.S.O. for the City of London Museum, 1963); John Fisher, 'Introductory notes', in *The A to Z of Elizabethan London* (London Topographical Society, Publication 122, 1979); and Laurence Worms, 'The London map trade to 1640', in David Woodward (ed.), *History of Cartography*, vol. 3, *Cartography in the European Renaissance* (Chicago and London, University of Chicago Press, in preparation). We are greatly indebted to Laurence Worms for his advice.

42 Fisher, 'Introductory Notes' (see note 41).

43 See Worms, 'The London Map Trade' (see note 41).

44 By Stephen Marks, in his London Museum lecture (see note 41).

45 An 8-sheet facsimile was published by the London Topographical Society in 1905 (Publication 17). For the literature on this map, see note 41. Only three extant copies are known, in the Guildhall Library, London; the Public Record Office; and Magdalene College, Cambridge (Pepys Collection).

46 A third version of the 'lost copperplate' map was engraved in pewter, probably by George Vertue and possibly from the original rather than the woodcut copy. It is dated 1737.

47 Worms, 'London map-trade' (see note 41).

48 Reproduced in facsimile by the Oxford Historical Society in 1884 with, in a companion volume, *Oxford Topography: An Essay* (Oxford, Oxford Historical Society, Publication 39 1899) by Herbert Hurst.

49 For a facsimile of the much-damaged sole extant copy of Hammond's map, see Clark and Gray, *Old Plans of Cambridge* (see note 27), Map no. 3, about 120×88 cm.

50 Lobel, 'The value of early maps' (see note 36), pp. 54–5; Clark and Gray, *Old Plans of Cambridge* (see note 27).

51 The drawing table was valued at 10 shillings: Catherine Delano-Smith, 'Son of Rudd: Edmund, another Tudor mapmaker?', *The Map Collector*, 64 (1993), p. 38. The implication of such a high value for an unfinished map is that it is sufficiently far advanced, and of a subject of sufficient interest, for the probate commissioners to recognise its potential value. A map of the university city of Cambridge, as opposed to some distant part of the world, was bound to impress them. For Hammond's details, see Clark and Gray, *Old Plans of Cambridge* (see note 27), Part 2, p. 27.

52 William Ravenhill and Margery Rowe, 'A decorated screen map of Exeter based on John Hooker's map of 1587', in Todd Gray, Margery Rowe and Audrey Erskine (eds), *Tudor and Stuart Devon: the Common Estate and Government* (Exeter, Exeter University Press, 1992), pp. 1–12. See below on the role of Hooker's plan.

53 James Elliot, *The City in Maps: Urban Mapping to 1900* (London, The British Library, 1987), pp. 26–37.

54 Skelton, 'Tudor town plans in John Speed's *Theatre*' (see note 28), pp. 109–20, quotation on pp. 109–10. One exception to Speed's 'rule' is his inclusion of a scale for Newcastle, the map of which

he attributes to William Mathew. We thank John Andrews for this point.

55 Paul Hindle, *Maps for Historians* (Chichester, Phillimore, 1998), p. 59.

56 A. J. Bird, 'John Speed's view of the urban hierarchy in Wales in the early seventeenth century', *Studia Celtica*, 10–11 (1975–6), pp. 401–11, quotation from p. 407.

57 Smith, 'Enduring image' (see note 35), p. 171; see also David Smith, 'Inset town plans on large-scale maps of Great Britain', *Cartographic Journal*, 29 (1992), pp. 118–36. This last is a contribution of the first importance, not least for the fact that it contains fundamental carto-bibliographical reference material.

58 The atlas examined was British Library, Maps C.10.d.8 (?1765), with 47 maps. For their probable dates, and an account of the atlas and its many compilations, see Donald Hodson, *County Atlases of the British Isles*, vol. 2 *Atlases Published 1743 to 1763 and their Subsequent Editions* (Tewin, Welwyn, Tewin Press, 1989), pp. 97–147.

59 British Library, Maps.25.a.2.

60 R. A. Skelton, 'Introduction', George Braun and Franz Hogenberg, · *Civitates Orbis Terrarum* (Amsterdam, Theatrum Orbis Terrarum, facsimile edition, 1966), p. vii.

61 Horst de la Croix, 'The literature on fortification in Italy', *Technology and Culture*, 4 (1963), pp. 30–50.

62 J. R. Kenyon, *Medieval Fortifications* (Leicester, Leicester University Press, 1990).

63 Keith Dawson, *Town Defences in Early Modern England* (unpublished University of Exeter Ph.D. thesis, 1995), especially pp. 222–4.

64 R. A. Skelton, 'The military surveyor's contribution to British cartography in the sixteenth century', *Imago Mundi*, 24 (1970), pp. 77–85; Marcus Merriman, 'Italian military engineers in Britain in the 1540s', in Sarah Tyacke (ed.), *English Map-Making 1500–1650* (London, The British Library, 1983), pp. 57–68. For an indispensable overview, see Peter Barber, 'England I. Pageantry, defense and government: maps at court to 1550' and 'England II: Monarchs, ministers, and maps, 1550–1625' in David Buisseret (ed.), *Monarchs, Ministers and Maps: The Emergence of Cartography as a Tool of Government in Early Modern Europe* (Chicago and London, University of Chicago Press, 1992), pp. 26–56 and 57–98. See also Stephen Johnston, 'Mathematical practitioners and instruments in Elizabethan England', *Annals of Science*, 48 (1991), pp. 319–44, on the harbour surveying work of Thomas Bedwell and Thomas Hood.

65 P. D. A. Harvey, in his 'The Portsmouth map of 1545 and the introduction of scale maps into England', in John Webb, Nigel Yates, and Sarah Peacock (eds), *Hampshire Studies* (Portsmouth, Portsmouth City Records Office, 1981), pp. 33–49, reports that all the known early maps which are

drawn to scale are the work of military engineers (p. 45).

66 Unless otherwise indicated, maps discussed below are illustrated in P. D. A. Harvey, *Maps in Tudor England* (London, Public Record Office and The British Library; Chicago, University of Chicago Press, 1993).

67 J. A. Bennett and S. A. Johnson, *The Geometry of War, 1500–1750* (Oxford: Museum of the History of Science, 1996). See also Ann Payne, *Views of the Past: Topographical Drawings in the British Library* (London, The British Library, 1987), p. 19, and Harvey, *Maps in Tudor England* (see note 66), p. 35.

68 Illustrated by Barber, 'England I' (see note 64), p. 38; L. R. Shelby, *John Rogers, Tudor Military Engineer* (Oxford, Clarendon Press, 1967), pp. 174–5; Barber, in 'England I' (see note 64), p. 37. See also Catherine Delano-Smith, 'Cartographic signs on European maps and their explanation before 1700', *Imago Mundi*, 37 (1985), pp. 9–29.

69 Harvey, *Maps in Tudor England* (see note 66), Fig. 27.

70 In the late fifteenth and early sixteenth centuries, one of the arguments put forward for the provision of chapels-of-ease in sparsely populated rural districts in southern coastal districts was the need to reduce the time villagers left their homes unprotected whilst attending a distant parish church.

71 P. D. A. Harvey, 'The Portsmouth map of 1545' (see note 65); Martin Biddle and John Summerson, 'Portsmouth and the Isle of Wight', in H. M. Colvin (ed.), *The History of the King's Works, vol. 4, 1485–1660 (Part II)* (London, H.M.S.O., 1982), pp. 488–569. Paul Harvey reports that English mapmakers had earlier mapped the English territory around Calais to scale, 1539–41. See also John A. Pinto, 'Origins and development of the ichnographic city plan', *Journal of the Society of Architectural Historians*, 35 (1976), pp. 35–50.

72 Biddle and Summerson, 'Portsmouth and the Isle of Wight' (see note 71), p. 506.

73 Donald Hodson, *Maps of Portsmouth before 1801* (Portsmouth, Portsmouth Record Series, 1978), p. xix.

74 Harvey, *Maps in Tudor England* (see note 66), Fig. 32.

75 British Library, Royal MS 18.D.viii.

76 Clark and Gray, *Old Plans of Cambridge* (see note 27) p. 2. We are indebted to Laurence Worms for drawing our attention to this reference.

77 Clark and Gray, *Old Plans of Cambridge* (see note 27) p. 3.

78 Ralph Agas, *Preparative to Platting of Landes* (1596), p. 18. In fact, few English towns embarked on systematic street paving before the late eighteenth century.

79 Jan Mokre, 'The environs map: Vienna and its surroundings *c*.1600–*c*.1800', *Imago Mundi*, 49 (1997), pp. 90–103.

80 British Library, Maps C.27.g.6 (64). Herman Moll produced a smaller version of Morden's map in about 1710: *A New Map Containing the Towns, Gentlemen's Houses, Villages and Other Remarks round London*; British Library, Maps 3479 (14). Such maps were by no means confined to London, although some, like an anonymous *Map of the Country Twenty Miles round Cambridge* [?1726] would seem to have been produced for the leisured market rather than for administration; British Library, Maps 1658 (5).

81 Described in Sir Herbert George Fordham, *John Cary, Engraver, Map, Chart and Print-Seller and Globe-Maker, 1754 to 1835* (Cambridge, Cambridge University Press, 1927; reprinted Folkestone, Dawson, 1976), pp. 22–3.

82 Peter Barber, 'Liberties and immunities', in Peter Barber and Christopher Board (eds), *Tales from the Map Room: Fact and Fiction about Maps and their Makers* (London, BBC Books, 1993), pp. 132–3. Barber tells us that other examples survive from London, York, and Norwich and that he suspects there may have originally been many more.

83 Data taken from Darlington and Howgego, *Printed Maps of London* (see note 37), though this work is (necessarily) an incomplete record of what is known today. Foreign maps and derivatives are excluded. Norden's map of Westminster (not mentioned by Darlington and Howgego) has been included.

84 Elliot, *The City in Maps* (see note 53), p. 44.

85 British Library, Add. MS 38,065. See Beresford, *History on the Ground* (see note 32), p. 154. For Saxton's only two town plans, see Ifor M. Evans and Heather Lawrence, *Christopher Saxton, Elizabethan Map-Maker* (Wakefield, Wakefield Publications and Holland Press, 1979), p. 100 and 111–2. The map (if indeed that is what it was, not a written survey) of Manchester has not survived; the map of Dewsbury is now in the Dewsbury Public Library.

86 British Library, Harley MS 3749.

87 W. L. D. Ravenhill, 'Maps for the landlord', in Barber and Board, *Tales from the Map Room* (see note 82), pp. 96–7. Norden's map of Toddington is British Library, Add. MS 38,065.

88 'A map of Worksop Manor together with the town of Worksop . . . belonging to the most Noble Edward Duke of Norfolk. Survey'd by George Kelk 1775'. Six sheets. Scale: 4 chains to one inch. (Nottinghamshire Archives, WS 2 L/1–6); 'A certain parcels of crown lands within the manor of Newark . . . in lease to his Grace the Duke of Newcastle and others from a survey made in the year 1788 with a sketch of the town of Newark'. Scale: 35 chains [to the inch] (Public Record Office, MR 275); 'Map of the town of Newark . . . from a survey taken in 1790' by W. Attenborough. Scale: 2 chains to one inch. (Nottingham Archives, NE 1R.).

89 John Schofield (ed.), *The London Surveys of Ralph Treswell* (London, London Topographical Society, Publication 135, 1987), p. 8. All known surveys by Treswell, including a map of Brittany, are listed here by Ralph Hyde. See also John Schofield, 'Ralph Treswell's surveys of London houses', in Tyacke (ed.), *English Map-Making 1500–1650* (see note 64), pp. 85–92. For Treswell's work in Brittany, see Michael Jones, 'Les Anglais à Crozon à la fin du XVIe siècle: le témoignage des cartes', *Mémoires de la Société d'Histoire et d'Archéologie de Bretagne*, 75 (1997), pp. 11–35.

90 Judith Etherton, 'New evidence: Ralph Treswell's association with St Bartholomew's Hospital', *London Topographical Record*, 27 (1995), pp. 103–17 and colour Fig. B; Peter Barber, 'A city for merchants', in Barber and Board, *Tales from the Map Room* (see note 82), pp. 134–5. See also John Schofield, 'Ralph Treswell's surveys of London houses *c*.1612', in Tyacke (ed.), *English Map-Making 1500–1600* (see note 64), pp. 85–92.

91 On Leybourne in general, see C. E. Kenney, 'William Leybourne, 1626–1716', *The Library*, 5th series, 5 (1951), pp. 159–71, E. G. R. Taylor, *The Mathematical Practitioners of Tudor and Stuart England* (Cambridge, Cambridge University Press, 1954), pp. 230–1, and A. W. Richeson, *English Land Measuring to 1800: Instruments and Practices* (Cambridge, Mass., M. I. T. Press and Society of the History of Technology, 1966), pp. 130–1.

92 Betty R. Masters, *The Public Markets of the City of London Surveyed by William Leybourne in 1677* (London, London Topographical Society Publication 117, 1974), p. 11.

93 Reproduced in monochrome in Masters, *Public Markets* (see note 92). The original of Leybourne's Book of Surveys, 1677, is in the Corporation of London Records Office (Plans, 92C).

94 The engraver was John Pine and the assembled map measured 2×4 m: Ralph Hyde, 'Introductory notes', in *The A to Z of Georgian London* (Lympne, Harry Margary for the London Topographical Society, Publication 126, 1982). Darlington and Howgego, *Printed Maps of London* (see note 37), no. 96.

95 Cited by Hyde, 'Introductory Notes' (see note 36), p. vii.

96 Darlington and Howgego, *Printed Maps of London* (see note 37), no. 193.

97 Hyde, 'Introductory Notes' (see note 36); Darlington and Howgego, *Printed Maps of London* (see note 37), no. 94.

98 Scale 1.25 inches to the mile. Darlington and Howgego, *Printed Maps of London* (see note 37), no. 101.

99 Peter Barber, 'Controlling civil unrest', in Barber and Board, *Tales from the Map Room* (see note 82), pp. 118–19.

100 Paul Laxon, 'Introduction', *The A to Z of Regency London* (Lympne, Harry Margary for the London Topographical Society, Publication 131, 1985).

101 Scale: 26 inches to one mile. The huge undertaking was financed through subscription: see Laxton, 'Introduction' (see note 100), p. vii, for a social analysis of the 1,116 names listed. Darlington and Howgego, *Printed Maps of London* (see note 37), no. 200.

102 Laxton, 'Introduction' (see note 100). Darlington and Howgego, *Printed Maps of London* (see note 37), no. 200 (2, 3, 4,).

103 David Smith, 'Public health and the large-scale mapping of British towns', *Journal of the International Map Collectors' Society*, 56 (1994), pp. 28–46.

104 Richard Oliver, *Ordnance Survey Maps: a Concise Guide for Historians* (London, Charles Close Society, 1993).

105 J. B. Harley, 'The Ordnance Survey 1:528 Board of Health town plans in Warwickshire 1848–1854', in T. R. Slater and P. J. Jarvis (eds), *Field and Forest: An Historical Geography of Warwickshire and Worcestershire* (Norwich, Geo Books, 1982), pp. 347–84.

106 Cited by Ralph Hyde in *Printed Maps of Victorian London* (Folkestone, Dawson, 1975), p. 12.

107 The boundaries were of the counties, county court districts, City and parliamentary boroughs, metropolitan local management districts, parishes, poor law districts, rural deaneries, postal districts, police court divisions, registry districts, and registry subdivisions: Hyde, *Printed Maps of Victorian London* (see note 106), no. 91, pp. 12–13 and 112–14. The map continued to be reissued until about 1901.

108 Gwyn Rowley, *British Fire Insurance Plans* (Hatfield, Charles E. Goad, 1984); Gwyn Rowley, 'British fire insurance plans: the Goad productions, *c.*1885–*c.*1970', *Archives*, 17 (1985), pp. 67–78.

109 Keith D. Lilley 'Geometry, urban planning and town design in the High Middle Ages', *Planning History* 20 (1998), pp. 7–15. Modern thinking, based on new evidence, runs counter to the traditional idea of the typical medieval town as organic, something that grew 'naturally'.

110 Maurice Beresford, *New Towns of the Middle Ages: Town Plantation in England, Wales and Gascony*, 2nd ed. (Gloucester, Alan Sutton, 1988) and J. K. St Joseph, *Medieval England. An Aerial Survey*, 2nd ed. (Cambridge, Cambridge University Press, 1979), Chapter 9.

111 Beresford, *New Towns* (see note 110), pp. 145–6.

112 Marcus Vitruvius Pollo, *De Architectura* (compiled first century BC); Leon Battista Alberti, *De Re Aedificatore* (written *c.*1450, printed in Latin in 1485), and Antonio Filarete, *Trattato d'Architettura* (1457–64). Filarete's treatise describes his plan for the radial-concentric ideal city of Sforzinda. See Helen Rosenau, *The Ideal City: Its Architectural Evolution in Europe* (London, Methuen, 3rd edition 1983), pp. 42–50.

113 Hyde, 'Introductory Notes' (see note 36).

114 Hyde, 'Introductory Notes' (see note 36).

115 Hyde, 'Introductory Notes' (see note 36). A single-sheet version of the original 6-sheet draft (which appears not to have survived) is in the British Library, Add MS 5415.E.1.

116 Hyde, 'Introductory Notes' (see note 36). Darlington and Howgego, *Printed Maps of London* (see note 37), no. 21; London Topographical Society Publication 104. A number of copies and derivatives followed, for example, Darlington and Howgego nos 16–20. Leake's map is sometimes referred to, confusingly, as Hollar's.

117 The two maps are listed separately in Darlington and Howgego, *Printed Maps of London* (see note 37), nos 22 and 23, although printed on the same sheet. A copy is in the British Library, Maps K.Top.20.18.

118 Wren's various drawings were later engraved: see British Library, Maps K.Top. 19 1–4. A reproduction of his final plan from the original drawing is printed in S. D. Adshead, 'Sir Christopher Wren and his plan for London', in Royal Institute of British Architects, *Sir Christopher Wren, A.D. 1632–1723* (London, Hodder and Stoughton, 1923), pp. 161–74; a slightly modified version of Wren's plan was published in 1760 in John Gwynn, *London and Westminster Improved*. The influence of the London rebuilding plans, legislation and procedures are assessed in M. Turner, *The Nature of Urban Renewal after Fires in Seven English Provincial Towns, circa 1675–1810* (unpublished University of Exeter Ph.D. thesis, 1985), especially pp. 114–18; the classic study of rebuilding is T. F. Reddaway, *The Rebuilding of London after the Great Fire* (London, Cape, 1940, 2nd edition 1951); also important is Steen Eiler Rasmussen, *London the Unique City*, revised abridged edition (Harmondsworth, Penguin Books, 1960), pp. 93–114.

119 Reproductions in Felix Barker and Peter Jackson, *The History of London in Maps* (London, Barrie and Jenkins, 1990), pp. 36–7.

120 Harcourt's layout was geometrically rigid, and Knight's, which provided for a canal running from the Fleet around the north of the city to Billingsgate, offended the King for the inappropriateness of seeking commercial gain out of tragedy: Barker and Jackson, *History of London* (see note 119), p. 36.

121 Even though never implemented in England, the post-Fire plans for a new London were not without influence. For instance, some twenty years later William Penn who had personally witnessed the Great Fire, based his plan for the new capital of his North American colony (Philadelphia, Pennsylvania) on Richard Newcourt's design for the rebuilding of London: John W. Reps, *The Making of Urban*

America (Princeton, N.J., Princeton University Press, 1965), pp. 161–3.

122 The fair copies of some of the surveys made between March 1666 and October 1668, gathered into four volumes by Peter Mills and John Oliver, have been published in reduced facsimile from the manuscript of 'Mills survey of ground staked out after the Fire of London' (Guildhall Library, MS 84) by the London Topographical Society (Publications 79 and 80, 1946) with an introduction by Walter M. Godfrey.

123 Hyde, 'Introductory Notes' (see note 36), p. v.

124 Walter Ison, The Georgian Buildings of Bath from 1700 to 1830 (London, Faber and Faber, 1948); Bryan Little, Bath Portrait: The Story of Bath, Its Life and Its Buildings, 3rd edition (Bristol, The Burleigh Press 1972); Barry Cunliffe, The City of Bath (Gloucester, Alan Sutton, 1986).

125 Tim Mowl and Brian Earnshaw, John Wood, Architect of Obsession (Bath, Millstream Books, 1988), p. 19.

126 Mowl and Earnshaw, John Wood (see note 125), p. 71; their comments on the 1735 map are on pp. 81–2.

127 Ebenezer Howard, Garden Cities of To-Morrow, edited by F. J. Osborn (London, Faber and Faber, 1965).

128 Cited in Lynam, 'English maps and map-makers of the sixteenth century' (see note 27), p. 22.

129 David Smith, Maps and Plans for the Local Historian and Collector (London, Batsford, 1988), pp. 154–6.

130 Thomas Frangenberg, 'Chorographies of Florence: the use of city views and plans in the sixteenth century', Imago Mundi, 46 (1994), pp. 41–64.

131 For example, A Pocket Map of the Citties of London and Westminster and the Suburbs thereof . . . by Thomas Bowles, followed in 1738 by The City Guide, or a Pocket Map of London, Westminster and Southwark: Darlington and Howgego, Printed Maps of London (see note 37), nos 69 and 82. On maps for way-finding in London in the nineteenth century, see Ralph Hyde, Printed Maps of Victorian London 1851–1900 (Folkestone, Dawson, 1975), especially p. 51, and Ralph Hyde, 'Introductory Notes', The A to Z of Victorian London (Lympne, Harry Margary for the Guildhall Library, 1987).

132 Jane E. Norton, Guide to the National and Provincial Directories of England and Wales, excluding London, published before 1856 (London, Royal Historical Society Guides and Handbooks No. 5, 1950, reprinted with some corrections, 1984), pp. 85–7.

133 S. De Beer, 'The development of the guide book until the early nineteenth century', Journal of the British Archaeological Association, 3rd series, 15 (1952), pp. 35–46, quotation from p. 36. We are indebted to Laurence Worms for the early use of the term 'handbook'.

134 Cited in David Smith, Victorian Maps of the British Isles (London, Batsford, 1985), p. 74.

135 Gareth Shaw and Allison Tipper, British Directories: A Bibliographic Guide to Directories Published in England and Wales (1850–1950) and Scotland (1773–1950) (Cassell, London, 2nd ed. 1996), pp. 24–31.

136 P. J. Atkins, The Directories of London, 1677–1977 (London, Mansell, 1990), p. 74.

137 Appendix to Reference Index of Patents of Invention . . ., 15 & 16 Vict.Cap.83.Sec.XXXII. We are grateful to Sarah Tyacke and Laurence Worms for telling us about this.

138 Ravenhill and Rowe, 'A decorated screen map' (see note 52).

139 Ravenhill and Rowe, 'A decorated screen map' (see note 52), p. 8. This non-landscape detail is present only on the larger, leather screen version of Hooker's map.

CHAPTER 7

1 Richard R. Oliver, Ordnance Survey Maps: A Concise Guide for Historians (London, Charles Close Society, 1993), p. 9; W. A. Seymour (ed.), A History of the Ordnance Survey (Folkestone, Dawson, 1980); T. Owen and E. Pilbeam, Ordnance Survey: Map Makers to Britain since 1791 (Southampton, Ordnance Survey, 1992), pp. 3–8.

2 C. and J. Greenwood's map of Yorkshire is titled: Map of the County of York, made on the basis of Triangles in the County, determined by Lieut. Coll. W. Mudge . . . and Captn. Thos. Colby. & surveyed in the years 1815, 1816 and 1817; see Chapter 3 above.

3 J. B. Harley, 'The re-mapping of England, 1750–1800', Imago Mundi, 19 (1965), pp. 56–67; R. A. Skelton, The Military Survey of Scotland (Edinburgh, Royal Scottish Geographical Society, 1967); I. D. Whyte and K. A. Whyte, Sources for Scottish Historical Geography: An Introductory Guide (Norwich, Geo Abstracts for Historical Geography Research Series No. 6, 1981); G. Whittington and A. J. S. Gibson, The Military Survey of Scotland 1747–1755: A Critique (Norwich, Geo Abstracts for Historical Geography Research Series, No. 18, 1986).

4 Cited in Owen and Pilbeam, Ordnance Survey (see note 1), p. 5.

5 Yolande Hodson, Ordnance Surveyors' Drawings 1789–c.1840 (Reading, Research Publications, 1989), p. 12.

6 Oliver, Ordnance Survey Maps (see note 1), pp. 9–10, 35, 40–1.

7 A townland was the lowest administrative unit of Ireland, each averaging about 100 hectares.

8 J. H. Andrews, History in the Ordnance Map: An Introduction for Irish Readers (Dublin, Ordnance

Survey, 1974; 2nd ed., Kerry, Montgomeryshire, David Archer, 1993) and *A Paper Landscape: the Ordnance Survey in Nineteenth-Century Ireland* (Oxford, Clarendon Press, 1975).

9 Oliver, *Ordnance Survey Maps* (see note 1), pp. 11–30; J. B. Harley, *The Ordnance Survey and Land-Use Mapping* (Norwich, Geo Abstracts for Historical Geography Research Series No. 2, 1979).

10 R. Oliver, *An Introduction to the Ordnance Survey One-Inch Seventh Series Map with a List of Editions* (Southampton, Ordnance Survey, 1991).

11 A. H. W. Robinson, *Marine Cartography in Britain. A History of the Sea Chart to 1855* (Leicester, Leicester University Press, 1962). Late seventeenth-century practice is discussed in considerable detail with respect to the Ouse and Humber rivers and estuaries by Paul Hughes, 'Thomas Surbey's 1699 survey of the Rivers Ouse and Humber', *Yorkshire Archaeological Journal*, 66 (1994), pp. 149–90.

12 Robinson, *Marine Cartography in Britain* (see note 11), p. 97.

13 Trinity House was the body responsible for erecting and maintaining navigational marks at sea; see Robinson, *Marine Cartography in Britain*, (note 11), p. 72 and pp. 75–84 where he discusses the work of one of the most celebrated private hydrographers, Lewis Morris. The work of the military engineer John Desmaretz at Portsmouth and Plymouth is reviewed on pp. 92–3. The charting of one section of the English coast is detailed in William Ravenhill, 'The marine cartography of south-west England from Elizabethan to modern times', in Michael Duffy *et al*, *The New Maritime History of Devon: from Early Times to the late Eighteenth Century*, vol. 1 (London, Conway Maritime Press; Exeter, University of Exeter Press, 1992), pp. 155–63. The history of charting Plymouth Sound and harbour is fully documented and illustrated in Elisabeth Stuart, *Lost Landscapes of Plymouth: Maps, Charts and Plans to 1800* (Stroud, Alan Sutton; Tring, Map Collector Publications, 1991), especially pp. 17–23.

14 J. B. Harley, *Maps for the Local Historian: A Guide to the British Sources* (London, National Council of Social Service for the Standing Conference for Local History, 1972), pp. 51–62.

15 Cited in Robinson, *Marine Cartography in Britain*, (see note 11) p. 102.

16 Dr Andrew Cook of the British Library Oriental and India Office Collections is writing a major study of Alexander Dalrymple's career at the East India Office and at the Hydrographic Office of the Admiralty.

17 Information on Dalrymple's *modus operandi* is kindly provided by Dr Andrew Cook.

18 Contemporary authorities and some modern historians have neither recognised nor credited Dalrymple's signal achievements. Our view of the founder of the Hydrographic Office will be revised by Andrew Cook and see also: A. C. F. David,

'Alexander Dalrymple and the emergence of the Admiralty chart', in D. Howse (ed.), *Five Hundred Years of Nautical Science* (Greenwich, National Maritime Museum, 1981), pp. 153–64. H. S. Edwards lists nearly 200 publications by Dalrymple in 'Alexander Dalrymple, FRS (1737–1808): first Hydrographer to the Admiralty', *The Map Collector*, 4 (1978), pp. 19–29, cited in T. Campbell, 'Episodes from the early history of British Admiralty charting', *The Map Collector*, 25 (1983), pp. 28–33.

19 Robinson, *Marine Cartography in Britain*, (see note 11) p. 127.

20 A. Day, *The Admiralty Hydrographic Service, 1795–1919* (London, HMSO, 1967), p. 43.

21 Cited in Ravenhill, 'Marine cartography of south-west England', (see note 13) p. 160. Useful though the maps of the military survey were, their accuracy was not uniform. For example, the island of Lundy off the north Devon coast was misplaced by the Trigonometrical Survey as it had not been tied in to the general triangulation of this area.

22 G. S. Ritchie, *The Admiralty Chart. British Naval Hydrography in the Nineteenth Century* (London, Hollis and Carter, 1967), pp. 238–48. In south-west England, for example, there was only one stretch of coast from Start Point in south Devon west to Gribbin Head by Fowey in east Cornwall that Beaufort accepted as having been surveyed accurately enough already (by White in 1828); see Ravenhill, 'Marine cartography of south-west England', (see note 13) p. 161.

23 Cited in Day, *The Admiralty Hydrographic Service*, (see note 20) pp. 334–5.

24 Cited in Robinson, *Marine Cartography in Britain*, (see note 11) p. 136.

25 Work in the coastal waters of Britain was, of course, paralleled simultaneously by increasing charting overseas.

26 Ritchie, *The Admiralty Chart*, (see note 22) p. 248.

27 The content of marine charts and particularly the land information that they contain is succinctly described in David Smith, *Maps and Plans for the Local Historian and Collector* (London, Batsford, 1988), pp. 139–47.

28 Harley, *Maps for the Local Historian* (see note 14), pp. 52–3 and see T. Campbell and A. David, 'Bibliographical notes on nineteenth-century British Admiralty charts', *The Map Collector*, 26 (1984), pp. 9–14. See also, A. David, *A Catalogue of Charts, Plans and Views, Printed at the Admiralty Office, for the Use of His Majesty's Navy in 1814* (Taunton, Andrew David, 1991) which is based mainly on a manuscript list presented to the British Museum in 1814 (Maps C.21.c.15) and entitled 'Admiralty chart catalogue, 1814'.

29 Arthur H. Robinson, *Early Thematic Mapping in the History of Cartography* (Chicago and London, University of Chicago Press, 1982), p. x. Robinson's

book is the standard work on the subject; see review by Elizabeth Clutton, in *The Map Collector*, 22 (1983), pp. 42–3.

30 Early in the development of the idea of thematic mapping four principal techniques for reducing area-located data to values independent of the size of area were developed. These are: (i) to divide the totals for each area by some arbitrary unit value and then place a number of uniform dots in each area, the number being related to the total value divided by the unit value (dot maps); (ii) to place in each area a symbol, usually a circle, the size of which is made proportional to the total value (proportional symbol/circle maps); (iii) to calculate percentages or the density of the variable being mapped (numbers per unit of area), divide this into a number of classes and assign a tone of shading to each class, usually the higher the value, the darker the shading (density or choropleth maps); (iv) to calculate the geographical coordinates of the centroid of the areal unit from which the data are drawn, make the assumption that the data values relate to these points, and then calculate isopleths (lines of equal value) from the point-located data (contour or isopleth maps).

31 See, for example, Peter Haggett, *The Geographer's Art* (Oxford, Blackwell, 1990), especially 'The art of the mappable', pp. 46–9.

32 A.-M. Guerry, *Essai sur la statistique morale de la France précédé d'un rapport à l'Académie des Sciences par MM. Lacroix, Silvestre et Girard* (Paris, 1833).

33 Günter Schilder, 'An unrecorded set of thematic maps by Hondius', *The Map Collector*, 59 (1992), pp. 44–7.

34 William Lambarde, *A Perambulation of Kent*, 2nd ed. (London, 1596), pp. 68–70; see Retha M. Warnicke, *William Lambarde. Elizabethan Antiquary 1536–1601* (Chichester, Phillimore, 1973), especially p. 31.

35 Cited in Robinson, *Early Thematic Mapping* (see note 29), p. 69.

36 Alan Cook, *Edmond Halley: Charting the Heavens and the Seas* (Oxford, Clarendon Press 1998), pp. 281–90.

37 See Chapter 8.

38 Robinson, *Early Thematic Mapping* (see note 29), pp. 68–188.

39 Martin J. S. Rudwick, 'The emergence of a visual language for geological science 1760–1840', *History of Science*, 14 (1976), pp. 149–95.

40 Rudwick, 'Emergence of a visual language', (see note 39) pp. 149–95.

41 Cited by Keith Needell, 'A comparison between chalk and cheese', *IMCOS* [International Map Collectors' Society] *Journal*, 63 (1995), p. 44.

42 H. A. Ireland, 'History of the development of geologic maps', *Bulletin of the Geological Society of America*, 54 (1943), pp. 1227–80; Roy Porter, *The Making of Geology: Earth Science in Britain* (Cambridge, Cambridge University Press, 1977).

43 Eila M. J. Campbell, 'An English philosophico-chorographical chart', *Imago Mundi*, 6 (1949), pp. 79–84; Michael Charlesworth, 'Mapping the body and desire: Christopher Packe's chorography of Kent', in Denis Cosgrove (ed.), *Mappings* (London, Reaktion Books, 1999), pp. 109–24.

44 R. C. Boud, 'The early development of British geological maps', *Imago Mundi*, 27 (1975), pp. 73–96; W. Maton, *Observations Relative Chiefly to the Natural History, Picturesque Scenery and Antiquities, of the Western Counties of England, Made in the Years 1794 and 1796*, 2 vols (Salisbury, J. Eaton, 1797).

45 Rudwick, 'Emergence of a visual language' (see note 39), pp. 162–3; Ireland, 'History of the development of geologic maps' (see note 42), which sets English developments into their international context.

46 Karen Severud Cook, 'From false starts to firm beginnings: early colour printing of geological maps', *Imago Mundi*, 47 (1995), pp. 155–72, quotation from p. 156.

47 J. B. Harley, 'The Ordnance Survey and the origins of official geological mapping in Devon and Cornwall', in K. J. Gregory and W. L. D. Ravenhill (eds), *Exeter Essays in Geography in Honour of Arthur Davies* (Exeter, University of Exeter, 1971), pp. 105–23.

48 Harley, 'The Ordnance Survey and the origins of official geological mapping' (see note 47), pp. 112–13; F. J. North, 'Geology's debt to Henry de la Beche', *Endeavour*, 3 (1944), pp. 15–19; North, 'From the geological map to the Geological Survey', *Transactions of the Cardiff Naturalists' Society*, 65 (1932), pp. 41–115.

49 H. E. Wilson, *Down to Earth: One Hundred Years of the British Geological Survey* (Edinburgh, Scottish Academy Press, 1985), pp. 3–32; Sir John Smith Flett, *The First Hundred Years of the Geological Survey of Great Britain* (London, H.M.S.O., 1937), pp. 60–4, 84–91 and 106–7.

50 Ralph Hyde, *Printed Maps of Victorian London* (Folkestone, Dawson, 1975), pp. 31–6.

51 We are grateful to Rodney Shirley for drawing our attention to the contribution of Alexander Keith Johnston to the thematic mapping of natural phenomena.

52 Cited in David Smith, *Victorian Maps of the British Isles* (London, Batsford, 1985), p. 61.

53 G. B. G. Bull, *Thomas Milne's Land Use Map of London and Environs in 1800* (London Topographical Society Publications 118 and 119, 1975–1976); C. Board, 'Town and Country', in P. Barber and C. Board (eds), *Tales from the Map Room: Fact and Fiction about Maps and their Makers* (London, BBC Books, 1993), pp. 142–3.

54 L. D. Stamp, *The Land of Britain; Its Use and Misuse* (London, Longmans, Green, 1962).

55 J. M. Hooke and R. J. P. Kain, *Historical Change in the Physical Environment* (London, Butterworths, 1982), pp. 19–21; R. M. Fuller, J. Sheail, and C. J. Barr, 'The land of Britain, 1930–1990: a comparative study of field mapping and remote sensing techniques', *Geographical Journal*, 160 (1994), pp. 173–84.

56 Simon Rycroft and Denis Cosgrove, 'Mapping the modern nation: Dudley Stamp and the Land Utilisation Survey', *History Workshop Journal*, 40 (1995), pp. 91–105.

57 C. Dupin, *Forces productives et commerciales de la France* (Paris, Bachelier, 1827).

58 On the development of the moral statistics movement in England see: M. J. Cullen, *The Statistical Movement in Early Victorian Britain: the Foundations of Empirical Social Research* (Hassocks, Sussex, The Harvester Press; New York, Barnes and Noble Books, 1975), pp. 65–74.

59 A. H. W. Robinson, 'The 1837 maps of Henry Dury Harness', *Geographical Journal*, 121 (1955), pp. 440–50 and T. W. Freeman, 'Some Irish maps, 1837–49', *Geographical Journal*, 122 (1956), pp. 129–31. We thank John Andrews for drawing our attention to the work of Harness and Larcom in Ireland.

60 Joseph Fletcher, 'Moral and educational statistics of England and Wales', *Journal of the Statistical Society of London*, 12 (1849), pp. 151–335; map entitled 'Ignorance in England and Wales as indicated by the men's signatures as marks in the marriage registers, 1844'.

61 Fletcher, 'Moral and educational statistics' (see note 60) map entitled, 'Dispersion of the population in England and Wales, 1841'.

62 Robinson, *Early Thematic Mapping* (see note 29), pp. 121–5.

63 E. W. Gilbert, 'Pioneer maps of health and disease in England', *Geographical Journal*, 124 (1958), pp. 172–83.

64 Thomas Shapter, *The History of Cholera in Exeter in 1832* (London, Churchill, 1849).

65 Gilbert, 'Pioneer maps of health and disease' (see note 63).

66 Robinson, *Early Thematic Mapping* (see note 29), p. 179.

67 In *Census of Ireland*, 1841, part 4, and see Robinson, *Early Thematic Mapping* (see note 29), pp. 184–6.

68 See Hyde, *Printed Maps of Victorian London* (see note 50), p. 24.

69 D. Smith, 'The social mapping of Henry Mayhew', *The Map Collector*, 30 (1986), pp. 2–3.

70 Kevin Bales, 'Charles Booth's survey of life and labour of the people in London 1889–1903', in Martin Bulmer, Kevin Bales and Kathryn Kish Sklar (eds), *The Social Survey in Historical Perspective* (Cambridge, Cambridge University Press, 1991),

pp. 66–110. Bales provides a detailed discussion of Booth's research methods.

71 David A. Reeder, 'Introduction' to *Charles Booth's Descriptive Map of London Poverty, 1889* (London Topographical Society Publication 130, 1984).

72 A brief review of Charles Booth's thematic mapping is in Hyde, *Printed Maps of Victorian London* (see note 50), pp. 28–31; Booth's data were overprinted on Edward Stanford's *Library Map of London*.

73 Reeder, 'Introduction' (see note 71).

74 J. Shepherd, J. Westaway and T. Lee, *Social Atlas of London* (Oxford, Clarendon Press, 1974).

CHAPTER 8

1 R. A. Skelton, *Maps: A Historical Survey of Their Study and Collecting. First Kenneth Nebenzahl Lecture at the Newberry Library, 1966* (Chicago, University of Chicago Press, 1972), p. 28.

2 James Westfall Thompson, *The Medieval Library* (Chicago, University of Chicago Press, 1939), p. 104.

3 Raymond Irwin, *The English Library. Its Sources and History* (London, George Allen and Unwin, 1966), p. 124, quoting John Stow on the destruction of the Guildhall library. For some consequences of such destruction, see C. R. Gillett, *Burned Books: Neglected Chapters in British History and Literature*, 2 vols (Port Washington, N.Y., Kennikat Press, 1932); Philip Gaskell, *Trinity College Library: The First 150 Years* (Cambridge, Cambridge University Press, 1980), p. 7, n. 3; J. C. T. Oates, *Cambridge University Library: A History* (Cambridge, Cambridge University Press, 1986), vol. 1, p. 77 ff.; Franklin B. Williams, 'Lost books of Tudor England', *The Library*, 5th series 33 (1978), pp. 1–14.

4 J. C. T. Oates, 'The libraries of Cambridge 1570–1700', in F. Wormald and C. E. Wright (eds) *The English Library before 1700. Studies in Its History* (London, Athlone Press, 1958), pp. 213–35.

5 Irwin, *The English Library* (see note 3), p. 125.

6 On the loss of music, see Iain Fenlon, *Music, Print and Culture in Early Sixteenth-Century Italy. The Panizzi Lectures, 1994.* (London, The British Library, 1995), pp. 2–4.

7 *The First Printed Catalogue of the Bodleian Library. A Facsimile* (Oxford, Clarendon Press, 1986).

8 Elisabeth S. Leedham-Green, *Books in Cambridge Inventories: Booklists from Vice-Chancellor's Court Probate Inventories in the Tudor and Stuart Periods*, 2 vols, (Cambridge, Cambridge University Press, 1986); Elisabeth Leedham-Green, 'Perne's wills', in Patrick Collinson, David McKitterick and Elisabeth Leedham-Green, *Andrew Perne: Quatercentenary Studies* (Cambridge, Cambridge University Library

for the Cambridge Bibliographical Society, 1991), pp. 79–119.

9 John Evans, 'Extracts from the private account book of Sir William More, of Loseley in Surrey, in the time of Queen Mary and Queen Elizabeth', *Archaeologia*, 35 (1855), pp. 284–310.

10 Many of these printed maps, especially those owned by William More in mid-century, would have been imported. From its description, the map of London at Cockesden is likely to have been either the 'lost copperplate' map of *c.*1558–9 or its woodcut copy (see chapter 6). The Cockesden inventory is recorded in James Orchard Halliwell, *Ancient Inventories of Furniture, Pictures, Tapestry, Plate etc Illustrative of the Domestic Manners of the English in the Sixteenth and Seventeenth Centuries* (London, printed for private circulation, 1845), pp. 73 and 81 (British Library, 741.k.19). For an interesting view of books and maps amongst the upper and land-owning classes from the end of the sixteenth century onwards, see A. Sarah Bendall, *Maps, Land and Society: A History, with a Carto-Bibliography of Cambridgeshire Estate Maps,* c.1600–1863 (Cambridge, Cambridge University Press, 1992), pp. 143–50. See also Catherine Delano-Smith, 'Map ownership in sixteenth-century Cambridge; the evidence of probate inventories', *Imago Mundi*, 47 (1995), pp. 67–93.

11 The two pioneering studies are Margaret Spufford's *Small Books and Pleasant Histories. Popular Fiction and its Readership in Seventeenth-century England* (Cambridge, Cambridge University Press, 1981) and Tessa Watt's *Cheap Print and Popular Piety 1550–1640* (Cambridge, Cambridge University Press, 1991). Margaret Spufford discusses road information in almanacs but see especially Bernard Capp, *English Almanacs 1500–1800. Astrology and the Popular Press* (London, Faber, 1979). For a continental perspective on the role of printing in popular culture, see R. W. Scribner, *For the Sake of Simple Folk. Popular Propaganda for the German Reformation* (Oxford, Clarendon Press, 1981).

12 Spufford, *Small Books* (see note 11), p. 113, citing Capp, *English Almanacs* (see note 11), p. 355.

13 In 1556, Leonard Digges gave a list of fairs in his almanac, and road directions appeared in the edition translated from G. Gossene's French almanac: Capp, *English Almanacs* (see note 11), p. 30.

14 John White, *A prognostication for . . . 1647* (London, Stationers' Company, 1647). The map (14 × 9 cm) occupies most of the page facing the table. It carries a scale bar with dividers and a com-pass rose. Earlier editions lack the map, having only a diagram of 'The Anatomy of Man' for medico-astrological use. Printed almanacs were early used to popularise new ideas and explain the use of new instruments, in the sixteenth century (by people such as Leonard and Thomas Digges and William

Cuningham) as in the eighteenth century: Capp, *English Almanacs* (see note 11), especially pp. 31, 180 and 200–3. Some contained diagrams of instruments to cut out and paste on to card.

15 Watt, *Cheap Print* (see note 11), Chapter 6, 'Godly tables for good householders', illustrates, among others 'The Map of Mortality' (1604) (Fig. 47) and '[Some f]yne gloves devised . . . to teche yonge peo[ple] good from evyll' (1559–67) (Fig. 49), both of which employ diagrams as well as text to emphasise the message.

16 Roger Chartier, *Cultural Uses of Print in Early Modern France*, translated by Lydia G. Cochrane (Princeton, Princeton University Press, 1987), p. 235, describes such *tours de cheminée* and *tours de lit*, and Watt, *Cheap Print* (see note 11), p. 131, notes that English cottages were so decorated.

17 Donald Hodson, *County Atlases of the British Isles. A Bibliography. vol. 1 1704–1742* (Tewin, Tewin Press, 1984), pp. 67–8.

18 The Westerham landslip was described, with a sketch-map, in John Chapman, *A Most True Report of the Moving and Sinking of Ground . . .* (1597; Bodleian Library, MS Wood, D. 28 (1)).

19 The *briefe relatione* contains a page with three maps (a regional map and the town plans of Orsoy and Calcar) and a picture of a whale stranded on the beach at Berckhey the previous year. At the time, the stranded whale had been taken as a portent but on the news-sheet it served as a political allegory. On the Berckhey whale, see Simon Schama, *The Embarrassment of Riches. An Interpretation of Dutch Culture in the Golden Age* (Berkeley and Los Angeles, University of California Press, 1988), pp. 130–3.

20 See Ruth Samson Luborsky and Elizabeth Morely Ingram, *A Guide to English Illustrated Books, 1536–1603*, 2 vols (Tempe, Arizona Center for Medieval and Renaissance Studies, 1998). The *Guide* documents over 5,000 pictures and diagrams from 1,800 publications, amongst which are some 300 maps (including re-uses), of which a score or so come from pamphlets, news-sheets and books on current political issues (not including voyages of discoveries).

21 Some of the 'lytle maps' in the Cambridge inventories, especially those of least value, may have been bought at the market or from pedlars: Delano-Smith 'Map ownership' (see note 10), pp. 67–93.

22 We are indebted to Tony Campbell for drawing our attention to these socially important double-hemispheric maps and their place in map history.

23 We are grateful to Geoffrey Armitage for access to his manuscript list of '17th and 18th century English large two-sheet double-hemisphere world maps, 1680–*c*.1800'.

24 British Library, Maps C.45.f.5. The map is described by Roderick Barron, *Blaeu* (Catalogue 26, Summer 1996), Item 22.

25 Herman Moll had already shown both Dampier's voyages and Halley's winds, the former in William

Dampier, *A New Voyage Round the World* (London, 1697) and both in the expanded edition of Dampier's book (1699). Examples of Moll's and Marshall's double-hemispheric world map are in the British Library (Maps K.Tab.17 (1); Maps MT. 11.h (17) respectively).

26 Barron, *Blaeu* (see note 24).

27 Armitage, 'World Maps' (see note 23).

28 Rotz intended his *Boke of Idrography* (British Library, Royal MS 20. E. IX) for Francis I of France but dedicated it instead to Henry VIII when he entered the English king's service in 1542: Helen M. Wallis (ed.) *The Boke of Idrography presented in 1542 by Jean Rotz to King Henry VIII, now in the British Library*, facsimile (Oxford, The Roxburgh Club, 1981), p. 9.

29 Carlo F. Capello, *Il mappamondo medioevale di Vercelli (1191–1218?)* (Turin, Università di Torino, 1976, Memorie e Studi Geografici, 10), pp. 9–11.

30 The Duchy of Cornwall *mappamundi* could have been a post-foundation gift to the College of Bonshommes at Ashridge, founded from St Albans in about 1350. See Graham Haslam, 'The Duchy of Cornwall map fragment', in Monique Pelletier (ed.), *Géographie du monde au Moyen Age et à la Renaissance* (Paris, Editions de Comité des Travaux Historiques et Scientifiques, 1989), pp. 33–44.

31 Colin G. C. Tite, *The Manuscript Library of Sir Robert Cotton. The Panizzi Lectures 1993* (London, The British Library, 1994), p. 11, referring to observations made by R. A. Skelton and John Summerson, *A Description of Maps and Architectural Drawings in the Collection made by William Cecil first Baron Burghley now at Hatfield House* (Oxford, The Roxburgh Club, 1971), pp. 10, 16, 20.

32 So many books were taken from the cathedral library by Kilwardby that the library's records start only with entries made by his successor: Irwin, *The English Library* (see note 3), p. 175. For the plundered Gospel, see James Westfall Thompson, *The Medieval Library* (Chicago, University of Chicago Press, 1939), p. 119.

33 Alfred Hessel, *A History of Libraries*, translated by Reuben Peiss (New Brunswick, Scarecrow Press, 1955), p. 32. Bedford bought them from the young Charles VII.

34 Many of these exiles were academics from the universities of Oxford and Cambridge, like John Ramidge himself, Thomas Harding, and Henry Jolliffe. Jolliffe had left many of his books in Stratford under the care of a neighbour of John Shakespeare, William Shakespeare's father: Christian Coppens, *Reading in Exile: The Libraries of John Ramidge (d. 1568), Thomas Harding (d. 1572) and Henry Jolliffe (d. 1578): Recusants in Louvain* (Cambridge, Libri Pertinentes, 1993).

35 Map historians should not rely on the indexes in modern studies of book or library history; for example, neither Irwin, *The English Library* (see note 3) nor Hessel, *History of Libraries* (see note 33), has an entry for maps although both refer to the map-rich libraries of John Dee and Samuel Pepys.

36 For example: as the present book was going to press, Evelyn Edson drew our attention to an unrecorded map in a fifteenth-century English manuscript of John Mandeville's Travels (British Library, Royal 17. C. XXXVIII, fol. 41v).

37 For medieval interest in the Anglo-Saxon period see Antonia Gransden, 'Realistic observation in twelfth-century England', *Speculum* 47 (1972), pp. 29–51, especially pp. 33–7. See also the same author's 'Prologues in the historiography of twelfth century England', in Daniel Williams (ed.) *England in the Twelfth Century. Proceedings of the 1988 Harlaxton Symposium* (Woodbridge, Suffolk, Boydell Press, 1990), pp. 55–81, and her 'Antiquarian studies in fifteenth-century England', *Antiquaries Journal* 60 (1980), pp. 75–97.

38 On allegorical maps in England, see Franz Reitinger, 'Mapping relationships: allegory, gender and the cartographical image in eighteenth-century France and England', *Imago Mundi*, 51 (1999), pp. 106–30 and plate 12.

Bibliography of Printed Works Cited in the Text

— *History or Narrative of the Great Level of the Fenns, called the Bedford Level (London, 1685).*

Abel, W. *Agricultural Fluctuations in Europe from the Thirteenth to the Twentieth Centuries* (London, Methuen, 1980).

Adams, I. H. 'Large-scale manuscript plans in Scotland', *Journal of the Society of Archivists*, 3 (1967), pp. 286–90.

Adams, I. H. 'Economic processes and the Scottish land surveyor', *Imago Mundi*, 27 (1975), pp. 13–18.

Adams, I. H. 'Estate plans', *The Local Historian*, 12 (1976), pp. 26–30.

Adams, I. H. 'The agents of agricultural change', in M. L. Parry and T. R. Slater (eds), *The Making of the Scottish Countryside* (London, Croom Helm, 1980).

Addison, Sir William. *The Old Roads of England* (London, Batsford, 1980).

Adshead, S. D. 'Sir Christopher Wren and his plan for London', in Royal Institute of British Architects, *Sir Christopher Wren, A.D. 1632–1723* (London, Hodder and Stoughton, 1923), pp. 161–74.

Agas, Ralph. *A Preparative to the Platting of Landes and Tenements for Surveigh* (London, 1596).

Aldridge, R. Henry. *The Case for Town Planning: a Practical Manual for the Use of Councillors, Officers, and Others Engaged in the Preparation of Town Planning Schemes* (London, National Housing and Town Planning Council, 1915).

Alexander, Jonathan J. G. *A Survey of Manuscripts Illuminated in the British Isles. vol. 1: Insular Manuscripts from the Sixth to the Ninth Century* (London, Harvey Miller, 1978).

Alexander, Jonathan J. G. and Otto Pächt. *Illuminated Manuscripts in the Bodleian Library, Oxford, vol. 3: British, Irish and Icelandic Schools* (Oxford, Clarendon Press, 1973).

Alington, Gabriel and Dominic Harbour. *The Hereford Mappamundi. A Medieval View of the World* (Leominster, Fowler Wright Books, 1996).

Andrews, J. H. *Ireland in Maps* (Dublin, Geographical Society of Ireland with the Ordnance Survey of Ireland, 1961).

Andrews, J. H. 'The Irish surveys of Robert Lythe', *Imago Mundi*, 19 (1965), pp. 22–37.

Andrews, J. H. *History in the Ordnance Map: An Introduction for Irish Readers* (Dublin, Ordnance Survey, 1974; 2nd edition, Kerry, Montgomeryshire, David Archer, 1993).

Andrews, J. H. *A Paper Landscape: the Ordnance Survey in Nineteenth-Century Ireland* (Oxford, Clarendon Press, 1975).

Andrews, J. H. *Plantation Acres. An Historical Study of the Irish Land Surveyor and his Maps* (Belfast, Ulster Historical Foundation, 1985).

Andrews, J. H. *Shapes of Ireland* (Dublin, Geography Publications, 1997).

Andrews, Kenneth R. *Trade, Plunder and Settlement. Maritime Enterprise and the Genesis of the British Empire, 1480–1630* (Cambridge, Cambridge University Press, 1984).

Anglo, Sydney. *Spectacle, Pageantry and Early Tudor Policy* (Oxford, Clarendon Press, Warburg Studies, 1996).

Armitage, Geoffrey. 'County carto-bibliographies of England and Wales. A selected list', *The Map Collector*, 52 (1990), pp. 49–50.

Armstrong, Charles A. J. 'Some examples of the distribution and speed of news in England at the time of the wars of the Roses', in R. W. Hunt, W. A. Pantin and R. W. Southern (eds), *Studies Presented to Frederick Maurice Powicke* (Oxford, Clarendon Press, 1948), pp. 429–59.

Ashley, Anthony. *The Mariners Mirrour*, facsimile ed. R. A. Skelton (Amsterdam, Theatrum Orbis Terrarum, 1966).

Aston, Margaret. *The Fifteenth Century. The Prospect of Europe* (London, Thames and Hudson, 1979).

Atkins, P. J. *The Directories of London, 1677–1977* (London, Mansell, 1990).

Baldwin, Robert. 'Colonial cartography and the role of the British chartered companies', in David Woodward (ed.), *History of Cartography*, Vol. 3, *Cartography in the European Renaissance* (Chicago and London, University of Chicago Press, forthcoming).

Bales, Kevin. 'Charles Booth's survey of life and labour of the people in London 1889–1903', in Martin Bulmer, Kevin Bales and Kathryn Kish Sklar (eds), *The Social Survey in Historical Perspective* (Cambridge, Cambridge University Press, 1991), pp. 66–110.

Barber, Peter. 'Old encounters new: the Aslake world map', in Monique Pelletier (ed.), *Géographie du monde au Moyen Age et à la Renaissance* (Paris, Editions du Comité des Travaux Historiques et Scientifiques, 1989), pp. 69–88.

Barber, Peter. 'British cartography', in Robert P. Maccubbin and Martha Hamilton-Phillips (eds), *The Ages of William III and Mary II: Power, Politics and Patronage 1688–1702. A Reference Encyclopaedia and Exhibition Catalogue* (Williamsburg, College of William and Mary in Virginia, 1989), pp. 92–104.

Barber, Peter. 'Necessary and ornamental: map use in England under the later Stuarts, 1660–1714', *Eighteenth Century Life*, 14 (1990), pp. 1–28.

Barber, Peter. 'England I. Pageantry, defense and government: maps at court to 1550' and 'England II: Monarchs, ministers, and maps, 1550–1625', in David Buisseret (ed.), *Monarchs, Ministers and Maps: The Emergence of Cartography as a Tool of Government in Early Modern Europe* (Chicago and London, University of Chicago Press, 1992), pp. 26–56 and 57–98.

Barber, Peter. 'The Minister leaves his mark', in Peter Barber and Christopher Board (eds), *Tales from the Map Room. Fact and Fiction about Maps and their Makers* (London, BBC Books, 1993), pp. 88–9.

Barber, Peter. 'Controlling civil unrest', in Peter Barber and Christopher Board (eds), *Tales from the Map Room: Fact and Fiction about Maps and their Makers* (London, BBC Books, 1993), pp. 118–19.

Barber, Peter. 'Liberties and immunities', in Peter Barber and Christopher Board (eds), *Tales from the Map Room: Fact and Fiction about Maps and their Makers* (London, BBC Books, 1993), pp. 132–3.

Barber, Peter. 'A city for merchants', in Peter Barber and Christopher Board (eds), *Tales from the Map Room: Fact and Fiction about Maps and their Makers* (London, BBC Books, 1993), pp. 134–5.

Barber, Peter. 'Finance and Flattery' in Peter Barber and Christopher Board (eds), *Tales from the Map Room. Fact and Fiction about Maps and their Makers* (London, BBC Books, 1993), pp. 140–1.

Barber, Peter. 'A glimpse of the earliest map-view of London?', *London Topographical Record*, 27 (1995), pp. 91–102.

Barber, Peter. 'Les îles Britanniques', in Marcel Watelet (ed.), *Geraldi Mercatoris. Atlas Europa* (Antwerp, Bibliothèque des amis du Fonds Mercatoris, 1994), pp. 70–1.

Barber, Peter. 'The Evesham world map: a late medieval English view of God and the world', *Imago Mundi*, 47 (1995), pp. 13–33.

Barber, Peter. 'Maps and monarchs in Europe 1550–1800', in Robert Oresko, G. C. Gibbs and H. M. Scott (eds), *Royal and Republican Sovereignty in Early Modern Europe* (Cambridge, Cambridge University Press, 1997), pp. 75–124.

Barber, Peter. 'The British Isles', in Marcel Watelet (ed.) *The Mercator Atlas of Europe circa 1570–1572*, English edition (Pleasant Hill, Oregon, Walking Tree Press, 1998), pp. 43–77 and 90–2 (notes).

Barber, Peter. 'Cartography, topography and history paintings', in David Starkey (ed.), *The Inventory of King Henry VIII*, vol. 2, *Text* (London, Society of Antiquaries, in preparation).

Barber, Peter and Michelle P. Brown. 'The Aslake world map', *Imago Mundi*, 44 (1992), pp. 24–44.

Barker, Felix and Peter Jackson. *The History of London in Maps* (London, Barrie and Jenkins, 1990).

Barley, M. W. 'Sherwood Forest, Nottinghamshire late 14th or early 15th century', in R. A. Skelton and P. D. A. Harvey (eds), *Local Maps and Plans from Medieval England* (Oxford, Clarendon Press, 1986), pp. 131–9.

Barron, Caroline and Nigel Saul (eds). *England and the Low Countries in the Late Middle Ages* (Stroud, Alan Sutton; New York, St Martin's Press, 1995).

Barron, Roderick (ed.). *The County Maps of Old England: Thomas Moule* (London, Studio Editions, 1990).

Barron, Roderick. *Blaeu* (Catalogue 26, Summer 1996, Item 22).

Bartlett, Robert. *Gerald of Wales, 1146–1223* (Oxford, Clarendon Press, 1982).

Batho, G. R. 'The finances of an Elizabethan nobleman: Henry Percy, Ninth Earl of Northumberland (1564–1632)', *Economic History Review*, 2nd series, 9 (1957), pp. 433–50.

Batho, G. R. 'Two newly discovered maps by Christopher Saxton', *Geographical Journal*, 125 (1959), pp. 70–4.

Beale, Philip. *A History of the Post in England from the Romans to the Stuarts* (Aldershot, Ashgate, 1998).

Beckett, John V. *A History of Laxton. England's Last Open-field Village* (Oxford, Blackwell, 1989).

Bede. 'Vita sanctorum abbatum monasterii' [Lives of the Abbots], in *Bede, Historical Works* (Cambridge, Mass., Harvard University Press, Loeb Classical Library, 1930), vol. 2, pp. 392–445.

Beech, Geraldine. 'Tithe maps', *The Map Collector*, 3 (1985), pp. 20–5.

de Beer, S. 'The development of the guide book until the early nineteenth century', *Journal of the British Archaeological Association*, 3rd series, 15 (1952), pp. 35–46.

Bendall, A. Sarah. *Maps, Land and Society: A History with a Carto-Bibliography of Cambridgeshire Estate Maps c.1600–1836* (Cambridge, Cambridge University Press, 1992).

Bendall, A. Sarah. 'Interpreting maps of the rural landscape: an example from late sixteenth-century Buckinghamshire', *Rural History*, 4 (1993), pp. 107–21.

Bendall, A. Sarah. 'Enquire "When the same platte was made and by whome and to what intent": sixteenth-century maps of Romney Marsh', *Imago Mundi*, 47 (1995), pp. 34–48.

Bendall, A. Sarah. 'Estate maps of an English county: Cambridgeshire, 1600–1836', in David Buisseret (ed.), *Rural Images: Estate Maps in the Old and New Worlds* (Chicago and London, Chicago University Press, 1996), pp. 63–90.

Bendall, A. Sarah. *Dictionary of Land Surveyors and Local Map-Makers of Great Britain and Ireland 1530–1850*, 2 vols (London, The British Library, 1997).

Bendall, A. Sarah. 'Mapping the English forests: Needwood 1598–1834', in John Langton (ed.), *Where History and Geography Met: English Forests from the Seventeenth to the Nineteenth Century* (forthcoming).

Bennett Durand, Dana. *The Vienna-Klosterneuburg Map Corpus of the Fifteenth Century. A Study in the Transition from Medieval to Modern Science* (Leiden, E. J. Brill, 1952).

Bennett, J. A. *The Divided Circle* (Oxford, Phaidon, 1987).

Bennett, J. A. 'Geometry and surveying in early-seventeenth century England', *Annals of Science*, 48 (1991), pp. 345–54.

Bennett, J. A. and S. A. Johnson. *The Geometry of War, 1500–1750* (Oxford, Museum of the History of Science, 1996).

Beresford, Maurice W. *History on the Ground: Six Studies in Maps and Landscapes* (London, Lutterworth Press, 1957; rev. ed., London, Methuen, 1971).

Beresford, Maurice W. *New Towns of the Middle Ages: Town Plantation in England, Wales and Gascony* (London, Lutterworth Press, 1967; 2nd ed. Gloucester, Alan Sutton, 1988).

Beresford, Maurice W. 'The decree rolls of Chancery as a source for economic history', *Economic History Review*, 2nd series, 32 (1979), pp. 1–10.

Beresford, Maurice W. 'Inclesmoor, West Riding of Yorkshire. *Circa* 1407', in R. A. Skelton and P. D. A. Harvey (eds), *Local Maps and Plans from Medieval England* (Oxford, Clarendon Press, 1986), pp. 147–61.

Beresford, Maurice W. 'Sherwood Forest, Nottinghamshire, late 14th or early 15th Century', in R. A. Skelton and P. D. A. Harvey (eds), *Local Maps and Plans from Medieval England* (Oxford, Clarendon Press, 1986), pp. 131–9.

Berlaimont, Noël. *Colloquia et dictionariolum septem linguarum, Belgicae, Anglicae, Teutonicae, Latinae, Italicae, Hispanicae, Gallicae* (Antwerp, 1586).

Bettey, J. H. 'Late-medieval Bristol: from town to city', *The Local Historian*, 28 (1998), pp. 3–15.

Biddle, Martin and John Summerson. 'Portsmouth and the Isle of Wight', in H. M. Colvin (ed.), *The History of the King's Works, vol. 4, 1485–1660 (Part II)* (London, H.M.S.O., 1982), pp. 488–569.

Binski, Paul. *The Painted Chamber at Westminster* (London, Society of Antiquaries, Occasional Paper n.s. 9, 1986).

Birch, Debra J. *Pilgrimage to Rome in the Middle Ages* (Woodbridge, Suffolk, Boydell and Brewer, 1998).

Bird, A. J. 'John Speed's view of the urban hierarchy in Wales in the early seventeenth century', *Studia Celtica*, 10–11 (1975–6), pp. 401–11.

Blackman, Herbert. 'Gun founding at Heathfield in the XVIII century', *Sussex Archaeological Collections*, 67 (1926), pp. 25–59.

Blagrave, John. *Baculum familliare* (London, 1590).

Blakemore, M. J. and J. B. Harley. 'Concepts in the History of Cartography. A Review and Perspective', *Cartographica*, 17:4 (1980).

Blome, Richard. *The Gentlemans Recreation* (London, 1686).

Blundeville, Thomas. *The true order and Method of wryting and reading Hystories* (London, 1574).

Blundeville, Thomas. *Exercises . . . [for] all Young Gentlemen . . .* (London, 1594).

Board, C. 'Town and Country', in Peter Barber and Christopher Board (eds), *Tales from the Map Room:*

Fact and Fiction about Maps and their Makers (London, BBC Books, 1993), pp. 142–3.

Boase, C. W. *Oxford* (London, Longmans, Green & Co., 1887).

de Boer, G. and R. A. Skelton. 'The earliest English chart with soundings', *Imago Mundi*, 23 (1969), pp. 9–16.

Booth, Charles. *Life and Labour of the People in London*, 2 vols, (London, Williams and Norgate, 1889, 1891, 2nd edition London, Macmillan & Co. 1902–3, 17 vols).

Borlase, Williams. *Observations on the Antiquities Historical and Monumental, of the County of Cornwall* (Oxford, W. Jackson, 1754).

Borlase, William. *Natural History of Cornwall* (Oxford, W. Jackson for the author, 1758).

Borough, William. *A Discourse of the Variation of the Compasse; or Magneticall Needle* (London, dated 1581 but printed in 1596).

Boud, R. C. 'The early development of British geological maps', *Imago Mundi*, 27 (1975), pp. 73–96.

Bourne, William. *A Regiment for the Sea* (London, 1576).

Bowden, P. 'Agricultural prices, farm profits and rents', in Joan Thirsk (ed.), *Agrarian History of England and Wales*, vol. 4, *1500–1640* (Cambridge, Cambridge University Press, 1967), pp. 593–695.

Brewer, J. S. (ed.). *Letters and Papers of Henry VIII* (London, Longmans, Green & Co., 1864).

British Parliamentary Papers (House of Commons, Session 1837), 'Copy of Papers Respecting the Proposed Survey of Lands under the Tithe Act', vol. 41 (1837), pp. 11–16.

British Parliamentary Papers (House of Commons), 'Report from the Select Committee on Survey of Parishes (Tithe Commutation Act); with the Minutes of Evidence', vol. 6 (1837).

Britnell, Richard H. *The Commercialisation of English Society 1000–1500*, 2nd ed. (Manchester and New York, Manchester University Press, 1996).

Broderson, Kai. *Terra Cognita. Studien zur römischen Raumerfassung*, Spudasmata 59 (Hildesheim, G. Olms, 1995).

Brotton, Jerry. *Trading Territories: Mapping the Early Modern World* (London, Reaktion, 1997).

Bruce-Mitford, R. S. L. 'The art of the Codex Amiatinus', *Journal of the British Archaeological Association*, 32 (1969), pp. 1–32.

Buisseret, David. *Rural Images: The Estate Plan in the Old and New Worlds* (Chicago, The Newberry Library, 1988).

Buisseret, David (ed.). *Monarchs, Ministers and Maps: The Emergence of Cartography as a Tool of Government in Early Modern Europe* (Chicago and London, University of Chicago Press, 1992).

Buisseret, David. 'Monarchs, ministers, and maps in France before the accession of Louis XIV', in David Buisseret (ed.), *Monarchs, Ministers and Maps: The Emergence of Cartography as a Tool of Government in Early Modern Europe* (Chicago and London, University of Chicago Press, 1992), pp. 99–123.

Buisseret, David (ed.), *Rural Images: Estate Maps in the Old and New Worlds* (Chicago and London, Chicago University Press, 1996).

Buisseret, David. 'Defining the estate map' in David Buisseret (ed.), *Rural Images: Estate Maps in the Old and New Worlds* (Chicago and London, Chicago University Press, 1996), pp. 1–4.

Buisseret, David (ed.). *Envisioning the City: Six Studies in Urban Cartography* (Chicago and London, University of Chicago Press, 1998).

Bull, G. B. G. *Thomas Milne's Land Use Map of London and Environs in 1800* (London Topographical Society Publications 118 and 119, 1975–1976).

Burden, Eugene. *The Printed Maps of Berkshire 1574–1900*, 3 parts (Ascot, Berkshire, the author, 1988–95).

Burke, Gerald. *Towns in the Making* (London, Edward Arnold, 1983).

Burnett, Charles. *The Introduction of Arabic Learning into England. The Panizzi Lectures 1996* (London, The British Library, 1997).

Butlin, R. A. 'Regions in England and Wales *circa* 1500–1914', in R. A. Dodgshon and R. A. Butlin (eds), *An Historical Geography of England and Wales*, 2nd ed. (London and New York, Academic Press, 1990), pp. 223–54

Campbell, Eila M. J. 'An English philosophico-chorographical chart', *Imago Mundi*, 6 (1949), pp. 79–84.

Campbell, Tony. *Maps and Plans in the Public Record Office, vol. 1, British Isles c.1410–1860* (London, H.M.S.O., 1967).

Campbell, Tony. 'The Drapers' Company and its school of seventeenth-century chart-makers', in Helen Wallis and Sarah Tyacke (eds), *My Head is a Map. Essays and Memoirs in Honour of R. V. Tooley* (London, Francis Edwards and Carta Press, 1973), pp. 81–106.

Campbell, Tony. 'Episodes from the early history of British Admiralty charting', *The Map Collector*, 25 (1983), pp. 28–33.

Campbell, Tony. 'The original monthly numbers of Moule's "English Counties" ', *The Map Collector* 31 (1985), pp. 26–39.

Campbell, Tony. *The Earliest Printed Maps, 1472–1500* (London, The British Library; Berkeley, University of California Press, 1987).

Campbell, Tony. 'Portolan charts from the late thirteenth century to 1500' in J. B. Harley and David Woodward (eds), *The History of Cartography* vol. 1 *Cartography in Prehistoric, Ancient, and Medieval Europe and the Mediterranean* (Chicago and London, University of Chicago Press, 1987), pp. 371–463.

Campbell, Tony. 'Cutting costs', in Peter Barber and Christopher Board (eds), *Tales from the Map Room. Fact and Fiction about Maps and their Makers* (London, BBC Books, 1993), pp. 34–5.

Campbell, Tony. 'For those in peril on the sea', in Peter Barber and Christopher Board (eds), *Tales from the Map Room. Fact and Fiction about Maps and their Makers* (London, BBC Books, 1993), pp. 164–5.

Campbell, Tony. 'Finding a safe haven when at sea', in Peter Barber and Christopher Board (eds), *Tales from the Map Room. Fact and Fiction about Maps and their Makers* (London, BBC Books, 1993), pp. 168–9.

Campbell, Tony. 'Four-fold map screen', *National Art Collections Fund Review for 1997*, p. 84.

Campbell, Tony, and A. David. 'Bibliographical notes on nineteenth-century British Admiralty charts', *The Map Collector*, 26 (1984), pp. 9–14.

Capello, Carlo F. *Il mappamondo medioevale di Vercelli (1191–1218?)* (Turin, Università di Torino, Memorie e Studi Geografici, 10, 1976).

Capp, Bernard. *English Almanacs 1500–1800. Astrology and the Popular Press* (London, Faber, 1979).

Carroll, R. A. *Printed Maps of Lincolnshire, 1576–1900: A Carto-Bibliography. With an Appendix on Road-Books* (Woodbridge, Suffolk, Boydell Press for the Lincoln Record Society, vol. 84, 1996).

Carruthers, Mary. *The Book of Memory. A Study of Memory in Medieval Culture* (Cambridge, Cambridge University Press, 1990).

Cary, John. *Traveller's Companion or a Delineation of the Turnpike Roads of England and Wales* (London, 1790).

Castiglione, Baldassare. *Il libro del cortegiano* (Florence, 1528).

Caxton, William. *Mirrour of the World* (London, 1490).

Chambers, John D. and Gordon E. Mingay. *The Agricultural Revolution, 1770–1860* (London, Batsford, 1966).

Chapman, John. *A Most True Report of the Moving and Sinking of Ground . . .* (London, 1597).

Chapman, John. 'Some problems in the interpretation of enclosure awards', *Agricultural History Review*, 26 (1978), pp. 108–14.

Chapman, John. 'The chronology of English enclosure', *Economic History Review*, 37 (1984), pp. 557–9.

Chapman, John. 'The extent and nature of Parliamentary enclosure', *Agricultural History Review*, 35 (1987), pp. 25–35.

Chapman, John. 'The interpretation of enclosure maps and awards', in K. Barker and R. J. P. Kain (eds), *Maps and History in South-West England* (Exeter, Exeter University Press, 1991), pp. 72–88.

Charlesworth, Andrew. *Social Protest in a Rural Society* (Norwich, Historical Geography Research Group, Research Series No. 1, 1979).

Charlesworth, Andrew (ed.). *An Atlas of Rural Protest in Britain, 1548–1900* (London, Croom Helm, 1983).

Charlesworth, Michael. 'Mapping the body and desire: Christopher Packe's chorography of Kent', in Denis Cosgrove (ed.), *Mappings* (London, Reaktion Books, 1999), pp. 109–24.

Chartier, Roger. *Cultural Uses of Print in Early Modern France*, translated by Lydia G. Cochrane (Princeton, Princeton University Press, 1987).

Chartres, John. 'Lincoln's inns', *Historian*, 7 (1983), pp. 16–19.

Chekin, Leonid. 'Easter tables and the pseudo-Isidorean Vatican map', *Imago Mundi*, 51 (1999), pp. 13–23.

Chilton, D. 'Land measurement in the sixteenth century', *Transactions of the Newcomen Society*, 31 (1957–9), pp. 111–29.

Clanchy, M. T. *From Memory to Written Record: England 1066–1307* (London, Edward Arnold, 1979; Oxford and Cambridge, Mass., Blackwell, 1993).

Clark, J. Willis and Arthur Gray. *Old Plans of Cambridge, 1574–1798, Reproduced in Facsimile with Descriptive Text*, 2 vols (Cambridge, Bowes and Bowes, 1921).

Clutton, Elizabeth. 'Review of A. H. Robinson. *Early Thematic Mapping in the History of Cartography* (Chicago, University of Chicago Press, 1982)', *The Map Collector*, 22 (1983), pp. 42–3.

Cobbett, William. *Rural Rides* (London, Dent, 1912).

Coleman, Janet. *English Literature in History: 1350–1400, Medieval Readers and Writers* (London, Hutchinson, 1981).

Coleman, Joyce. *Public Reading and the Reading Public in Late Medieval England and France* (Cambridge, Cambridge University Press, 1996).

Conley, Tom. *The Self-Made Map. Cartographic Writing in Early Modern France* (Minneapolis and London, University of Minnesota Press, 1997).

Connolly, Seán and Jennifer O'Reilly. *Bede: On the Temple* (Liverpool, Liverpool University Press, 1995).

Cook, Alan. *Edmond Halley: Charting the Heavens and the Seas* (Oxford, Clarendon Press, 1998).

Cook, Karen Severud. 'From false starts to firm beginnings: early colour printing of geological maps', *Imago Mundi*, 47 (1995), pp. 155–72.

Copeland, Richard. *The Rutter of the Sea* (London, 1528).

Coppens, Christian. *Reading in Exile: The Libraries of John Ramidge (d. 1568), Thomas Harding (d. 1572) and Henry Jolliffe (d. 1578): Recusants in Louvain* (Cambridge, Libri Pertinentes, 1993).

Cormack, Lesley B. *Charting an Empire. Geography at the English Universities, 1580–1620* (Chicago, University of Chicago Press, 1997).

Cosgrove, Denis (ed.). *Mappings* (London, Reaktion Books, 1999).

Courcelle, Pierre. *Late Latin Writers and their Greek Sources* (Cambridge, Mass., Harvard University Press, 1969).

de la Croix, Horst. 'The literature on fortification in Italy', *Technology and Culture*, 4 (1963), pp. 30–50.

Crone, G. R. *Early Maps of the British Isles, AD 1000 – AD 1579* (London, Royal Geographical Society, Reproductions of Early Maps 7, 1961).

Crone, G. R., E. M. J. Campbell and R. A. Skelton. 'Landmarks in British cartography. I. Early cartographic activity in Britain', *Geographical Journal*, 128 (1962), pp. 406–10.

Crosby, A. W. *The Measure of Reality. Quantification and Western Society 1250–1600* (Cambridge, Cambridge University Press, 1997).

Cullen, M. J. *The Statistical Movement in Early Victorian Britain: the Foundations of Empirical Social Research* (Hassocks, Sussex, The Harvester Press; New York, Barnes and Noble Books, 1975).

Cuningham, William. *The Cosmographical Glasse* (London, 1589).

Cunliffe, Barry. *The City of Bath* (Gloucester, Alan Sutton, 1986).

Cuvier, Georges and Alexandre Brongniart. *Essai sur la géographie minéralogique des environs de Paris* (Paris, 1811).

Dampier, William. *A New Voyage Round the World* (London, 1697, expanded 1699).

Darby, H. C. 'The agrarian contribution to surveying in England', *Geographical Journal*, 82 (1933), pp. 529–35.

Darby, H. C. *The Draining of the Fens* (Cambridge, Cambridge University Press, 1940).

Darby, H. C. 'The age of the improver: 1600–1800', in H. C. Darby (ed.), *The New Historical Geography of England* (Cambridge, Cambridge University Press, 1973), pp. 302–88.

Darby, H. C. *The Changing Fenland* (Cambridge, Cambridge University Press, 1983).

Darlington, Ida and James Howgego. *Printed Maps of London, circa 1553–1850* (London, George Philip, 1964).

David, A. C. F. 'Alexander Dalrymple and the emergence of the Admiralty chart', in D. Howse (ed.), *Five Hundred Years of Nautical Science* (Greenwich, National Maritime Museum, 1981), pp. 153–64.

David, A. *A Catalogue of Charts, Plans and Views, Printed at the Admiralty Office, for the Use of His Majesty's Navy in 1814* (Taunton, Andrew David, 1991).

Davies, R. (ed.), *Estate Maps of Wales, 1600–1836* (Aberystwyth, National Library of Wales, 1982).

Dawson, Keith. *Town Defences in Early Modern England* (unpublished University of Exeter Ph.D. thesis, 1995).

Day, A. *The Admiralty Hydrographic Service, 1795–1919* (London, H.M.S.O., 1967).

Dee, John. 'Mathematical preface' to *The Elements of Geometrie* (London, 1570).

Defoe, Daniel. *Tour of Great Britain* (London, 1724–6).

Delano-Smith, Catherine. 'Cartographic signs on European maps and their explanation before 1700', *Imago Mundi*, 37 (1985), pp. 9–29.

Delano-Smith, Catherine. 'Cartography in the prehistoric period in the Old World: Europe, the Middle East, and North Africa', in J. B. Harley and

David Woodword (eds), *The History of Cartography*, vol. 1, *Cartography in Prehistoric, Ancient, and Medieval Europe and the Mediterranean* (Chicago and London, University of Chicago Press, 1987), pp. 54–101.

Delano-Smith, Catherine. 'Maps as art and science: maps in sixteenth-century Bibles', *Imago Mundi*, 42 (1990), pp. 65–83.

Delano-Smith, Catherine. 'Son of Rudd: Edmund, another Tudor mapmaker?', *The Map Collector*, 64 (1993), p. 38.

Delano-Smith, Catherine. 'Map ownership in sixteenth-century Cambridge; the evidence of probate inventories', *Imago Mundi*, 47 (1995), pp. 67–93.

Delano-Smith, Catherine. 'Milieus of mobility: itineraries, road maps, and route maps', in James R. Akerman (ed.), *Maps for Mobility. The Twelfth Nebenzhal Lectures in the History of Cartography, October 1996, at The Newberry Library* (Chicago and London, University of Chicago Press, forthcoming).

Delano-Smith, Catherine and Mayer Gruber. 'Rashi's legacy: maps of the Holy Land', *The Map Collector*, 59 (1992), pp. 30–5.

Delano-Smith, Catherine and Elizabeth M. Ingram. *Maps in Bibles 1500–1600. An Illustrated Catalogue* (Geneva, Droz, 1991).

Delano-Smith, Catherine and Roger J. P. Kain. *La Cartogràfia Anglesa* (Barcelona, Institut Cartogràfic de Catalunya, 1997).

Dickens, Bruce. 'Premonstratensian itineraries from Titchfield Abbey MS at Welbeck', *Proceedings of Leeds Philosophical and Literary Society* (Literary and Historical Section), 4 (1938), pp. 349–61.

Digges, Leonard. *A boke named Tectonicon . . .* (London, 1556).

Digges, Leonard and Thomas. *A geometrical practise, named Pantometria* (London, 1571).

Dilke, O. A. W. 'Maps in the treatises of Roman land surveyors', *Geographical Journal*, 127 (1967), pp. 417–26.

Dilke, O. A. W. 'Illustrations from Roman surveyors' manuals', *Imago Mundi*, 21 (1967), pp. 9–29.

Dilke, O. A. W. *The Roman Land Surveyors: An Introduction to the Agrimensores* (Newton Abbot, David and Charles, 1971).

Dilke, O. A. W. 'Ground survey and measurement in Roman towns', in Francis Grew and Brian Hobley (eds), *Roman Urban Topography in Britain and the Western Empire* (London, The Council for British Archaeology, Research Report No. 59, 1975), pp. 6–13.

Dilke, O. A. W. 'The culmination of Greek cartography in Ptolemy', in J. B. Harley and David Woodward (eds), *The History of Cartography*, vol. 1, *Cartography in Prehistoric, Ancient, and Medieval Europe and the Mediterranean* (Chicago and London, University of Chicago Press, 1987), pp. 177–200.

Dilke, O. A. W. 'Maps in the service of the state: Roman cartography to the end of the Augustan Era', in J. B. Harley and David Woodward (eds), *The History of Cartography*, vol. 1, *Cartography in Prehistoric, Ancient, and Medieval Europe and the Mediterranean* (Chicago and London, University of Chicago Press, 1987), pp. 201–11.

Dilke, O. A. W. 'Roman large-scale mapping in the early Empire', in J. B. Harley and David Woodward (eds), *The History of Cartography*, vol. 1, *Cartography in Prehistoric, Ancient, and Medieval Europe and the Mediterranean* (Chicago and London, Chicago University Press, 1987), pp. 212–33.

Dilke, O. A. W. 'Itineraries and geographical maps in the early and late Roman Empires', in J. B. Harley and David Woodward (eds), *The History of Cartography*, vol. 1 *Cartography in Prehistoric, Ancient, and Medieval Europe and the Mediterranean* (Chicago and London, University of Chicago Press, 1987), pp. 234–8.

Donkin, R. A. 'Changes in the early Middle Ages', in H. C. Darby (ed.), *A New Historical Geography of England* (Cambridge, Cambridge University Press, 1973), pp. 75–135.

Dodgshon, R. A. and Butlin, R. A. (eds). *An Historical Geography of England and Wales* (London and New York, Academic Press, 1978; 2nd ed. 1990).

Donn, Benjamin. *The Mathematical Essays* (London, 1758).

Donn, Benjamin. *The Accountant and Geometrician* (London, 1765).

Dovey, Zillah. *An Elizabethan Progress. The Queen's Journey into East Anglia, 1578* (Stroud, Alan Sutton, 1996).

Drayton, Michael. *Poly-Olbion, or a Chorographical Description of Great Britain* (London, 1612–22).

Dujardin-Troadec, Louis. *Les cartographes bretons du Conquet. La navigation en images, 1543–1650* (Brest, Imprimerie commerciale et administrative, 1966).

Duke Humfrey's Library and The Divinity School, 1488–1988. Catalogue to an exhibition at The Bodleian Library, June-August 1988 (Oxford, Bodleian Library, 1988).

Dupin, C. *Forces productives et commerciales de la France* (Paris, Bachelier, 1827).

Eccles, M. (ed.). *The Macro Plays* (London, Early English Text Society, Extra Series No. 262, 1969).

Eden, Peter (ed.). *Dictionary of Land Surveyors and Local Cartographers of Great Britain and Ireland 1550–1850*, 3 vols and supplement (Folkestone, Dawson, 1975–9).

Eden, Peter. 'Three Elizabethan estate surveyors: Peter Kempe, Thomas Clerke and Thomas Langdon', in Sarah Tyacke (ed.), *English Map-Making 1500–1650. Historical Essays* (London, The British Library, 1983), pp. 68–84.

Eden, Richard. *The Arte of Navigation* (London, 1596).

Edney, Matthew. *Mapping of an Empire. The Geographical Construction of British India, 1765–1843* (Chicago and London, University of Chicago Press, 1997).

Edson, Evelyn. 'Matthew Paris' "other" map of Palestine', *The Map Collector*, 66 (1994), pp. 18–22.

Edson, Evelyn. 'World maps and Easter tables: computus maps in context', *Imago Mundi*, 48 (1996), pp. 25–42.

Edson, Evelyn. *Mapping Time and Space. How Medieval Mapmakers Viewed their World* (London, The British Library, 1997, 1999).

Edwards, H. S. 'Alexander Dalrymple, FRS (1737–1808): first Hydrographer to the Admiralty', *The Map Collector*, 4 (1978), pp. 19–29.

Van Eerde, Katherine S. *John Ogilby and the Taste of his Times* (Folkestone, Dawson, 1976).

Elliot, James. *The City in Maps: Urban Mapping to 1900* (London, The British Library, 1987).

Elvey, Elizabeth M. *A Handlist of Buckinghamshire Estate Maps* (Buckingham, Buckinghamshire Record Society, Lists and Indexes No. 2, 1963).

Elyot, Thomas. *The Boke named the Governour* (London, 1531).

Emmison, F. G. *Types of Open Field Parishes in the Midlands* (London, Historical Association, Pamphlet No. 108, 1937).

Emmison, F. G. *Catalogue of Maps in the Essex Record Office* (Chelmsford, Essex County Council, 1947).

Emmison, F. G. 'Estate maps and surveys', *History*, 48 (1963), pp. 34–7.

Englisch, Brigitte. 'Erhard Etzlaub's projection and method of mapping', *Imago Mundi*, 48 (1996), pp. 103–23.

English, Barbara. *Yorkshire Enclosure Awards* (Hull, University of Hull Studies in Regional and Local History No. 5, 1985).

Etherton, Judith. 'New evidence: Ralph Treswell's association with St Bartholomew's Hospital', *London Topographical Record*, 27 (1995), pp. 103–17.

Evans, Eric J. *The Contentious Tithe: The Tithe Problem and English Agriculture, 1750–1850* (London, Routledge and Kegan Paul, 1976).

Evans, Eric J. *Tithes: Maps, Apportionments and the 1836 Act* (Chichester, British Association for Local History, 1993).

Evans, Ifor M. and Heather Lawrence. *Christopher Saxton, Elizabethan Map-Maker* (Wakefield, Wakefield Historical Publications; London, The Holland Press, 1979).

Evans, John. 'Extracts from the private account book of Sir William More, of Loseley in Surrey, in the time of Queen Mary and Queen Elizabeth', *Archaeologia*, 35 (1855), pp. 284–310.

Ewing Duncan, David. *The Calendar* (London, Fourth Estate, 1998).

Fenlon, Iain. *Music, Print and Culture in Early Sixteenth-Century Italy. The Panizzi Lectures 1994* (London, The British Library, 1995).

Fischer, D. H. *The Great Wave. Price Revolutions and the Rhythm of History* (New York and Oxford, Oxford University Press, 1996).

Fisher, John. 'Introductory notes', in *The A to Z of Elizabethan London* (London Topographical Society, Publication 122, 1979).

Fitzherbert, John. *Here begynneth a right frutefull mater: and hath to name the boke of surveying and improveme[n]ts* (London, 1523).

Fletcher, David H. *Estate Maps of Christ Church, Oxford, The Emergence of Map Consciousness, c.1600–c.1840* (unpublished University of Exeter Ph.D. thesis, 1990).

Fletcher, David H. *The Emergence of Estate Maps: Christ Church, Oxford 1600–1840* (Oxford, Oxford University Press, 1995).

Fletcher, David H. 'Map or terrier? The example of Christ Church Oxford estate management 1600–1840', *Transactions of the Institute of British Geographers*, 23 (1998), pp. 221–37.

Fletcher, Joseph. 'Moral and educational statistics of England and Wales', *Journal of the Statistical Society of London*, 12 (1849), pp. 151–335.

Flint, Valerie I. *Honorius Augustodunensis of Regensburg* (Aldershot, Variorum, Authors of the Middle Ages 6, 1995).

Flint, Valerie I. 'The Hereford map; its authors, two scenes and a border', *Transactions of the Royal Historical Society*, 6th Series, 8 (1998), pp. 19–44.

Ford, Brian J. *Images of Science. A History of Scientific Illustration* (London, The British Library; New York, Oxford University Press, 1992).

Fordham, H. G. 'Descriptive list of maps of the Great Level of the Fens', *Studies in Carto-Bibliography* (Oxford, Clarendon Press, 1914), pp. 65–8.

Fordham, H. G. *Studies in Carto-Bibliography, British and French, and in the Bibliography of Itineraries and Road-Books* (Oxford, Clarendon Press, 1914).

Fordham, H. G. *John Cary, Engraver, Map, Chart and Print-Seller and Globe-Maker, 1754 to 1853* (Cambridge, Cambridge University Press, 1925; reprinted Folkestone, Dawson, 1976).

Fordham, H. G. 'Earliest tables of highways of England and Wales, 1541–61', *The Library*, 4th ser. 8 (1927–8), pp. 349–54.

Fox, H. S. A. 'Exeter, Devonshire, *circa* 1420', in R. A. Skelton and P. D. A. Harvey (eds), *Local Maps and Plans from Medieval England* (Oxford, Clarendon Press, 1986), pp. 163–9.

Frangenberg, Thomas. 'Chorographies of Florence. The use of city views and city plans in the sixteenth century', *Imago Mundi*, 46 (1994), pp. 41–64.

Freeman, T. W. 'Some Irish maps, 1837–49', *Geographical Journal*, 122 (1956), pp. 129–31.

French, Peter J. *John Dee: The World of an Elizabethan Magus* (London, Routledge and Kegan Paul, 1972).

Frere, S. S. and J. K. St Joseph. *Roman Britain from the Air* (Cambridge, Cambridge University Press, 1983).

Frisius, Gemma. *Libellus de locorum discribendorum ratione* (Antwerp, 1533).

Fuller, G. Joan. 'The development of roads in the Surrey-Sussex Weald and coastlands between 1700 and 1900', *Transactions of the Institute of British Geographers*, 19 (1953), pp. 37–49.

Fuller, R. M., J. Sheail, and C. J. Barr. 'The land of Britain, 1930–1990: a comparative study of field mapping and remote sensing techniques', *Geographical Journal*, 160 (1994), pp. 173–84.

Furnivall, F. J. and A. W. Pollard (eds). *The Macro Plays*, 2nd ed. (London, Early English Text Society, Extra Series No. 91, 1924).

Garland, Ken. *Mr Beck's Underground Map* (Harrow Weald, Capitol Transport Publishing, 1994).

Gatti Perer, M. L. (ed). *La Gerusalem celeste. Catalogo della mostra, Milano, Università Cattolica del S. Cuore, 20 maggio – 5 guigno 1983* (Milan, Università Cattolica, 1983).

Gautier Dalché, Patrick. *Géographie et culture. La représentation de l'espace du VIe au XIIe siècle* (Aldershot, Ashgate Publishing, Variorum Collected Studies Series, 1997).

Geoffrey of Monmouth, *The History of the Kings of Britain*, translated by Lewis Thorpe (London, Folio Society, 1969).

Gilbert, E. W. 'Pioneer maps of health and disease in England', *Geographical Journal*, 124 (1958), pp. 172–83.

Gillett, C. R. *Burned Books: Neglected Chapters in British History and Literature*, 2 vols (Port Washington, N.Y., Kennikat Press, 1932).

Gillies, John. *Shakespeare and the Geography of Difference* (Cambridge, Cambridge University Press, 1994).

Gillow, Joseph. *Lord Burghley's Map of Lancashire in 1590* (London, Arden Press for the Catholic Record Society, 1907).

Glanville, Philippa. *London in Maps* (London, The Connoisseur, 1972).

Glunz, H. H. *History of the Vulgate in England, from Alcuin to Roger Bacon* (Cambridge, Cambridge University Press, 1933).

Gough, Richard. *British Topography*, 2 vols (London, 1780).

Grafton, Richard. *An Abridgement of the Chronicle of England* (London, 1571).

Gransden, Antonia. 'Realistic observation in twelfth-century England', *Speculum*, 47 (1972), pp. 29–51, reprinted in Antonia Gransden, *Legends, Traditions and History in Medieval England* (London and Rio Grande, The Hambledon Press, 1992), pp. 175–98.

Gransden, Antonia. 'Antiquarian studies in fifteenth-century England', *Antiquaries Journal*, 60 (1980), pp. 75–97.

Gransden, Antonia. 'Prologues in the historiography of twelfth-century England', in Daniel Williams (ed.), *England in the Twelfth Century. Proceedings of the 1988 Harlaxton Symposium* (Woodbridge, Suffolk, Boydell Press, 1990), pp. 55–81.

Gransden, Antonia. *Legends, Traditions and History in Medieval England* (London and Rio Grande, The Hambledon Press, 1992).

Guerry, A.-M. *Essai sur la statistique morale de la France précédé d'un rapport à l'Académie des Sciences par MM. Lacroix, Silvestre et Girard* (Paris, 1833).

Haggett, Peter. *The Geographer's Art* (Oxford, Blackwell, 1990).

Hallam, H. E. 'Wildmore Fen, Lincolnshire, 1224 × 1249', in R. A. Skelton and P. D. A. Harvey (eds), *Local Maps and Plans from Medieval England* (Oxford, Clarendon Press, 1986), pp. 71–81.

Halliwell, James Orchard. *Rara Mathematica. A Collection of Treatises on the Mathematics* (London?, 1834).

Halliwell, James Orchard. *Ancient Inventories of Furniture, Pictures, Tapestry, Plate etc Illustrative of the Domestic Manners of the English in the Sixteenth and Seventeenth Centuries* (London, printed for private circulation, 1845).

Harley, J. B. *Christopher Greenwood, County Mapmaker and his Worcestershire Map of 1822* (Worcester, Worcestershire Historical Society, 1962).

Harley, J. B. 'The Society of Arts and the surveys of English counties 1759–1809', *Journal of the Royal Society of Arts*, 112 (1963), pp. 43–6, 119–24, 269–75, 538–43.

Harley, J. B. 'The re-mapping of England 1750–1800', *Imago Mundi* 19 (1965), pp. 56–67.

Harley, J. B. 'A proposed survey of Lancashire by Francis and Netlam Giles', *Transactions of the Lancashire and Cheshire Historic Society*, 116 (1965), pp. 197–206.

Harley, J. B. 'From Saxton to Speed', *Cheshire Round*, 1 (1966), pp. 174–88.

Harley, J. B. 'John Strachey of Somerset: an antiquarian cartographer of the early eighteenth century', *Cartographic Journal*, 3 (1966), pp. 2–5.

Harley, J. B. 'Maps of early Georgian Cheshire', *Cheshire Round* 1 (1966), pp. 256–69.

Harley, J. B. 'English county map-making in the early years of the Ordnance Survey: the map of Surrey by Joseph Lindley and William Crossley', *Geographical Journal*, 132 (1966), pp. 372–8.

Harley, J. B. 'The bankruptcy of Thomas Jefferys: an episode in the economic history of eighteenth century map making', *Imago Mundi*, 20 (1967), pp. 27–48.

Harley, J. B. 'Uncultivated fields in the history of cartography', *Cartographic Journal*, 4 (1967), pp. 7–11.

Harley, J. B. *William Yates's Map of the County of Lancashire, 1786* (Birkenhead, Historic Society of Lancashire and Cheshire, 1968).

Harley, J. B. *The County maps from William Camden's Britannia 1695 by Robert Morden. A Facsimile* (Exeter, David and Charles, 1970).

Harley, J. B. (ed.). *J. Ogilby, Britannia (1675)* (Amsterdam, Theatrum Orbis Terrarum, 1970).

Harley, J. B. 'The Ordnance Survey and the origins of official geological mapping in Devon and Cornwall', in K. J. Gregory and W. L. D. Ravenhill (eds), *Exeter Essays in Geography in Honour of Arthur Davies* (Exeter, University of Exeter, 1971), pp. 105–23.

Harley, J. B. *Maps for the Local Historian: A Guide to the British Sources* (London, National Council for Social Service for the Standing Conference for Local History, 1972).

Harley, J. B. 'The map collection of William Cecil, First Baron Burghley', *The Map Collector*, 3 (1978), pp. 12–19.

Harley, J. B. 'Christopher Saxton and the first atlas of England and Wales 1579–1979', *The Map Collector*, 8 (1979), pp. 3–11.

Harley, J. B. *The Ordnance Survey and Land-Use Mapping* (Norwich, Geo Abstracts for the Historical Geography Research Series No. 2, 1979).

Harley, J. B. 'The Ordnance Survey 1:528 Board of Health town plans in Warwickshire 1848–1854', in T. R. Slater and P. J. Jarvis (eds), *Field and Forest: An Historical Geography of Warwickshire and Worcestershire* (Norwich, Geo Books, 1982), pp. 347–84.

Harley, J. B. 'Meaning and ambiguity in Tudor cartography' in Sarah Tyacke (ed.), *English Map-Making 1500–1650* (London, The British Library, 1983), pp. 22–45.

Harley, J. B. 'Maps, knowledge and power', in D. Cosgrove and S. Daniels (eds), *The Iconography of Landscape: Essays on the Symbolic Representation, Design and Use of Past Environments* (Cambridge, Cambridge University Press, 1988), pp. 277–312.

Harley, J. B., D. V. Fowkes and J. C. Harvey, *P. P. Burdett's Map of Derbyshire, with an Explanatory Introduction* (Matlock, Derbyshire Archaeological Society, 1975).

Harley, J. B. and J. C. Harvey. 'Introduction' to the facsimile edition of *A Survey of the County of Yorkshire by Thomas Jefferys, 1775* (Lympne, Harry Margary, 1973).

Harley, J. B. and Paul Laxton, *A Survey of the County Palatine of Chester by P. P. Burdett 1777* (Liverpool, Lund Press for the Historic Society of Lancashire and Cheshire, Occasional Series 1, 1974).

Harley, J. B. and E. A. Stuart, 'George Withiell: a west country surveyor of the late-seventeenth century', *Devon and Cornwall Notes and Queries*, 35 (1982), pp. 45–58.

Harley, J. B. and David Woodward (eds). *The History of Cartography*. Vol. 1, *Cartography in Prehistoric, Ancient, and Medieval Europe and the Mediterranean* (Chicago and London, Chicago

University Press, 1987); vol. 2 (in 3 books) *Cartography in the Traditional Islamic and South Asian Societies* (Chicago and London, Chicago University Press, 1992–98).

Harrison, Philip and Mark Brayshay. 'Post-horse routes, royal progresses and government communications in the reign of James I', *Journal of Transport History*, n.s. 18 (1997), pp. 116–33.

Hartzell, K. D. 'Diagrams for liturgical ceremonies, late 14th Century', in R. A. Skelton and P. D. A. Harvey (eds), *Local Maps and Plans from Medieval England* (Oxford, Clarendon Press, 1986), pp. 339–43.

Harvey, John H. (ed.). *William Worcestre, Itineraries* (Oxford, Clarendon Press, 1969).

Harvey, John H. 'Symbolic plan of a city, early 15th Century', in R. A. Skelton and P. D. A. Harvey (eds), *Local Maps and Plans from Medieval England* (Oxford, Clarendon Press, 1986), pp. 342–3.

Harvey, P. D. A. 'An Elizabethan map of manors in north Dorset', *British Museum Quarterly*, 29 (1965), pp. 82–4.

Harvey, P. D. A. *The History of Topographical Maps: Symbols, Pictures and Surveys* (London, Thames and Hudson, 1980).

Harvey, P. D. A. 'The Portsmouth map of 1545 and the introduction of scale maps into England', in John Webb, Nigel Yates, and Sarah Peacock (eds), *Hampshire Studies* (Portsmouth, Portsmouth City Records Office, 1981), pp. 33–49.

Harvey, P. D. A. 'A 13th-century plan from Waltham Abbey, Essex', *Imago Mundi*, 22 (1986), pp. 10–12.

Harvey, P. D. A. 'Boarstall, Buckinghamshire. 1444 × 1446', in Skelton and Harvey, *Local Maps and Plans from Medieval England* (Oxford, Clarendon Press, 1986), pp. 211–19 and Plate 18.

Harvey, P. D. A. 'Wormley, Hertfordshire, 1220 × 1230', in R. A. Skelton and P. D. A. Harvey (eds), *Local Maps and Plans from Medieval England* (Oxford, Clarendon Press, 1986), pp. 59–70.

Harvey, P. D. A. 'Shouldham, Norfolk, 1440 × 1441', in R. A. Skelton and P. D. A. Harvey (eds), *Local Maps and Plans from Medieval England* (Oxford, Clarendon Press, 1986), pp. 195–201.

Harvey, P. D. A. 'Local and Regional Cartography in Medieval Europe', in J. B. Harley and David Woodward (eds), *History of Cartography*, vol. 1, *Cartography in Prehistoric, Ancient, and Medieval Europe and the Mediterranean* (Chicago and London, University of Chicago Press, 1987), pp. 464–501.

Harvey, P. D. A. *Medieval Maps* (London, The British Library; Toronto, University of Toronto Press, 1991).

Harvey, P. D. A. 'Matthew Paris's maps of Britain', in P. R. Coss and D. S. Lloyd (eds), *Thirteenth Century England IV: Proceedings of the Newcastle upon Tyne Conference, 1992* (Woodbridge, Suffolk, The Boydell Press, 1992), pp. 118–21.

Harvey, P. D. A. *Maps in Tudor England* (London, The Public Record Office and The British Library; Chicago, University of Chicago Press, 1993).

Harvey, P. D. A. *Mappa Mundi. The Hereford World Map* (London, Hereford Cathedral and The British Library; Toronto, University of Toronto Press, 1996).

Harvey, P. D. A. 'English estate maps: their early history and their use as historical evidence', in David Buisseret (ed.), *Rural Images: Estate Maps in the Old and New Worlds* (Chicago and London, Chicago University Press, 1996), pp. 27–61.

Harvey, P. D. A. 'The Sawley (Henry of Mainz) map', *Imago Mundi*, 49 (1997), pp. 33–42.

Harvey, P. D. A. and Harry Thorpe. *The Printed Maps of Warwickshire 1576–1900* (Warwick, Warwick County Council in association with University of Birmingham, 1959).

Haslam, Graham. 'The Duchy of Cornwall map fragment', in Monique Pelletier (ed.), *Géographie du monde au Moyen Age et à la Renaissance* (Paris, Editions du Comité des Travaux Historiques et Scientifiques, 1989), pp. 33–44.

Heawood, Edward. *English County Maps in the Collection of the Royal Geographical Society* (London, Royal Geographical Society, Reproductions of early engraved maps, 1932, 2).

Helgerson, Richard. 'The land speaks: cartography, chorography, and subversion in Renaissance England', *Representations*, 16 (1986), pp. 50–85.

Hessel, Alfred. *A History of Libraries* (translated by Reuben Peiss, New Brunswick, Scarecrow Press, 1955).

Hieronymus, De Sancto Marco. *Opusculum de universali mundi machina* (London, 1505).

Hindle, Brian Paul. 'The towns and roads of the Gough map (*c.*1360)', *Manchester Geographer*, 1 (1980), pp. 35–49.

Hindle, Brian Paul. *Medieval Roads* (Princes Risborough, Shire Publications, 1982).

Hindle, Brian Paul. *Maps for Local Historians* (London, Batsford, 1988).

Hindle, Brian Paul. *Maps for Historians* (Chichester, Phillimore, 1998).

Hitchcock, Robert. *A Politique Platt for the Honour of the Prince . . .* (London, 1580).

Hobsbawm, Eric J. and George F. E. Rudé. *Captain Swing* (London, Lawrence and Wishart, 1969).

Hodnett, Edward. *English Woodcuts 1480–1535* (Oxford, Oxford University Press, 1973).

Hodson, Donald. *Printed Maps of Hertfordshire 1577–1900* (Folkestone, Dawson, 1974).

Hodson, Donald. *Maps of Portsmouth before 1801* (Portsmouth, Portsmouth Record Series, 1978).

Hodson, Donald. *County Atlases of the British Isles*, vol. 1, *Atlases Published 1704 to 1742*, vol. 2, *Atlases Published 1743–1763* (Tewin, Tewin Press, 1984–9) and vol. 3, *Atlases Published 1764–1789* (London, The British Library, 1997).

Hodson, Yolande. *Ordnance Surveyors' Drawings 1789–c.1840* (Reading, Research Publications, 1989).

Holder, Arthur G. *Bede: On the Tabernacle* (Liverpool, Liverpool University Press, 1994).

Holinshed, Raphael. *The First Volume of the Chronicles of England, Scotland and Ireland* (London, 1577).

Holmes, M. A. *Moorfields in 1559. An Engraved Copper Plate from the Earliest Known Map of London* (London, H.M.S.O. for the City of London Museum, 1963).

Holwell, John. *A Sure Guide to the Practical Surveyor, in Two Parts* (London, 1678).

Homer, Henry S. *An Essay on the Nature and Method of Ascertaining the Specifick Shares of Proprietors upon the Enclosure of Common Fields* (Oxford, S. Parker, 1766).

Hood, Thomas. *The Marriners guide set forth in the forme of a dialogue* (London, dated 1581 but printed in 1596).

Hooke, Della. *Worcestershire Anglo-Saxon Charter-Bounds* (Woodbridge, Suffolk, Boydell Press, 1990).

Hooke, Della. *Pre-Conquest Charter-Bounds of Devon and Cornwall* (Woodbridge, Suffolk, Boydell Press, 1994).

Hooke, J. M. and R. J. P. Kain. *Historical Change in the Physical Environment* (London, Butterworths, 1982).

Howard, Ebenezer. *Garden Cities of To-Morrow*, edited by F. J. Osborn (London, Faber and Faber, 1965).

Hughes, Paul. 'Thomas Surbey's 1699 survey of the Rivers Ouse and Humber', *Yorkshire Archaeological Journal*, 66 (1994), pp. 149–90.

Hull, F. 'Isle of Thanet, Kent, late 14th Century × 1414', in R. A. Skelton and P. D. A. Harvey (eds), *Local Maps and Plans from Medieval England* (Oxford, Clarendon Press, 1986), pp. 119–26.

Humphreys, John. *Elizabethan Sheldon Tapestries* (London, Oxford University Press, 1929).

Hurst, Herbert. *Oxford Topography: An Essay* (Oxford, Oxford Historical Society, Publication 39, 1899).

Hyde, Ralph. *Printed Maps of Victorian London 1851–1900* (Folkestone, Dawson, 1975).

Hyde, Ralph. 'Introductory notes', in *The A to Z of Georgian London* (Lympne, Harry Margary for the London Topographical Society, Publication 126, 1982).

Hyde, Ralph. 'A "handy" map', *The Map Collector*, 35 (June 1986), p. 47.

Hyde, Ralph. 'Introductory Notes', in *The A to Z of Restoration London* (The City of London, 1676) (London, London Topographical Society, Publication 145, 1992).

Ingram, Elizabeth M. 'Maps as readers' aids: maps and plans in Geneva bibles', *Imago Mundi*, 45 (1993), pp. 29–44.

Ingram, Elizabeth M. 'The map of the Holy Land in the Coverdale Bible: a map by Holbein?', *The Map Collector*, 64 (1993), pp. 26–31.

Ireland, H. A. 'History of the development of geologic maps', *Bulletin of the Geological Society of America*, 54 (1943), pp. 1227–80.

Irwin, Raymond. *The English Library. Its Sources and History* (London, George Allen and Unwin, 1966).

Ison, Walter. *The Georgian Buildings of Bath from 1700 to 1830* (London, Faber and Faber, 1948).

Jardine, Lisa. *Worldly Goods. A New History of the Renaissance* (London, Macmillan, 1996).

Johnston, Stephen. 'Mathematical practitioners and instruments in Elizabethan England', *Annals of Science*, 48 (1991), pp. 319–44.

Jolly, David C. *Maps in British Periodicals* (Brookline, Mass., David C. Jolly, 1992), vol. 1.

Jones, Michael. 'Les Anglais à Crozon à la fin du XVIe siècle: le témoignage des cartes', *Mémoires de la Société d'Histoire et d'Archéologie de Bretagne*, 75 (1997), pp. 11–35.

Jones, Philip E. 'Deptford, Kent and Surrey; Lambeth, Surrey; London', in R. A. Skelton and P. D. A. Harvey (eds), *Local Maps and Plans from Medieval England* (Oxford, Clarendon Press, 1986), pp. 251–62.

Jusserand, J. J. *Les Anglais au Moyen Age: la vie nomade et les routes d'Angleterre au IXe siècle* (Paris, 1884); 4th edition translated by L. T. Smith as *English Wayfaring Life in the Middle Ages* (London, Ernest Benn, 1950).

Kain, Roger J. P. 'R. K. Dawson's proposals in 1836 for a cadastral survey of England and Wales', *Cartographic Journal*, 12 (1975), pp. 81–8.

Kain, Roger J. P. *An Atlas and Index of the Tithe Files of Mid-Nineteenth-Century England and Wales* (Cambridge, Cambridge University Press, 1986).

Kain, Roger J. P. and E. Baigent. *The Cadastral Map in the Service of the State: A History of Property Mapping* (Chicago and London, University of Chicago Press, 1992).

Kain, Roger J. P. and Richard R. Oliver. *The Tithe Maps of England and Wales: A Cartographic Analysis and County-by-County Catalogue* (Cambridge, Cambridge University Press, 1995).

Kain, Roger J. P. and Richard R. Oliver. *The Enclosure Maps of England and Wales* (Cambridge, Cambridge University Press, forthcoming).

Kain, Roger J. P. and Hugh C. Prince, *The Tithe Surveys of England and Wales* (Cambridge, Cambridge University Press, 1985).

Kamal, Youssouf. *Monumenta cartographica Africae et Aegypti*, 5 vols in 16 parts (Cairo, 1926–51).

Karrow, Robert W. *Mapmakers of the Sixteenth Century and Their Maps* (Winnetka, Illinois, Speculum Press for The Newberry Library, 1993).

Kennedy, Edward S. and Mary Helen Kennedy. *Geographical Coordinates of Localities from Islamic Sources* (Frankfurt, Institut für Geschichte der Arabisch-Islamischen Wissenschaften, 1987).

Kenney, C. E. 'William Leybourne, 1626–1716', *The Library*, 5th series, 5 (1951), pp. 159–71.

Kenyon, J. R. *Medieval Fortifications* (Leicester, Leicester University Press, 1990).

Kerridge, Eric (ed.), *Surveys of the Manor of Philip, First Earl of Pembroke and Montgomery 1631–2* (Devizes, Wiltshire Archaeological and Natural History Society, Records Branch, No. 9, 1953).

Kingsley, D. *Printed Maps of Sussex 1575–1900* (Lewes, Sussex Record Society, 1982).

Kintgen, Eugene R. *Reading in Tudor England* (Pittsburgh, University of Pittsburgh Press, 1996).

Kitchen, Frank. *Cosmo-choro-poly-grapher: An Analytical Account of the Life and Work of John Norden* (unpublished D.Phil. thesis, University of Sussex, 1993).

Kitchen, Frank. 'John Norden (*c.*1547–1625): estate surveyor, topographer, county mapmaker and devotional writer', *Imago Mundi*, 49 (1997), pp. 43–61.

Klein, Christopher M. *Maps in Eighteenth-Century British Magazines. A Checklist* (Chicago, The Newberry Library, Herman Dunlap Smith Center, Publication No. 3, 1987).

Knowles, M. D. 'Clerkenwell and Islington, Middlesex mid–15th century', in R. A. Skelton and P. D. A. Harvey (eds), *Local Maps and Plans from Medieval England* (Oxford, Clarendon Press, 1986), pp. 221–8

Koeman, Cornelis. *The History of Lucas Jansoon Waghenaer and his 'Spieghel der Zeevaerdt'* (Lausanne, Elseviers, 1964).

Kupfer, Marcia. 'The lost *mappamundi* at Chalivoy-Milon', *Speculum*, 66 (1991), pp. 540–71.

Lamb, H. H. *Climate, History and the Modern World* (London and New York, Methuen, 1982).

Lambarde, William. *A Perambulation of Kent* (2nd ed., London, 1596).

Latham, Robert and William Matthews (eds). *The Diary of Samuel Pepys*, 11 vols (London, Bell, 1970–83).

Laughton, J. K. (ed.). *State Papers Relating to the Defeat of the Spanish Armada, Anno 1588*, 2 vols (London, Navy Records Society, 1894).

Lawton, R. 'Population and society, 1730–1914', in R. A. Dodgshon and R. A. Butlin (eds), *An Historical Geography of England and Wales*, 2nd ed. (London and New York, Academic Press, 1990), pp. 285–321.

Laxton, Paul. 'John Rocque's survey of Berkshire', *Journal, Durham University Geographical Society*, 8 (1966), pp. 41–9.

Laxton, Paul. *John Rocque's Map of Berkshire, 1761* (Lympne, Harry Margary, 1973), Introduction.

Laxton, Paul. *250 Years of Map-Making in the County of Hampshire* (Lympne, Harry Margary, 1976), Introduction.

Laxton, Paul. 'The geodetic and topographical evaluation of English county maps, 1740–1840', *Cartographic Journal*, 13:1 (1976), pp. 37–54.

Laxton, Paul. 'Introduction', *The A to Z of Regency London* (Lympne, Harry Margary for the London Topographical Society, Publication 131, 1985).

Laxton, Paul. 'Wind and water power' in J. Langton and R. J. Morris (eds), *Atlas of Industrializing Britain* (London, Methuen, 1986), pp. 69–71.

Laxton, Paul. 'Greenwood, Christopher', in *New Dictionary of National Biography* (Oxford, Oxford University Press, forthcoming).

Laxton, Paul. 'Burdett, Peter Perez', in C. S. Nicholls (ed.) *The Dictionary of National Biography: Missing Persons* (Oxford, Oxford University Press, 1993), p. 102.

Laxton, Paul. 'Rocque, John', in C. S. Nicholls (ed.) *The Dictionary of National Biography: Missing Persons* (Oxford, Oxford University Press, 1993), pp. 563–4.

Laxton, Paul and R. I. Hodgson. 'Commonfield and enclosure – two case studies', *Journal, Durham University Geographical Society*, 9 (1967), pp. 41–74.

Lecoq, Danielle. 'La "mappemonde" du *De arca Noe mystica* de Hughes de Saint-Victor (1128–1129)', in Monique Pelletier (ed.), *Géographie du monde au Moyen Age et à la Renaissance* (Paris, Editions du Comité des Travaux Historiques et Scientifiques, 1989), pp. 9–31.

Leedham-Green, Elizabeth. *Books in Cambridge Inventories: Booklists from Vice-Chancellors' Court Probate Inventories in the Tudor and Stuart Periods*, 2 vols (Cambridge, Cambridge University Press, 1986).

Leedham-Green, Elisabeth. 'Perne's wills', in Patrick Collinson, David McKitterick and Elisabeth Leedham-Green, *Andrew Perne: Quartercentenary Studies* (Cambridge, Cambridge University Press for the Cambridge Bibliographical Society, 1991), pp. 79–119.

Leland, John. *Itinerary* (1530–40); first printed as *The Itinerary of John Leland the Antiquary* by T. Hearne, (Oxford, 1710).

Lester, G. A. *Sir John Patson's 'Grete Book'. A Descriptive Catalogue, with an Introduction, of British Library MS Lansdowne 285* (Cambridge, D. S. Brewer, 1984)

Lewin, H. G. *Early British Railways 1801–44* (London, Locomotive Publishing Company, 1925).

Lewis, Susanne. *The Art of Matthew Paris in the 'Chronica majora'* (Berkeley, University of California Press; Aldershot, Scolar Press with Corpus Christi College, Cambridge, 1987).

Leybourne, William. *The Compleat Surveyor . . .* (London, 1653).

Lilley, Keith D. 'Geometry, urban planning and town design in the High Middle Ages', *Planning History*, 20 (1998), pp. 7–15.

Little, Bryan. *Bath Portrait: The Story of Bath, Its Life and Its Buildings*, 3rd ed. (Bristol, The Burleigh Press, 1972).

Lobel, M. D. 'The value of early maps as evidence for the topography of English towns', *Imago Mundi*, 22 (1968), pp. 50–61.

Luborsky, Ruth S. and Elizabeth M. Ingram. *A Guide to English Illustrated Books, 1536–1603*, 2 vols (Tempe, Arizona Center for Medieval and Renaissance Studies, 1998).

Ludwig, Friedrich. *Untersuchungen über die Reise- und Marschgeschwindigkeit im XII und XIII Jahrhundert* (Berlin, Ernst Siegried Mittler & Son, 1897).

Lynam, Edward. *British Maps and Map-makers*, 3rd impression, rev. (London, Collins, 1947).

Lynam, Edward. 'English maps and map-makers of the sixteenth century', *Geographical Journal*, 116 (1950), pp. 7–28.

Lynam, Edward. *The Mapmaker's Art. Essays in the History of Maps* (London, Batchworth Press, 1953).

MacEachren, Alan M. and Gregory B. Johnson. 'The evolution, application and implications of strip format travel maps', *Cartographic Journal*, 24 (1987), pp. 147–58.

Machiavelli, Niccolo. *Arte della guerra* (Florence, 1521).

Mangani, Giorgio. 'Abraham Ortelius and the hermetic meaning of the cordiform map', *Imago Mundi*, 50 (1998), pp. 59–83.

Manley, Gordon. *Climate and the British Scene* (London, Collins, 1952).

Marcombe, David. 'Saxton's apprenticeship: John Rudd, a Yorkshire cartographer', *The Yorkshire Archaeological Journal*, 50 (1978), pp. 171–5.

Marcombe, David. 'John Rudd: a forgotten Tudor mapmaker?' *The Map Collector*, 64 (1993), pp. 34–7.

Marks, Stephen Powys. *The Map of Mid Sixteenth Century London: an Investigation into the Relationship between a Copper Engraved Map and its Derivatives* (London, London Topographical Society, Publication 100, 1964).

Marks, Stephen Powys. 'The copperplate map – a lecture and an exhibition', *London Topographical Society Newsletter*, 47 (November 1998), pp. 5–6.

Marshall, William. *Rural Economy of the West of England*, 2 vols (London, 1976).

Marshall, William. *On the Appropriation and Inclosure of Commonable and Intermixed Lands* (London, W. Bulmer, 1801).

Marshall, William. *On the Landed Property of England* (London, G. and W. Nichol, 1804).

Masters, Betty R. *The Public Markets of the City of London Surveyed by William Leybourne in 1677* (London, London Topographical Society Publication 117, 1974).

Mastoris, Stephanos and Sue Groves (eds). *Sherwood Forest in 1609: a Crown Survey by Richard Bankes* (Nottingham, Thoroton Society, Thoroton Society Record Series, Vol. 40, 1997).

Mastoris, Steph[anos]. 'A newly-discovered perambulation map of Sherwood Forest in the early seventeenth century' *Transactions of the Thoroton Society of Nottinghamshire*, 102 (1998), pp. 79–92

Maton, W. *Observations Relative Chiefly to the Natural History, Picturesque Scenery and Antiquities, of the Western Counties of England, Made in the Years 1794 and 1796*, 2 vols (Salisbury, J. Eaton, 1797).

Mayhew, Henry. *London Labour and the London Poor*, 4 vols (London, 1861–2; enlarged photographic reprint, London, Frank Cass, 1967).

McKendrick, Scot. 'Tapestries from the Low Countries during the fifteenth century', in Caroline Barron and Nigel Saul (eds), *England and the Low Countries in the Late Middle Ages* (Stroud, Alan Sutton; New York, St Martin's Press, 1995), pp. 43–60.

McKenzie, D. F. *Bibliography and the Sociology of Texts. The Panizzi Lectures 1985* (London, The British Library, 1986).

McRae, Andrew. 'To know one's own: estate surveying and the representation of the land in early modern England', *Huntington Library Quarterly*, 56 (1993), pp. 332–57.

McRae, Andrew. *God Speed the Plough: The Representation of Agrarian England, 1500–1660* (Cambridge, Cambridge University Press, 1996).

McLean, Antonia. *Humanism and the Rise of Science in Tudor England* (London, Heinemann, 1972).

Meehan, Denis. *Adamnan's De Locis Sanctis* (Dublin, Dublin Institute for Advanced Studies, Scriptores Latini Hiberniae 3, 1958).

Mercier, Raymond P. 'Geodesy', in J. B. Harley and David Woodward (eds), *The History of Cartography*, vol. 2, book 1, *Cartography in the Traditional Islamic and South Asian Societies* (Chicago and London, University of Chicago Press, 1992), pp. 175–88.

Merriman, Marcus. 'Italian military engineers in Britain in the 1540s', in Sarah Tyacke (ed.), *English*

Map-Making 1500–1650 (London, The British Library, 1983), pp. 57–68.

Miller, Konrad. *Mappaemundi. Die ältesten Weltkarten*, 6 vols (Stuttgart. J. Roth, 1895–8).

Miller, Konrad (ed.). *Die Peutingersche Tafel* (1887–8; reproduced Stuttgart, F. A. Brockhaus, 1962).

Mills, David. 'Diagrams for staging plays, early or mid–15th Century', in R. A. Skelton and P. D. A. Harvey (eds), *Local Maps and Plans from Medieval England* (Oxford, Clarendon Press, 1986), pp. 344–5.

Minet, William. 'Some unpublished plans of Dover Harbour', *Archaeologia* 72 (1921–22), pp. 185–224.

Mitchell, Joan B. 'Early maps of Great Britain: I. The Matthew Paris maps', *The Geographical Journal*, 80 (1933), pp. 27–34.

Mitchell, Rose and David Crook. 'A fifteenth-century map of the Lincolnshire Fenlands', *Imago Mundi*, 51 (1999), pp. 40–50 and Plates 3 and 4.

Mokre, Jan. 'The environs map; Vienna and its surroundings *c.*1600–*c.*1800', *Imago Mundi*, 49 (1997), pp. 90–103.

Mone, Franz Joseph. *Anzeiger fur kunde der teutschen Weltkarten*, vol. 5 (Karlsruhe, 1836).

Morgan, N. J. *Early Gothic Manuscripts 1250–1285. A Survey of Manuscripts Illustrated in the British Isles*, IV, vol. 2 (London, Harvey Miller, 1988).

Morgan, Victor. 'The cartographic image of "The Country" in early modern England', *Transactions of the Royal Historical Society*, 5th series, 29 (1979), pp. 129–54.

Morison, Fynes. *An Itinerary written by Fynes Moryson first in the Latin Toungue containing his ten yeeres travell through the twelve dominions of Germany, Bohemialand, Sweitzerland, Netherland, Denmarke, Poland, Italy, Turkey, France, England, Scotland <u>and</u> Ireland* (London, 1617).

Morris, A. E. J. *History of Urban Form before the Industrial Revolutions*, 3rd ed. (London, Longman, 1994).

Motzo, B. R. (ed.) *Il Compasso di Navigare* (Cagliari, Università di Cagliari, 1947).

Mowl, Tim and Brian Earnshaw. *John Wood, Architect of Obsession* (Bath, Millstream Books, 1988).

Moyes, A. 'Transport 1730–1900' in R. A. Dodgshon and R. A. Butlin (eds), *An Historical Geography of England and Wales*, 1st ed. (London and New York, Academic Press, 1978), pp. 401–29.

Munby, J. T. *People and Fields in Medieval Portchester* (Winchester, Hampshire County Council, Portsmouth Record Series, forthcoming).

Münster, Sebastian. *Erlerung des newen Instruments der Sunnen* (Oppenheim, 1528).

Murdoch, John E. *Album of Science. Antiquity and the Middle Ages* (New York, Charles Scribner's Sons, 1984).

Needell, Keith. 'A comparison between chalk and cheese', *IMCOS* [International Map Collectors' Society] *Journal*, 63 (1995), p. 44.

Nichols, Harold. *Local Maps of Derbyshire. An Inventory and Introduction* (Matlock, Derbyshire Library Service, 1980).

Nichols, Harold. *Local Maps of Nottinghamshire to 1800. An Inventory* (Nottingham, Nottingham County Council Leisure Services, 1987).

Nicholson, Timothy R. *Wheels on the Road. Maps of Britain for the Cyclist and Motorist 1870–1940* (Norwich, Geo-Books, 1983).

Norden, John. *Speculum Britanniae. The first parte. An historicall and chorographical discription of Middlesex* (London, 1593).

Norden, John. *Speculi Britanniae Pars Topographicall & Historical Description of Cornwall* (*c.*1597; printed London, 1728).

Norden, John. *Speculi Britanniae pars The Description of Hartfordshire* (London, 1598).

Norden, John. *The Surveyors Dialogue.* (London, 1607).

Norden, John. *England. An Intended Guyde* (London, 1625).

Norman, Robert, *The Safeguard of Sailors, or the Great Rutter. London 1584* (Amsterdam, Theatrum Orbis Terrarum; Norwood N. J., Walter J. Johnson, 1976).

Norman, Robert. *The New Attractive* (London, dated 1581 but printed in 1596).

Norris, Edwin. *The Ancient Cornish Drama* (Oxford, Oxford University Press, 1859).

North, F. J. 'From the geological map to the Geological Survey', *Transactions of the Cardiff Naturalists' Society*, 65 (1932), pp. 41–115.

North, F. J. 'Geology's debt to Henry de la Beche', *Endeavour*, 3 (1944), pp. 15–19.

Norton, Jane E. *Guide to the National and Provincial Directories of England and Wales, excluding London, Published before 1856* (London, Royal Historical Society Guides and Handbooks No. 5, 1950, reprinted with some corrections, 1984).

Oates, J. C. T. 'The libraries of Cambridge 1570–1700', in F. Wormald and C. E. Wright (eds) *The English Library before 1700. Studies in its History* (London, Athlone Press, 1958), pp. 213–35.

Oates, J. C. T. *Cambridge University Library: A History* (Cambridge, Cambridge University Press, 1986).

Ogilby, John. *Britannia . . . An Illustration of the Kingdom of England and Dominion of Wales.* (London, 1675).

Oliver, Richard R. *An Introduction to the Ordnance Survey One-Inch Seventh Series Map with a List of Editions* (Southampton, Ordnance Survey, 1991).

Oliver, Richard. R. *Ordnance Survey Maps: a Concise Guide for Historians* (London, Charles Close Society, 1993).

Oliver, Richard R. and Roger J. P. Kain, 'Maps and the assessment of parish rates in nineteenth-century England and Wales, *Imago Mundi*, 50 (1998), pp. 156–73.

O'Loughlin, Thomas. 'The view from Iona; Adomnán's mental maps', *Peritia. Journal of the Medieval Academy of Ireland*, 10 (1996), pp. 98–123.

O'Loughlin, Thomas. 'Adomnán and Arculf: the case of an expert witness', *Journal of Medieval Latin*, 7 (1997), pp. 127–46.

O'Loughlin, Thomas. 'The map of Europe (N.L.I. 700): a window into the world of Giraldus Cambrensis', *Imago Mundi*, 51 (1999), pp. 24–39 and Plate 2.

Orteberg, Veronica. 'Archbishop Sigeric's journey to Rome in 990', *Anglo-Saxon England*, 19 (1990), pp. 197–276.

Owen, Dorothy M. 'Clenchwarton, Norfolk, late 14th or early 15th century', in R. A. Skelton and P. D. A. Harvey (eds), *Local Maps and Plans from Medieval England* (Oxford, Clarendon Press, 1986), pp. 127–30.

Owen, T. and E. Pilbeam. *Ordnance Survey: Map Makers to Britain since 1791* (Southampton, Ordnance Survey, 1992).

Parker, Geoffrey. *The Army of Flanders and the Spanish Road 1567–1659* (Cambridge, Cambridge University Press, 1972).

Parker, Matthew (ed.). *Matthew Paris Chronica Maiora* (Zurich, 1571).

Parry, M. L. *Climatic Change, Agriculture and Settlement* (Folkestone, Dawson, 1978).

Parsons, E. J. S. *The Map of Great Britain circa AD 1360 known as the Gough Map. An Introduction to* the Facsimile (Oxford, Oxford University Press for the Bodleian Library and the Royal Geographical Society, 1958).

Patten, William. The Expedicion into Scotla[n]d (London, 1547).

Pawson, Eric. Transport and Economy. The Turnpike Roads of Eighteenth Century Britain (London, Academic Press, 1977).

Pawson, E. 'The framework of industrial change 1730–1900', in R. A. Dodgshon and R. A. Butlin (eds), An Historical Geography of England and Wales, 1st ed. (London and New York, Academic Press, 1978), pp. 267–89.

Payne, Ann. Views of the Past: Topographical Drawings in the British Library (London, The British Library, 1987).

Peacham, Henry. The Complete Gentleman (London, 1622).

Pedley, Mary. 'Maps, war, and commerce: business correspondence with the London map firm of Thomas Jefferys and William Faden', Imago Mundi, 48 (1996), pp. 161–73.

Pegolotti, Francesco Balducci, La Pratica della mercatura, edited by Allan Evans (Cambridge, Mass., Medieval Academy of Arts, 24, 1934).

Pelletier, Monique (ed.). Géographie du monde au Moyen Age et à la Renaissance (Paris, Editions du Comité des Travaux Historiques et Scientifiques, 1989).

Phillips, A. D. M. 'The seventeenth century maps and surveys of William Fowler', Cartographic Journal, 17 (1980), pp. 100–10.

Pinto, John A. 'Origins and development of the ichnographic city plan', Journal of the Society of Architectural Historians, 35 (1976), pp. 35–50.

Porter, Roy. The Making of Geology: Earth Science in Britain (Cambridge, Cambridge University Press, 1977).

Postan, Michael M. 'The trade of medieval Europe' in Michael. M. Postan and Edward Miller (eds), The Cambridge Economic History of Europe, 2nd ed., vol. 2, Trade and Industry in the Middle Ages (Cambridge, Cambridge University Press, 1987), pp. 191–6.

Pottage, Alain. 'The measure of land', Modern Law Review, 57 (1994), pp. 361–84.

Potter, Jonathan, Ltd. Choice Items from Stock, Catalogue 13 (1998).

Pounds, N. J. G. An Historical Geography of Europe, 1500–1840 (Cambridge, Cambridge University Press, 1979).

Price, D. J. 'Medieval land surveying and topographical maps', Geographical Journal, 121 (1955), pp. 1–10.

Prior, Oliver H. Caxton's Mirrour of the World (London, Kegan Paul, Trench Trübner; New York, Humphrey Milford, Oxford University Press, Early English Text Society, Extra Series 110, 1913).

Proude, Richard. New Rutter of the Sea, for the North Parts (London, 1541).

Ralph, Elizabeth. 'Bristol circa 1480', in R. A. Skelton and P. D. A. Harvey (eds), Local Maps and Plans from Medieval England (Oxford, Clarendon Press, 1986), pp. 309–16.

Rasmussen, Steen Eiler. London the Unique City, revised abridged edition (Harmondsworth, Penguin Books, 1960).

Rathborne [Rathbone], Aaron. The Surveyor (London, 1616).

Ravenhill, William L. D. Benjamin Donn. A Map of the County of Devon, 1765. Facsimile and Introduction (Exeter, University of Exeter, 1965).

Ravenhill, William L. D. John Norden's Manuscript Maps of Cornwall and Its Nine Hundreds (Exeter, University of Exeter, 1972).

Ravenhill, William L. D. 'Joel Gascoyne: a pioneer of large-scale county mapping', Imago Mundi, 26 (1972), pp. 60–70.

Ravenhill, William L. D. Two Hundred and Fifty Years of Map-making in the County of Sussex (Lympne, Harry Margary, 1974).

Ravenhill, William L. D. 'John Adams, his map of England, its projection, and his Index Villaris of 1680', Geographical Journal, 144 (1978), pp. 427–37.

Ravenhill, William L. D. 'Christopher Saxton's surveying: an enigma', in Sarah Tyacke (ed.), English Map-Making 1500–1650 (London, The British Library, 1983), pp. 112–19.

Ravenhill, William L. D. 'The South West in the eighteenth-century re-mapping of England', in Katherine Barker and Roger Kain (eds), Maps and History in South-West England (Exeter, Exeter University Press, 1991), pp. 1–27.

Ravenhill, William L. D. 'The marine cartography of south-west England from Elizabethan to modern times', in Michael Duffy et al, The New Maritime History of Devon: from Early Times to the late Eighteenth Century, vol. 1 (London, Conway Maritime Press; Exeter, University of Exeter Press, 1992), pp. 155–63.

Ravenhill, William L. D. Christopher Saxton's 16th Century Maps: The Counties of England and Wales

(Shrewsbury, Chatsworth Library and Airlife Publishing, 1992).

Ravenhill, William L. D. 'Maps for the landlord', in Peter Barber and Christopher Board (eds), *Tales from the Map Room: Fact and Fiction about Maps and their Makers* (London, BBC Books, 1993), pp. 96–7.

Ravenhill, William L. D. and O. J. Padel. *A Map of the County of Cornwall newly Surveyed by Joel Gascoyne. Reprinted in Facsimile with an Introduction* (Exeter, Devon and Cornwall Record Society, n.s. 34, 1991).

Ravenhill, William L. D. and Margery Rowe. 'A decorated screen map of Exeter based on John Hooker's map of 1587', in Todd Gray, Margery Rowe and Audrey Erskine (eds), *Tudor and Stuart Devon. The Common Estate and Government. Essays Presented to Joyce Youings* (Exeter, University of Exeter Press, 1992), pp. 1–12.

Recorde, Robert. *The Ground of Artes, teaching the Worke and Practise of Arithmeticke* (London, 1543).

Reddaway, T. F. *The Rebuilding of London after the Great Fire* (London, Cape, 1940, second edition 1951).

Reeder, David A. 'Introduction' to *Charles Booth's Descriptive Map of London Poverty, 1889* (London Topographical Society Publication 130, 1984).

Reinhartz, Dennis. *The Cartographer and the Litterati–Herman Moll and his Intellectual Circle* (Lampeter, Edwin Mellon, 1997).

Reiss, Timothy J. *Knowledge Discovery and Imagination in Early Modern Europe* (Cambridge, Cambridge University Press, 1997).

Reitinger, Franz. 'Mapping relationships: allegory, gender and the cartographical image in eighteenth-century France and England', *Imago Mundi*, 51 (1999), pp. 106–30 and Plate 12.

Reps, John W. *The Making of Urban America* (Princeton, N.J., Princeton University Press, 1965).

Reynolds, L. D. and N. G. Wilson. *Scribes and Scholars: A Guide to the Transmission of Greek and Latin Literature* (Oxford University Press 1968, 3rd ed. Oxford, Clarendon Press, 1991).

Reynolds, Susan. 'Chertsey, Surrey, and Laleham, Middlesex, mid- or late-15th century's in R. A. Skelton and P. D. A. Harvey (eds), *Local Maps and Plans from Medieval England* (Oxford, Clarendon Press, 1986), pp. 237–43, and Plate 21.

Reynolds, Susan. 'Staines, Middlesex. 1469 × *circa* 1477', in R. A. Skelton and P. D. A. Harvey (eds), *Local Maps and Plans from Medieval England* (Oxford, Clarendon Press, 1986), pp. 245–50.

Richeson, A. W. *English Land Measuring to 1800: Instruments and Practices* (Cambridge, Mass., M.I.T. Press and Society of the History of Technology, 1966).

Ritchie, G. S. *The Admiralty Chart. British Naval Hydrography in the Nineteenth Century* (London, Hollis and Carter, 1967), pp. 238–48.

Roach, William. 'William Smith "A Description of the Cittie of Nuremberg" (Beschreibung der Reichsstadt Nürnburg)', *Mitteilungen des Vereins für Geschichte der Stadt Nürnberg*, 48 (1958), pp. 194–245.

Roberts, B. K. 'An early Tudor sketch map: its context and implications,' *History Studies*, 1 (1968), pp. 33–8.

Roberts, B. K. 'North-west Warwickshire; Tanworth in Arden, Warwickshire', in R. A. Skelton and P. D. A. Harvey (eds), *Local Maps and Plans from Medieval England* (Oxford, Clarendon Press, 1986), pp. 317–28.

Roberts, B. K. *The Making of the English Village* (Harlow, Longman, 1987).

Robinson, A. H. 'The 1837 maps of Henry Dury Harness', *Geographical Journal*, 121 (1955), pp. 440–50.

Robinson, A. H. *Early Thematic Mapping in the History of Cartography* (Chicago, University of Chicago Press, 1982).

Robinson, A. H. W. *Marine Cartography in Britain. A History of the Sea Chart to 1855* (Leicester, Leicester University Press, 1962).

Robinson, Henry W. and Walter Adams (eds). *The Diary of Robert Hooke 1672–1680* (London, Taylor and Francis, 1935).

Robinson, Howard. *The British Post Office. A History* (Princeton, NJ, Princeton University Press, 1948).

Rogers, James E. Thorold. *History of Agriculture and Prices in England*, 7 vols (Oxford, Clarendon Press, 1866–1902), vol. 4.

Rosenau, Helen. *The Ideal City: Its Architectural Evolution in Europe* , 3rd ed. (London, Methuen, 1983).

Rowley, Gwyn. *British Fire Insurance Plans* (Hatfield, Charles E. Goad, 1984).

Rowley, Gwyn. 'British fire insurance plans: the Goad productions, *c*.1885–*c*.1970', *Archives*, 17 (1985), pp. 67–78.

Rudé, George F. E. 'English rural disturbances on the eve of the first Reform Bill, 1830–1', *Past and Present*, 37 (1967), pp. 87–102.

Rudwick, Martin J. S. 'The emergence of a visual language for geological science 1760–1840', *History of Science*, 14 (1976), pp. 149–95.

Rycroft, Simon and Denis Cosgrove. 'Mapping the modern nation: Dudley Stamp and the Land Utilisation Survey', *History Workshop Journal*, 40 (1995), pp. 91–105.

Salisbury, C. R. 'An early Tudor map of the River Trent in Nottinghamshire', *Transactions of the Thoroton Society of Nottinghamshire*, 87 (1983), pp. 54–9.

Savage, Ernest. *Old English Libraries. The Making, Collection, and Use of Books during the Middle Ages* (London, Methuen, 1911).

Scafi, Alessandro. 'Mapping Eden', in Denis Cosgrove (ed.), *Mappings* (London, Reaktion Books, 1999), pp. 50–70 and pp. 273–6 (notes).

Schama, Simon. *The Embarrassment of Riches. An Interpretation of Dutch Culture in the Golden Age* (Berkeley and Los Angeles, University of California Press, 1988).

Schilder, Günter. 'A Dutch manuscript rutter: an unique portrait of the European coasts in the late sixteenth century', *Imago Mundi*, 43 (1991), pp. 59–71.

Schilder, Günter. 'An unrecorded set of thematic maps by Hondius', *The Map Collector*, 59 (1992), pp. 44–7.

Schofield, John. 'Ralph Treswell's surveys of London houses', in Sarah Tyacke (ed), *English Map-Making 1500–1650. Historical Essays* (London, The British Library, 1983), pp. 85–92.

Schofield, John (ed.). *The London Surveys of Ralph Treswell* (London, London Topographical Society, Publication 135, 1987).

Schultz, Jürgen. 'Jacopo de'Barbari's View of Venice: map-making, city views and moralized geography before the year 1500', *Art Bulletin*, 60 (1978), pp. 425–74.

Scott, Kathleen L. 'Design, decoration and illustration', in J. Griffiths and D. Pearsall (eds), *Book Production and Publishing in Britain, 1375–1475* (Cambridge, Cambridge University Press, 1989), pp. 31–64.

Scott, Kathleen L. *Later Gothic Manuscripts, 1390–1490. A Survey of Manuscripts Illuminated in the British Isles*, vol. VI, 2 vols (London, Harvey Miller, 1996).

Scribner, R. W. *For the Sake of Simple Folk. Popular Propaganda for the German Reformation* (Oxford, Clarendon Press, 1981).

Seymour, W. A. (ed.), *A History of the Ordnance Survey* (Folkestone, Dawson, 1980).

Shapter, Thomas. *The History of Cholera in Exeter in 1832* (London, Churchill, 1849).

Shaw, Gareth and Allison Tipper. *British Directories: A Bibliographic Guide to Directories Published in England and Wales (1850–1950) and Scotland (1773–1950)*, 2nd ed. (Cassell, London, 1996).

Shearer, Barbara (ed.). *Catalogue of Medieval and Renaissance Manuscripts, Beinecke Library, Yale*, 2 vols (Binghamton, N.Y., Medieval and Renaissance Texts and Studies Vol. 48, 1987).

Shelby, L. R. *John Rogers, Tudor Military Engineer* (Oxford, Clarendon Press, 1967).

Shepherd, J., J. Westaway and T. Lee. *Social Atlas of London* (Oxford, Clarendon Press, 1974).

Shirley, Rodney W. *Early Printed Maps of the British Isles. A Bibliography. 1477–1650* (London, Holland Press Cartographica 5, 1980).

Shirley, Rodney W. *The Mapping of the World. Early Printed World Maps 1472–1700*, 3rd ed. (London, Holland Press, 1993).

Shirley, Rodney W. *Printed Maps of the British Isles 1650–1750* (Tring, Map Collector Publications; London, The British Library, 1988).

Simmons, Matthew. *Direction for English Travaillers* (London, 1635).

Skelton, R. A. 'Tudor town plans in John Speed's *Theatre*', *Archaeological Journal*, 108 (1951), pp. 109–20.

Skelton, R. A. 'Four English county maps', *British Museum Quarterly*, 22 (1960), pp. 47–50.

Skelton, R. A. 'Introduction', in George Braun and Franz Hogenberg, *Civitates Orbis Terrarum*, facsimile edition (Amsterdam, Theatrum Orbis Terrarum, 1966), p. vii.

Skelton, R. A. *The Military Survey of Scotland* (Edinburgh, Royal Scottish Geographical Society, 1967).

Skelton, R. A. 'The military surveyor's contribution to British cartography in the sixteenth century', *Imago Mundi*, 24 (1970), pp. 77–85.

Skelton, R. A. *County Atlases of the British Isles 1579 to 1850. A Bibliography. 1579–1703* (London, Carta Press, 1970).

Skelton, R. A. *Maps: A Historical Survey of Their Study and Collecting*. First Kenneth Nebenzahl Lecture at the Newberry Library, 1966; Chicago (University of Chicago Press, 1972).

Skelton, R. A. *Saxton's Survey of England and Wales with a facsimile of Saxton's Wall-Map of 1583*, with a Preface by J. B. Harley (Amsterdam, N. Israel, *Imago Mundi* Supplement No. 6, 1974).

Skelton, R. A. and P. D. A. Harvey (eds). *Local Maps and Plans from Medieval England* (Oxford, Clarendon Press, 1986).

Skelton, R. A. and John Summerson. *A Description of the Maps and Architectural Drawings in the Collection Made by William Cecil, First Baron Burghley, Now at Hatfield House* (Oxford, for The Roxburghe Club, 1971).

Slicher van Bath, B. H. *Agrarian History of Western Europe, 1500–1850* (London, Arnold, 1963).

Smith, A. G. R. *Science and Society in the Sixteenth and Seventeenth Centuries* (London, Thames and Hudson, 1972).

Smith, David. *Victorian Maps of the British Isles* (London, Batsford, 1985).

Smith, David. 'The social mapping of Henry Mayhew', *The Map Collector*, 30 (1986), pp. 2–3.

Smith, David. *Maps and Plans for the Local Historian and Collector* (London, Batsford, 1988).

Smith, David. 'The representation of industry on large-scale county maps of England and Wales 1700–c.1840', *Industrial Archaeology Review*, 12 (1990), pp. 153–77.

Smith, David. 'The enduring image of early British townscapes', *Cartographic Journal*, 28 (1991), pp. 163–75.

Smith, David. 'Inset town plans on large-scale maps of Great Britain', *Cartographic Journal*, 29 (1992), pp. 118–36.

Smith, David. 'The demise of a nineteenth-century cartographic project: the publication of Lysons' *Magna Britannia*', IMCOS [International Map Collectors Society] *Journal*, 53 (1993), pp. 5–13.

Smith, David. 'Public health and the large-scale mapping of British towns', IMCOS [International Map Collectors Society] *Journal*, 56 (1994), pp. 28–46.

Smith, David. 'The preparation of the town plans for Lysons' *Magna Britannia*', *Cartographic Journal*, 32 (1995), pp. 11–17.

Smith, Sir John Flett. *The First Hundred Years of the Geological Survey of Great Britain* (London H.M.S.O., 1937).

Smith, Thomas R. 'Manuscript and printed sea-charts in seventeenth century London: the case of the Thames School', in Norman J. W. Thrower (ed.), *The Compleat Plattmaker. Essays on Chart, Map, and Globe Making in England in the Seventeenth and Eighteenth Centuries* (Berkeley, Los Angeles and London, University of California Press, 1978), pp. 45–100.

Snape, M. G. 'Durham 1439 × *circa* 1442', in R. A. Skelton and P. D. A. Harvey (eds), *Local Maps and Plans from Medieval England* (Oxford, Clarendon Press, 1986), pp. 189–94.

Snape, M. G. and B. K. Roberts, 'Tursdale Beck, County Durham, *circa* 1430 × *circa* 1442' in R. A. Skelton and P. D. A. Harvey (eds), *Local Maps and Plans from Medieval England* (Oxford, Clarendon Press, 1986), pp. 171–87.

Somers Cocks, J. V. 'Dartmoor, Devonshire. Late 15th or early 16th century', in R. A. Skelton and P. D. A. Harvey (eds), *Local Maps and Plans from Medieval England* (Oxford, Clarendon Press, 1986), pp. 293–302 and Plate 26.

Spadaforda, David. *The Idea of Progress in Eighteenth-Century Britain* (New Haven and London, Yale University Press, 1990).

Speed, John. *A Prospect of the most Famous Parts of the World* (London, 1627).

Sprent, R. P. 'Introduction', in Thomas Chubb, *The Printed Maps in the Atlases of Great Britain and Ireland. A Bibliography*, 1579–1870 (London, Ed. J. Burrow, 1927; reprinted in facsimile, London, Dawson, 1966).

Spufford, Margaret. *Small Books and Pleasant Histories. Popular Fiction and its Readership in Seventeenth-century England* (Cambridge, Cambridge University Press, 1981).

St Joseph, J. K. *Medieval England. An Aerial Survey*, 2nd ed. (Cambridge, Cambridge University Press, 1979).

Stagl, Justin. 'The methodising of travel in the sixteenth century. A tale of three cities', *History and Anthropology*, 4 (1990), pp. 303–38.

Stamp, L. D. *The Land of Britain: Its Use and Misuse* (London, Longmans, Green, 1962).

Stegena, Lajos. *Lazarus Secretarius: The First Hungarian Mapmaker and His Work* (Budapest, Akadémiai Kiadó, 1982).

Stenton, Sir Frank. 'The roads in the Gough map', in E. J. S. Parsons, *The Map of Great Britain circa AD 1360 known as the Gough Map. An Introduction to the Facsimile* (Oxford, Oxford University Press for the Bodleian Library and the Royal Geographical Society, 1958), pp. 16–20.

Stevens, Brian. *Plans of English Towns by G. Cole and J. Roper, 1810* (Monmouth, Historic Prints, 1970).

Stone, Jeffrey C. *A Short History of the Cartography of Africa* (Lampeter, Edward Mellon Press, 1995).

Stuart, Elisabeth. *Lost Landscapes of Plymouth: Maps, Charts and Plans to 1800* (Stroud, Alan Sutton; Tring, Map Collector Publications Ltd, 1991).

Stubbs, William (ed.). *The Historical Works of Gervase of Canterbury*, Rerum Britannicarum medii ævii scriptores, 2 vols (London, 1879–80).

Sturmy, Samuel. *The Mariners Magazine* (London, 1669).

Talbert, Richard. 'Rome's empire and beyond: the spatial aspect', in *Gouvernants et gouvernés dans l'Imperium Romanum*. Cahiers des Etudes Anciennes 26 (Montreal, Université Laval, Département d'Histoire, 1990), pp. 215–23.

Tate, W. E. *A Domesday of English Enclosure Acts and Awards*, Michael E. Turner, ed. (Reading, University of Reading, 1978).

Taub, Liba Chaia. *Ptolemy's Universe. The Natural Philosophy and Ethical Foundations of Ptolemy's Astronomy* (Chicago and La Salle, Illinois, Open Court, 1993).

Taub, Liba. 'The historical function of the *Forma urbis Romae*', *Imago Mundi*, 45 (1993), pp. 9–19.

Taylor, Christopher. *Roads and Tracks of Roman Britain*, 2nd ed. (London, Orion Books, 1994).

Taylor, E. G. R. *Tudor Geography, 1485–1583* (London, Methuen, 1930).

Taylor, E. G. R. *Late Tudor and Early Stuart Geography* (London, Methuen, 1934).

Taylor, E. G. R. 'Robert Hooke and the cartographical projects of the late seventeenth century', *Geographical Journal*, 90 (1937), pp. 529–40.

Taylor, E. G. R. 'Notes on John Adams and contemporary map makers', *Geographical Journal*, 97 (1941), pp. 182–4.

Taylor, E. G. R. 'The surveyor,' *Economic History Review*, 17 (1947), pp. 121–33.

Taylor, E. G. R. *The Mathematical Practitioners of Tudor and Stuart England* (Cambridge, Cambridge University Press, 1954).

Taylor, E. G. R. 'John Dee and the map of North-East Asia', *Imago Mundi*, 12 (1955), pp. 103–6.

Taylor, E. G. R. *The Haven-Finding Art: A History of Navigation from Odysseus to Captain Cook* (London, Hollis and Carter, 1956).

Taylor, E. G. R. *A Regiment for the Sea and Other Writings on Navigation by William Bourne of Gravesend, a gunner (c.1535–1582)* (Cambridge, Cambridge University Press for the Hakluyt Society, 1963).

Taylor, E. G. R. and M. W. Richey. *The Geometrical Seaman. A Book of Early Nautical Instruments* (London, Hollis and Carter for the Institute of Navigation, 1962).

Taylor, John. *The 'Universal Chronicle' of Ranulf Higden* (Oxford, Clarendon Press, 1966).

Terrell, Christopher. 'The seaman's view', *The Map Collector*, 31 (1985), pp. 2–7.

Thirsk, Joan. *Fenland Farming in the Sixteenth Century* (Leicester, University of Leicester Department of English Local History, Occasional Papers No. 3, 1953).

Thirsk, Joan (ed.). *The Agrarian History of England and Wales*, vol. 4 (Cambridge, Cambridge University Press, 1967).

Thomas, Keith. 'Numeracy in early modern England', *Transactions of the Royal Historical Society*, 5th series, 37 (1987), pp. 103–32.

Thompson, F. M. L. *Chartered Surveyors: The Growth of a Profession* (London, Routledge and Kegan Paul, 1968).

Thompson, James W. *The Medieval Library* (Chicago, University of Chicago Press, 1939).

Thorndike, Lynn. *The Sphere of Sacrobosco and Its Commentaries* (Chicago, Chicago University Press, 1949).

Tite, Colin G. C. *The Manuscript Library of Sir Robert Cotton. The Panizzi Lectures, 1993* (London, The British Library, 1994).

Trent Foley, W. and A. G. Holden. *Bede: A Biblical Miscellany* (Liverpool, Liverpool University Press, 1999).

Turner, Michael. *The Nature of Urban Renewal after Fires in Seven English Provincial Towns, circa 1675–1810* (unpublished University of Exeter Ph.D. thesis, 1985).

Turner, Michael E. *English Parliamentary Enclosure: Its Historical Geography and Economic History* (Folkestone, Dawson, 1980).

Turner, Michael E. 'The landscape of parliamentary enclosure', in Michael Reed (ed.), *Discovering Past Landscapes* (London, Croom Helm, 1984), pp. 132–66.

Turner, Michael E. *Enclosures in Britain, 1750–1830* (London, Macmillan, 1984).

Tyacke, Sarah. *London Map-Sellers 1660–1720* (Tring, Map Collector Publications, 1978).

Tyacke, Sarah, and John Huddy. *Christopher Saxton and Tudor Map-Making* (London, The British Library, 1980).

Tyacke, Sarah (ed.) *English Map-Making 1500–1650. Historical Essays* (London, The British Library, 1983).

Tyacke, Sarah. 'English overseas chartmaking', in David Woodward (ed.), *History of Cartography*, vol. 3, *Cartography in the European Renaissance* (Chicago and London, University of Chicago Press, forthcoming).

Tyson, Blake 'John Adams's cartographic correspondence to Sir Daniel Flemming of Rydal Hall, Cumbria, 1676–1687', *Geographical Journal*, 151 (1985), pp. 21–39.

Urry, William. 'Canterbury, Kent, *circa* 1153 × 1161', in R. A. Skelton and P. D. A. Harvey (eds), *Local Maps and Plans from Medieval England* (Oxford, Clarendon Press, 1986), pp. 43–58.

Varley, John. 'John Rocque. Engraver, surveyor, cartographer and map-seller', *Imago Mundi*, 5 (1948), pp. 83–91.

Vaughan, Richard. *Matthew Paris* (Cambridge, Cambridge University Press, 1958).

Verner, Coolie. 'John Seller and the chart trade in seventeenth-century England', in J. W. Thrower (ed.), *The Compleat Plattmaker. Essays on Chart, Map, and Globe Making in England in the Seventeenth and Eighteenth Centuries* (Berkeley, Los Angeles; London, University of California Press, 1978), pp. 127–57.

Vitruvius, *On Architecture*, translated by F. Granger, 2 vols (London, Heinemann, Loeb Classical Library, 1931–4).

Waghenaer, Lucas Jansoon. *Spieghel der Zeevaerdt* (Leiden, 1584); English translation, Anthony Ashley, *The Mariners Mirrour* (London 1588).

Wallis, Faith E. *MS Oxford, St John's 17: a Medieval Manuscript in Context* (Unpublished Ph.D. thesis, University of Toronto, 1985).

Wallis, Faith E. *Bede: The Reckoning of Time* (Liverpool, Liverpool University Press, 1999).

Wallis, Helen M. ' "Geographie is better than Divinitie": maps, globes and geography in the days of Samuel Pepys', in Norman J. W. Thrower (ed.), *The Compleat Plattmaker. Essays on Chart, Map and Globe Making in England in the Seventeenth and Eighteenth Centuries* (Berkeley, Los Angeles; London, University of California Press, 1978), pp. 1–43.

Wallis, Helen M. (ed.). *The Maps and Text of the Boke of Idrography Presented by Jean Rotz to Henry VIII now in the British Library* (Oxford, The Roxburghe Club, 1981).

Wallis, Helen M. and W. P. Cumming. 'Charts by John Friend preserved at Chatsworth House, Derbyshire, England', *Imago Mundi*, 25 (1971), p. 81.

Wallis, Helen M. and Sarah Tyacke (eds). *My Head is a Map: Essays and Memoirs in Honour of R. V. Tooley* (London, Francis Edwards and Carta Press, 1973).

Walton, J. R. 'Agriculture and rural society, 1730–1914', in R. A. Dodgshon and R. A. Butlin (eds), *An Historical Geography of England and Wales*, 2nd ed. (London, Academic Press, 1990), pp. 323–50.

Warnicke, Retha M. *William Lambarde. Elizabethan Antiquary 1536–1602* (London and Chichester, Phillimore, 1973).

Watelet, Marcel (ed.). *Gérard Mercator Cosmographe. Le Temps et l'Espace* (Antwerp, Fonds Mercator Paribus, 1995).

Waters, David W. *The Art of Navigation in Elizabethan and Early Stuart Times* (London, Hollis and Carter, 1958).

Waters, David W. *The Rutter of the Sea. The Sailing Directions of Pierre Garcie. A Study of the First English and French Printed Sailing Directions. With facsimile reproductions* (New Haven and London, Yale University Press, 1967).

Watt, Tessa. *Cheap Print and Popular Piety 1550–1640* (Cambridge, Cambridge University Press, 1991).

Whalley, George H. *The Tithe Act and the Tithe Act Amendment Act* (London, Shaw and Sons, 1838).

Wheatley, Henry B. and Edmund W. Ashbee (eds). *William Smith, 'The Particular Description of England, 1588, With Views of some of the Chief Towns and Armorial Bearings of Nobles and Bishops'* (London, [by subscription], 1879).

White, John. *A prognostication for . . . 1647* (London, Stationers' Company, 1647)

Whittington, G. and A. J. S. Gibson. *The Military Survey of Scotland 1747–1755: A Critique* (Norwich, Geo Abstracts for the Historical Geography Research Series, No. 18, 1986).

Whyte, I. D. and K. A. Whyte. *Sources for Scottish Historical Geography: An Introductory Guide* (Norwich, Geo Abstracts for the Historical Geography Research Series No. 6, 1981).

Wilkinson, John. *Jerusalem Pilgrims before the Crusades* (Warminister, Aris & Phillips, 1977).

Williams, Edward and William Mudge, 'An account of the trigonometrical survey carried on in 1791, 1792, 1793 and 1794', *Philosophical Transactions of the Royal Society*, 85 (1795), pp. 414–591.

Williams, John. *The Illustrated Beatus. A Corpus of the Illustrations of the Commentary on the Apocalypse* (London, Harvey Miller, 1994–).

Williams, John. 'Isidore, Orosius and the Beatus map' *Imago Mundi*, 49 (1997), pp. 7–32.

Willmoth, Francis. ' "The Genius of All Arts" and the use of instruments: Jonas Moore (1617–1679) as a mathematician, surveyor, and astronomer', *Annals of Science*, 48 (1991), pp. 355–65.

Willmoth, Francis. *Sir Jonas Moore* (Woodbridge, Suffolk, Boydell Press, 1993).

Wilson, H. E. *Down to Earth: One Hundred Years of the British Geological Survey* (Edinburgh, Scottish Academy Press, 1985).

Wiseman, Peter. 'Julius Caesar and the Hereford world map', *History Today*, 37 (1987), pp. 53–7.

Woodward, David. 'The study of the history of cartography: a suggested framework', *American Cartographer*, 1 (1974), pp. 101–15.

Woodward, David (ed.). *Five Centuries of Map Printing. The Kenneth Nebenzahl, Jr. Lectures in the History of Cartography* (Chicago, University of Chicago Press for The Newberry Library, 1975).

Woodward, David. 'Reality, symbolism, time and space in medieval world maps', *Annals of the American Association of Geographers*, 75 (1985), pp. 510–21.

Woodward, David. 'Medieval *mappaemundi*', in J. B. Harley and David Woodward (eds), *History of Cartography*, vol. 1, *Cartography in Prehistoric, Ancient, and Medieval Europe and the Mediterranean* (Chicago, University of Chicago Press, 1987), pp. 286–370.

Woodward, David. 'Roger Bacon's terrestrial co-ordinate system', *Annals of the Association of the American Geographers*, 80 (1990), pp. 109–22.

Woodward, David (ed.), *History of Cartography*, vol. 3, *Cartography in the European Renaissance* (Chicago, University of Chicago Press, forthcoming).

Woodward, David (ed.). *Approaches and Challenges in a World-wide History of Cartography Project* (Barcelona, Instituto Cartogràfic de Catalunya, in preparation).

Woodward, David with Herbert M. Howe. 'Roger Bacon on geography and cartography', in Jeremiah Hackett (ed.), *Roger Bacon and the Sciences. Commemorative Essays* (Leiden, New York and Cologne, Brill, 1997), pp. 198–222.

Woodward, David and G. Malcolm Lewis (eds). *History of Cartography*, vol. 2, book 3, *Cartography in the Traditional African, American, Arctic, Australian, and Pacific Societies* (Chicago and London, University of Chicago Press, 1998).

Woolley, R. M. *Catalogue of the Manuscripts of Lincoln Cathedral Chapter Library* (Oxford, Clarendon Press, 1927).

Wordie, J. R. 'The chronology of English enclosure, 1500–1914', *Economic History Review*, 36 (1983), pp. 483–505.

Worms, Laurence. *Some British Mapmakers* (London, Ash Rare Books, Catalogue [1992]).

Worms, Laurence. 'Thomas Kitchen's "Journey of Life": Hydrographer to George III, mapmaker and engraver', *The Map Collector*, 62 and 63 (1993), pp. 2–8, 14–20.

Worms, Laurence. 'Faden, William', in C. S. Nicholls (ed.), *The Dictionary of National Biography: Missing Persons* (Oxford, Oxford University Press, 1993), p. 218.

Worms, Laurence. 'The London map trade to 1640', in David Woodward (ed.), *History of Cartography*, vol. 3, *Cartography in the European Renaissance* (Chicago and London, University of Chicago Press, forthcoming).

Wright, J. K. *Geographical Lore of the Time of the Crusades* (New York, American Geographical Society, Research Series 15, 1925; republished with additions, New York, Dover Publications, 1965).

Yates, E. M. 'Dark Age and medieval settlement on the edge of wastes and forests', *Field Studies*, 2 (1965), pp. 133–53.

Yelling, J. A. *Common Field and Enclosure in England 1450–1850* (London, Macmillan, 1977).

Index

Page numbers in *italics* refer to illustrations